Biomolecular Simulations in Structure-Based Drug Discovery

Sippl, W., Jung, M. (Eds.)

Epigenetic Drug Discovery

2018

ISBN: 978-3-527-34314-0

Vol. 74

Giordanetto, F. (Ed.)

Early Drug Development

2018

ISBN: 978-3-527-34149-8

Vol. 73

Handler, N., Buschmann, H. (Eds.)

Drug Selectivity

2017

ISBN: 978-3-527-33538-1

Vol. 72

Vaughan, T., Osbourn, J., Jalla, B. (Eds.)

Protein Therapeutics

2017

ISBN: 978-3-527-34086-6

Vol. 71

Ecker, G. F., Clausen, R. P., and Sitte, H. H. (Eds.)

Transporters as Drug Targets

2017

ISBN: 978-3-527-33384-4

Vol. 70

Martic-Kehl, M. I., Schubiger, P.A. (Eds.)

Animal Models for Human Cancer

Discovery and Development of Novel Therapeutics

2017

ISBN: 978-3-527-33997-6

Vol. 69

Holenz, Jörg (Ed.)

Lead Generation

Methods and Strategies

2016

ISBN: 978-3-527-33329-5

Vol. 68

Erlanson, Daniel A. / Jahnke, Wolfgang (Eds.)

Fragment-based Drug Discovery

Lessons and Outlook

2015

ISBN: 978-3-527-33775-0

Vol. 67

Urbán, László / Patel, Vinod F. / Vaz, Roy J. (Eds.)

Antitargets and Drug Safety

2015

ISBN: 978-3-527-33511-4

Vol. 66

Keserü, György M. / Swinney, David C. (Eds.)

Kinetics and Thermodynamics of Drug Binding

2015

ISBN: 978-3-527-33582-4

Vol. 65

Pfannkuch, Friedlieb / Suter-Dick, Laura (Eds.)

Predictive Toxicology

From Vision to Reality

2014

ISBN: 978-3-527-33608-1

Vol. 64

Biomolecular Simulations in Structure-Based Drug Discovery

Edited by
Francesco L. Gervasio and Vojtech Spiwok

WILEY-VCH

Series Editors

Prof. Dr. Raimund Mannhold
Rosenweg 7
40489 Düsseldorf
Germany

Dr. Helmut Buschmann
Aachen, Germany
Sperberweg 15
52076 Aachen
Germany

Dr. Jörg Holenz
GSK
R&D Neurosciences TAU
1250 S. Collegeville Road, PA
United States

Volume Editors

Francesco L. Gervasio
University College London
Chair of Biomolecular Modelling
20 Gordon Street
WC1H 0AJ London
United Kingdom

Vojtech Spiwok
Univ. of Chemistry and Technology
Dept. of Biochemistry and Microbiology
Technická 3
166 28 Prague 6
Czech Republic

Library of Congress Card No.: applied for

British Library Cataloguing-in-Publication Data
A catalogue record for this book is available from the British Library.

Bibliographic information published by the Deutsche Nationalbibliothek
The Deutsche Nationalbibliothek lists this publication in the Deutsche Nationalbibliografie; detailed bibliographic data are available on the Internet at <http://dnb.d-nb.de>.

Print ISBN: 978-3-527-34265-5
ePDF ISBN: 978-3-527-80684-3
ePub ISBN: 978-3-527-80685-0
oBook ISBN: 978-3-527-80683-6

Cover Design SCHULZ Grafik-Design, Fußgönheim, Germany
Typesetting SPi Global, Chennai, India
Printing and Binding C.O.S. Printers Pte Ltd, Singapore

Printed on acid-free paper

10 9 8 7 6 5 4 3 2 1

Contents

Foreword *xiii*

Part I Principles *1*

1 **Predictive Power of Biomolecular Simulations** *3*
Vojtěch Spiwok
1.1 Design of Biomolecular Simulations *4*
1.2 Collective Variables and Trajectory Clustering *6*
1.3 Accuracy of Biomolecular Simulations *8*
1.4 Sampling *10*
1.5 Binding Free Energy *14*
1.6 Convergence of Free Energy Estimates *16*
1.7 Future Outlook *20*
References *21*

2 **Molecular Dynamics–Based Approaches Describing Protein Binding** *29*
Andrea Spitaleri and Walter Rocchia
2.1 Introduction *29*
2.1.1 Protein Binding: Molecular Dynamics Versus Docking *30*
2.1.2 Molecular Dynamics – The Current State of the Art *31*
2.2 Protein–Protein Binding *32*
2.3 Protein–Peptide Binding *34*
2.4 Protein–Ligand Binding *36*
2.5 Future Directions *38*
2.5.1 Modeling of Cation-p Interactions *38*
2.6 Grand Challenges *39*
References *39*

Part II Advanced Algorithms *43*

3 Modeling Ligand–Target Binding with Enhanced Sampling Simulations *45*
Federico Comitani and Francesco L. Gervasio
3.1 Introduction *45*
3.2 The Limits of Molecular Dynamics *46*
3.3 Tempering Methods *47*
3.4 Multiple Replica Methods *48*
3.5 Endpoint Methods *50*
3.5.1 Alchemical Methods *50*
3.6 Collective Variable-Based Methods *51*
3.6.1 Metadynamics *52*
3.7 Binding Kinetics *57*
3.8 Conclusions *59*
References *60*

4 Markov State Models in Drug Design *67*
Bettina G. Keller, Stevan Aleksić, and Luca Donati
4.1 Introduction *67*
4.2 Markov State Models *68*
4.2.1 MD Simulations *68*
4.2.2 The Molecular Ensemble *69*
4.2.3 The Propagator *69*
4.2.4 The Dominant Eigenspace *70*
4.2.5 The Markov State Model *72*
4.3 Microstates *75*
4.4 Long-Lived Conformations *77*
4.5 Transition Paths *79*
4.6 Outlook *81*
Acknowledgments *82*
References *82*

5 Monte Carlo Techniques for Drug Design: The Success Case of PELE *87*
Joan F. Gilabert, Daniel Lecina, Jorge Estrada, and Victor Guallar
5.1 Introduction *87*
5.1.1 First Applications *88*
5.1.2 Free Energy Calculations *88*
5.1.3 Optimization *88*
5.1.4 MC and MD Combinations *89*
5.2 The PELE Method *90*
5.2.1 MC Sampling Procedure *91*
5.2.2 Ligand Perturbation *91*
5.2.3 Receptor Perturbation *91*
5.2.4 Side-Chain Adjustment *93*

5.2.5 Minimization *93*
5.2.6 Coordinate Exploration *93*
5.2.7 Energy Function *94*
5.3 Examples of PELE's Applications *94*
5.3.1 Mapping Protein Ligand and Biomedical Studies *94*
5.3.2 Enzyme Characterization *96*
Acknowledgments *97*
References *97*

6 Understanding the Structure and Dynamics of Peptides and Proteins Through the Lens of Network Science *105*
Mathieu Fossépré, Laurence Leherte, Aatto Laaksonen, and Daniel P. Vercauteren
6.1 Insight into the Rise of Network Science *105*
6.2 Networks of Protein Structures: Topological Features and Applications *107*
6.2.1 Topological Features and Analysis of Networks: A Brief Overview *107*
6.2.2 Centrality Measures and Protein Structures *110*
6.2.3 Software *114*
6.3 Networks of Protein Dynamics: Merging Molecular Simulation Methods and Network Theory *117*
6.3.1 Molecular Simulations: A Brief Overview *117*
6.3.2 How Can Network Science Help in the Analysis of Molecular Simulations? *118*
6.3.3 Software *119*
6.4 Coarse-Graining and Elastic Network Models: Understanding Protein Dynamics with Networks *120*
6.4.1 Coarse-Graining: A Brief Overview *120*
6.4.2 Elastic Network Models: General Principles *123*
6.4.3 Elastic Network Models: The Design of Residue Interaction Networks *124*
6.5 Network Modularization to Understand Protein Structure and Function *128*
6.5.1 Modularization of Residue Interaction Networks *128*
6.5.2 Toward the Design of Mesoscale Protein Models with Network Modularization Techniques *130*
6.6 Laboratory Contributions in the Field of Network Science *131*
6.6.1 Graph Reduction of Three-Dimensional Molecular Fields of Peptides and Proteins *132*
6.6.2 Design of Multiscale Elastic Network Models to Study Protein Dynamics *135*
6.7 Conclusions and Perspectives *140*
Acknowledgments *142*
References *142*

Part III Applications and Success Stories *163*

7 From Computers to Bedside: Computational Chemistry Contributing to FDA Approval *165*
Christina Athanasiou and Zoe Cournia
7.1 Introduction *165*
7.2 Rationalizing the Drug Discovery Process: Early Days *166*
7.2.1 Captopril (Capoten®) *167*
7.2.2 Saquinavir (Invirase®) *167*
7.2.3 Ritonavir (Norvir®) *168*
7.3 Use of Computer-Aided Methods in the Drug Discovery Process *168*
7.3.1 Ligand-Based Methods *169*
7.3.1.1 Overlay of Structures *169*
7.3.1.2 Pharmacophore Modeling *171*
7.3.1.3 Quantitative Structure–Activity Relationships (QSAR) *172*
7.3.2 Structure-Based Methods *173*
7.3.2.1 Molecular Docking – Virtual Screening *175*
7.3.2.2 Flexible Receptor Molecular Docking *179*
7.3.2.3 Molecular Dynamics Simulations *179*
7.3.2.4 *De Novo* Drug Design *180*
7.3.2.5 Protein Structure Prediction *181*
7.3.2.6 Rucaparib (Zepatier®) *184*
7.3.3 *Ab Initio* Quantum Chemical Methods *185*
7.4 Future Outlook *186*
References *190*

8 Application of Biomolecular Simulations to G Protein–Coupled Receptors (GPCRs) *205*
Mariona Torrens-Fontanals, Tomasz M. Stepniewski, Ismael Rodríguez-Espigares, and Jana Selent
8.1 Introduction *205*
8.2 MD Simulations for Studying the Conformational Plasticity of GPCRs *207*
8.2.1 Challenges in GPCR Simulations: The Sampling Problem and Simulation Timescales *208*
8.2.2 Making Sense Out of Simulation Data *209*
8.3 Application of MD Simulations to GPCR Drug Design: Why Should We Use MD? *210*
8.4 Evolution of MD Timescales *214*
8.5 Sharing MD Data via a Public Database *216*
8.6 Conclusions and Perspectives *216*
Acknowledgments *217*
References *217*

9 Molecular Dynamics Applications to GPCR Ligand Design *225*
Andrea Bortolato, Francesca Deflorian, Giuseppe Deganutti, Davide Sabbadin, Stefano Moro, and Jonathan S. Mason
9.1 Introduction *225*

9.2 The Role of Water in GPCR Structure-Based Ligand Design *226*
9.2.1 WaterMap and WaterFLAP *228*
9.3 Ligand-Binding Free Energy *230*
9.4 Ligand-Binding Kinetics *233*
9.4.1 Supervised Molecular Dynamics (SuMD) *235*
9.4.2 Adiabatic Bias Metadynamics *238*
9.5 Conclusion *241*
References *242*

10 Ion Channel Simulations *247*
Saurabh Pandey, Daniel Bonhenry, and Rudiger H. Ettrich
10.1 Introduction *247*
10.2 Overview of Computational Methods Applied to Study Ion Channels *248*
10.2.1 Homology Modeling *248*
10.2.2 All-atom Molecular Dynamics Simulations *249*
10.2.2.1 Force Fields *250*
10.2.3 Methods for Calculation of Free Energy *251*
10.2.3.1 Free Energy Perturbation *251*
10.2.3.2 Umbrella Sampling *251*
10.2.3.3 Metadynamics *252*
10.2.3.4 Adaptive Biased Force Method *252*
10.3 Properties of Ion Channels Studied by Computational Modeling *253*
10.3.1 A Refined Atomic Scale Model of the *Saccharomyces cerevisiae* K^+-translocation Protein Trk1p *253*
10.3.2 Homology Modeling, Docking, and Mutagenesis Studies of Human Melatonin Receptors *254*
10.3.3 Selectivity and Permeation in Voltage-Gated Sodium (Na_V) Channels *254*
10.3.4 Study of Ion Conduction Mechanism, Favorable Translocation Path, and Ion Selectivity in KcsA Using Free Energy Perturbation and Umbrella Sampling *257*
10.3.5 Ion Conductance Calculations *260*
10.3.5.1 Voltage-Dependent Anion Channel (VDAC) *261*
10.3.5.2 Calculation of Ion Conduction in Low-Conductance GLIC Channel *261*
10.3.6 Transient Receptor Potential (TRP) Channels *263*
10.4 Free Energy Methods Applied to Channels Bearing Hydrophobic Gates *264*
10.5 Conclusion *270*
Acknowledgments *271*
References *271*

11 Understanding Allostery to Design New Drugs *281*
Giulia Morra and Giorgio Colombo
11.1 Introduction *281*
11.2 Protein Allostery: Basic Concepts and Theoretical Framework *282*
11.2.1 The Classic View of Allostery *283*

11.2.2 The Thermodynamic Two-State Model of Allostery 283
11.2.3 From Thermodynamics to Protein Structure and Dynamics 285
11.2.4 Entropy in Allostery: The Ensemble Allostery Model 287
11.3 Exploiting Allostery in Drug Discovery and Design 288
11.3.1 Computational Prediction of Allosteric Behavior and Application to Drug Discovery 288
11.3.2 Identification of Allosteric Binding Sites Through Structural and Dynamic Approaches 289
11.4 Chaperones 291
11.5 Kinases 293
11.6 GPCRs 294
11.7 Conclusions 296
References 296

12 Structure and Stability of Amyloid Protofibrils of Polyglutamine and Polyasparagine from Molecular Dynamics Simulations 301
Viet Hoang Man, Yuan Zhang, Christopher Roland, and Celeste Sagui
12.1 Introduction 301
12.2 Polyglutamine Protofibrils and Aggregates 303
12.2.1 Investigations of Oligomeric Q_8 Structures 303
12.2.2 Time Evolution, Steric Zippers, and Crystal Structures of 4×4 Q_8 Aggregates 306
12.2.3 Monomeric Q_{40} Protofibrils 308
12.3 Amyloid Models of Asparagine (N) and Glutamine(Q) 311
12.3.1 Initial Structures 313
12.3.2 Monomeric PolyQ β Hairpins Are More Stable than PolyN Hairpins 314
12.3.3 N-rich Oligomers Are Most Stable in Class 1 Steric Zippers with 2-by-2 Interdigitation 315
12.3.4 PolyQ Oligomers Are Most Stable in Antiparallel Stranded β Sheets with 1-by-1 Steric Zippers 316
12.3.5 PolyQ Structures Show Higher Stability than Most Stable PolyN Structures 317
12.3.6 Thermodynamic Considerations of Aggregate Formation 318
12.4 Summary 319
Acknowledgments 320
References 320

13 Using Biomolecular Simulations to Target Cdc34 in Cancer 325
Miriam Di Marco, Matteo Lambrughi, and Elena Papaleo
13.1 Background 325
13.2 Families of E2 Enzymes 327
13.3 Cdc34 Protein Sequence and Structure 328
13.4 Cdc34 Heterogeneous Conformational Ensemble in Solution 329

13.5 Long-Range Communication in Family 3 Enzymes: A Structural Path from the Ub-Binding Site to the E3 Recognition Site *330*
13.6 Cdc34 Modulation by Phosphorylation: From Phenotype to Structure *331*
13.7 The Dual Role of the Acidic Loop of Cdc34: Regulator of Activity and Interface for E3 Binding *332*
13.8 Different Strategies to Target Cdc34 with Small Molecules *333*
13.9 Conclusions and Perspectives *334*
Acknowledgments *336*
References *336*

Index *343*

Foreword

Computational chemistry tools, from quantum chemistry techniques to molecular modeling, have greatly contributed to a number of fields, ranging from geophysics and material chemistry to structural biology and drug design. Dangerous, expensive, and laborious experiments can be often replaced "in silico" by accurate calculations. In drug discovery, a number of techniques at various levels of accuracy and computational cost are in use. Methods on the more accurate end of the spectrum such as fully atomistic molecular simulations have been shown to be able to reliably predict a number of properties of interest, such as the binding pose or the binding free energy. However, they are computationally expensive. This fact has so far hampered the systematic application of simulation-based methods in drug discovery, while inexpensive heuristic molecular modeling methods, such as protein–ligand docking are routinely used.

However, things are rapidly changing and the potential of atomistic biomolecular simulations in academic and industrial drug discovery is becoming increasingly clear. The question is whether we can expect an evolution or a revolution in this field. There are examples of other areas of life sciences where a revolution took or is taking place. For example, sequencing of the human genome took a decade and was funded by governments of several countries. Today, sequencing of eukaryotic genomes has become a routine, and a million-genome project is on the way owing to highly efficient and inexpensive parallel sequencing technology. Similarly, genetic manipulations are becoming significantly easier and more efficient owing to CRISPR/Cas technology. At the same time, the deep learning revolution is having a deep impact on many fields. The open question is whether we can expect such a revolution in biomolecular simulations due to new groundbreaking technology and convergence with machine learning techniques or a stepwise evolution due to the availability of new hardware, of grid and cloud resources, as well as advances in force-field accuracy, enhanced sampling techniques, and other achievements.

The aim of this book is to report on the current state and promising future directions for biomolecular simulations in drug discovery. Although we personally believe that there is true potential for a simulation-based revolution in drug discovery, we will let the readers draw their own conclusions.

In the first part of the book, called *Principles*, we give an overview of biomolecular simulation techniques with focus on modeling protein–ligand interactions. When applying any molecular modeling method, we have to ask the question how accurate is the method in comparison with the experiment. There are three major factors influencing the overall accuracy of biomolecular simulations. First, the method itself is approximative. Second, we use a simplified structure–energy relationship (such as molecular mechanics force field), which is approximative, especially for new classes of molecules. And, finally third, the simulated system is an image of a single or few molecules observed for a short time in contrast to the experiment that typically provides observations averaged over a vast number of molecules and over a significantly longer time. In the other words, sampling of states in the simulation may be incomplete compared to sampling in the experiment. These issues are discussed in Chapter 1. Chapter 2 focuses on the "sampling problem," in contexts relevant to drug discovery, namely, in modeling of protein–protein, protein–peptide, and protein–ligand interactions.

The second part of the book is called *Advanced Algorithms.* It presents algorithms used to solve problems presented in the first part of the book, especially the sampling problem. It is possible to artificially force the system to sample more states than in a conventional molecular simulation. The dynamics in such simulations is biased, but it is possible to derive statistically meaningful long-timescale behavior and free energies from such simulations. These techniques, referred to as enhanced sampling techniques, are presented in Chapter 3. The methods include sampling enhancement obtained by raising the temperature (tempering methods), methods employing artificial potentials or forces acting on selected degrees of freedom, combined approaches, and other methods.

The traditional approach to evaluate protein–ligand interactions in drug discovery is based on thermodynamics, i.e. measurement or prediction of K_i, IC_{50}, binding ΔG, or similar parameters. However, recently it turned out that kinetics of protein–ligand binding and unbinding is highly important, often more important than the thermodynamics. Markov state models presented in Chapter 4 provide an elegant way to describe thermodynamics and kinetics of the studied process from various types of molecular simulations.

Other solutions to the sampling problem are based on a simplified representation of the studied system or of its dynamics. These approaches are covered in Chapters 5 and 6. Chapter 5 presents an alternative sampling approach based on a Monte Carlo method: PELE. The dynamics of the system is simplified to harmonic vibrations of a protein and translations and rotations of a ligand. This is used in each step to propose the new state of the system, which is either accepted or rejected in the spirit of the Monte Carlo method. The algorithm is highly efficient in exploring ligand and target dynamics, as demonstrated by a number of ligand design applications. Chapter 6 presents an overview of network models. It is possible to represent the structure of a protein as a network of interactions. This approach makes it possible to simplify (coarse grain) the studied system, study the system in terms of normal modes, and combine these coarse-grained models with fine-grained models.

The third part of the book is called *Applications and Success Stories.* Chapter 7 provides an overview of the applications of molecular modeling methods in drug discovery. It presents various molecular modeling methods, including quantitative structure–activity relationship (QSAR) and ligand-based models, pharmacophore modeling, protein–ligand docking, biomolecular simulations, and quantum chemistry methods. Each technique is presented together with its practical impact in drug development and with examples of approved drugs.

Chapters 8 and 9 focus on the largest group of drug targets – G protein–coupled receptors (GPCRs), one from the academic and one from industrial perspective. The issues covered by these chapters include sampling problem, the role of membrane and water, free energy predictions, ligand binding kinetics, and others. Simulation of GPCRs is challenging partially due to their membrane environment. Another important group of membrane-bound targets are ion channels covered in Chapter 10. Special topics related to ion channels, such as modeling of ion selectivity and ion conductance, are described in this chapter.

Allostery is a very important topic when studying protein–ligand interactions because many ligands bind to sites other than those expected and/or make an effect on sites other than the binding one. Allostery, its thermodynamics, ways of modeling, and application on various drug targets are described in Chapter 11.

The last two chapters are focused on specific topics of current relevance in drug discovery. Chapter 12 presents the way to address protein misfolding and aggregation by biomolecular simulations. This is illustrated on polyglutamine and polyasparagine protofibrils from simulations to thermodynamic models of aggregate formation. Chapter 13 targets the cell cycle and the role of ubiquitin-mediated proteolysis. In the example of Cdc34, it is illustrated how biomolecular simulations can be integrated with structural biology and other methods to elucidate the structure and dynamics of a drug target.

This book was realized thanks to the invitation from Prof. Gerd Folkers and thanks to support by him and other series editors. We gratefully acknowledge their support and patience. We also thank Dr. Frank Weinreich, Dr. Stefanie Volk, and Dr. Sujisha Karunakaran from Wiley-VCH for their support and pleasant collaboration on this volume.

We believe that the book can add more dynamics to drug design and more drug design to biomolecular simulations.

Prague and London, July 2018

Francesco L. Gervasio
Vojtěch Spiwok

Part I

Principles

1

Predictive Power of Biomolecular Simulations

Vojtěch Spiwok

University of Chemistry and Technology, Prague, Department of Biochemistry and Microbiology, Technická 3, 166 28 Prague 6, Czech Republic

Biomolecular simulations are becoming routine in structure-based drug design and related fields. This chapter briefly presents the history of molecular simulations, basic principles and approximations, and the most common designs of computational experiments. I also discuss statistical analysis of simulation results together with possible limits of accuracy.

The history of computational modeling of molecular structure and dynamics goes back to 1953, to the work of Rosenbluth and coworkers [1]. It introduced the Markov chain Monte Carlo as a method to study a simplified model of the fluid system. Atoms of the studied system were perfectly inelastic and the system was two-dimensional (2D) instead of three-dimensional (3D), so the analogy with real molecular systems was not perfect. The first molecular dynamics simulation (i.e. modeling of motions) on the same system was done by Alder and Wainwright in 1957 [2] using perfectly elastic collision between 2D particles. The first molecular simulation with specific atom types was done by Rahman in 1964 [3]. Rahman used a CDC 3600 computer to simulate dynamics of 864 argon atoms modeled using Lennard-Jones potential. The first simulation of liquid water was published by Rahman and Stillinger in 1971 [4].

Another big milestone was the first biomolecular simulation. McCammon, Gelin, and 2013 Nobel Prize winner Karplus simulated 9.2 ps of the life of the bovine pancreatic trypsin inhibitor (BPTI, also known as aprotinin) in vacuum [5]. The simulation was performed during the CECAM (Centre Européen de Calcul Atomic et Moléculaire) workshop "Models of Protein Dynamics" in Orsay, France on CECAM computer facilities [6]. It was one of the first works showing proteins as a dynamic species with fluid-like internal motions, even though in the native state.

Biomolecular simulations have undergone a huge progress in terms of accuracy, size of simulated systems, and simulated times since their pioneer times. However, the question arises whether this progress is enough for their practical application in drug discovery, protein engineering, and related applied fields. To address this issue, let me present here the concept of the hype cycle [7] developed

Biomolecular Simulations in Structure-Based Drug Discovery,
First Edition. Edited by Francesco L. Gervasio and Vojtech Spiwok.

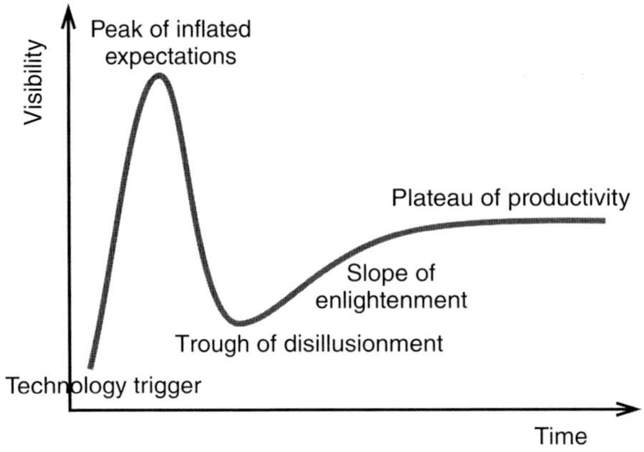

Figure 1.1 Gartner hype cycle of inventions.

by Gartner Inc. and depicted in Figure 1.1. According to this concept, every new invention starts by a *Technology Trigger*. Visibility of the invention grows until it reaches the *Peak of Inflated Expectations*. At this point, failures of the invention start to dominate over its benefits and the invention falls into the phase of *Trough of Disillusionment*. From this phase a new and slower progress starts in the phase of *Slope of Enlightenment* toward the *Plateau of Productivity*. Biomolecular simulation passed the *Technology Trigger* and *Peak of Inflated Expectations* as many expected that biomolecular simulation would become routine and an inexpensive alternative to experimental testing of compounds for biological activity. Now, in my opinion, biomolecular simulations are located on the *Slope of Enlightenment* with a slow but steady progress toward the *Plateau of Productivity*.

1.1 Design of Biomolecular Simulations

Biomolecular simulations can follow different designs. I use the term design to describe the setup of the simulation procedure chosen in order to answer the research hypothesis. There are three major designs of molecular simulation. The first design starts from a predicted structure of the molecular system, which we want to evaluate, for example, a protein–ligand complex predicted by a simple protein–ligand docking. I refer to this as the *evaluative design* (Figure 1.2). The research hypothesis is: Does the predicted structure represent real structure? The basic assumption behind this design is that an accurately predicted structure of the system, for example, an accurately modeled structure of the complex, is lower in free energy than an inaccurately predicted one. The system therefore tends to be stable in a simulation starting from an accurately modeled structure and tends to be unstable in a simulation starting from an inaccurate structure. The evaluative design can be represented by the study of Cavalli et al. [8]. This study was published in 2004, and simulated times are therefore significantly shorter (typically 2.5 ns) than those available today. Nevertheless, the same length of simulations can be used today with much higher throughput in terms of the number of tested compounds or their binding poses; therefore, the study is still highly actual. Docking of propidium into human acetylcholine esterase (Alzheimer disease target) by

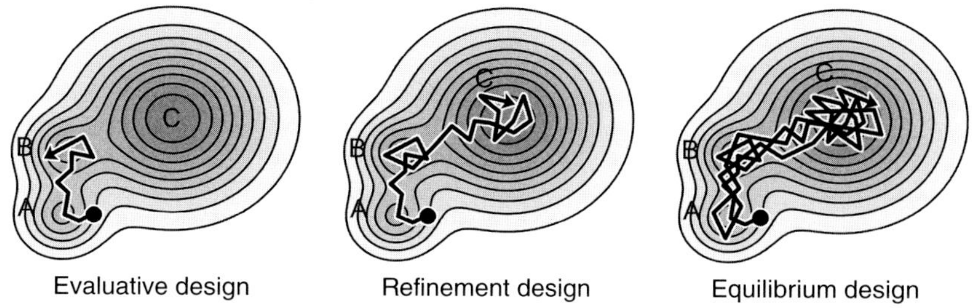

Figure 1.2 Schematic illustration of designs of biomolecular simulations. Horizontal dimensions correspond to coordinates of the system, and contours correspond to the free energy.

the program Dock resulted in the prediction of 36 possible binding poses (clusters of docked binding poses). Six of them were then subjected to 2.5-ns simulation. Evolution of these systems was analyzed in terms of root-mean-square deviation (RMSD). Binding poses with high stability in simulations were similar to experimentally determined binding poses for a homologous enzyme.

The second design is referred to as *refinement design* (Figure 1.2). It uses an assumption similar to the evaluative design, i.e. that molecular simulations tend to evolve from high-free energy states to low-free energy states. In the refinement design, it is hoped that the dynamics can drive the system from the predicted structure, even though incorrectly predicted, to global free energy minimum, the correct structure, or at least close to it. Naturally, shorter simulation times are necessary to demonstrate correctness or incorrectness of a model by the evaluative design. Longer simulation times are necessary to drive the system from the incorrect to the correct state by the refinement design. In the previous paragraph, I used the study of Cavalli et al. from 2004 [8] as an example of evaluative design. I can present the refinement design on the work published by the same author 11 years later [9]. They used unbiased simulation to predict the binding pose of picomolar inhibitor 4′-deaza-1′-aza-2′-deoxy-1′-(9-methylene)-immucillin-H in human purine nucleoside phosphorylase. They carried out 14 simulations (500 ns each) of the system containing the trimeric enzyme, 9 ligand molecules (to increase its concentration) placed outside the protein molecule, solvent, and ions. From these simulations, 11 evolved toward binding with a good agreement with the experimentally determined structure of the complex. RMSD from the experimentally determined structure of the complex dropped during these simulations from approximately 6 to 0.2–0.3 nm.

The last design introduced here is referred to as *equilibrium design* (Figure 1.2). In this design, we hope that the simulation is sufficiently long (or sampling is sufficiently enhanced) to explore all relevant free energy minima and to sample them according to their distribution in the real system. Naturally, the equilibrium design requires longest simulation times or highest sampling enhancement from all three simulation designs. As an example I can present the study by D.E. Shaw Research [10]. The authors simulated systems containing the protein FK506 binding protein (FKBP) with one of six fragment ligands, water, and ions. They carried out 10-μs simulations for each ligand. The dissociation constant of a complex can be calculated from its association kinetics as $K_D = k_{off}/k_{on}$. Weak binding (high K_D) together with reasonably fast binding kinetics therefore implies that unbinding is also sufficiently fast. For this reason, microsecond timescales were enough to observe multiple binding and unbinding events for millimolar ligands. The fragments identified by these simulations as relatively strong binders can be selected and combined into larger compounds with higher affinity in the manner of fragment-based drug design [11]. Fragment-based drug design and molecular dynamics simulation seem to be a good combination. Fragment-based design requires testing of a low number of weak ligands. This is good, since biomolecular simulations are computationally expensive. Reciprocally, weak binding enables to use molecular dynamics simulations in available timescales. Moreover, unlike some experimental methods of fragment-based drug design, molecular simulations provide binding pose prediction that can be used to combine fragments.

The three designs described are not without pitfalls. Most of these pitfalls are caused by limitations of simulated timescales. It is often difficult or impossible to simulate timescales long enough to destabilize the structure in the evaluation design, reach the global free energy minimum in the refinement design, or obtain the equilibrium distribution in the equilibrium design. This problem can be addressed by enhanced sampling techniques discussed later in this chapter.

The main problem of the evaluative design is that many correct structures of proteins or protein–ligand complexes are relatively flexible. It is therefore difficult to decide whether high flexibility (in terms of RMSD or ligand displacement) indicates a wrong model or not.

This is not the only problem of biomolecular simulation designs. Figure 1.2 shows three minima A, B, and C. Even an incorrect model A may be separated by a large energy barrier from the structure B and from the correct structure C. This can make A stable in the timescales of an evaluative simulation. Similarly, when a refinement simulation evolves from structure A to structure B and stays there, it is not guaranteed that B is the correct structure. Finally, even if a perfect equilibrium sampling is reached between A and B, the unexplored structure C can still exist.

1.2 Collective Variables and Trajectory Clustering

When the system is fully sampled and equilibrium distribution of states is achieved in the equilibrium design, it is possible to calculate a free energy profile of the studied system. For this it is necessary to classify states along the trajectory. In other than equilibrium design, it is necessary to monitor the progress of the simulation. These analyses often employ the concept of collective variables (CVs). A CV is a parameter that can be calculated from the atomic coordinates of the studied system. It can be calculated in every simulation snapshot, so it can be viewed as a function of time (i.e. $\mathbf{s}(t)$). It has to be chosen so that its value changes with the progress of the simulated process. Finally, CVs should be relevant to the experiment. There are simple CVs such as distances between atoms or geometrical centers or 3-point (valence) and 4-point (torsion) angles. RMSD from the reference structure often used to monitor stability during simulation is also an example of CV. Other more sophisticated CVs include those specifically developed for studying intermolecular interactions [12] and protein folding [13], principal component analysis (PCA), and related methods [14, 15], machine-learning-based CVs [16–18], and others.

Once values of some CV (or CVs) are calculated for all snapshots along the trajectory, it is possible to calculate one-dimensional (1D), 2D, or multidimensional histograms. These histograms can be expressed in energy units as estimated free energy surface:

$$F(\mathbf{s}) = -kT \ \log(P(\mathbf{s})) \tag{1.1}$$

where F is a (relative) free energy surface, $\mathbf{s}$ is a multidimensional vector of CVs, P is its probability distribution (histogram), k is the Boltzmann constant, and T is temperature. Calculation of an accurate free energy surface requires complete sampling of all relevant states of the simulated system. Its accuracy is addressed later.

A discontinuous alternative to CVs is trajectory clustering. Cluster analysis of simulation coordinates (usually preprocessed by fitting to a reference structure to remove translational and rotational motions) makes it possible to place each simulation snapshot to a certain cluster. Similar to CVs, it is possible to estimate free energy surface as

$$F_i = -kT \ \log(P_i) \tag{1.2}$$

where F_i and P_i are free energy and probability, respectively, of the ith cluster. Several clustering algorithms, general as well as tailored for molecular simulations, have been tested in the analysis of molecular simulations. Several packages and tools have been developed for trajectory clustering, namely, the gmx cluster from Gromacs package [19], Gromos tools [20], CPPTRAJ from Amber package [21], and stand-alone packages Bio3D (for R) [22], MDAnalysis (for Python) [23] and MDTraj (for Python) [24]. Many of these tools make it possible to analyze trajectories in terms of both clusters and CVs. Popular algorithms for trajectory clustering are nonhierarchical K-means [25], K-medoids [26], and Gromos algorithm by Daura and coworkers [27]. Hierarchical methods can be used for a tree-based representation of free energy surfaces [28], but they are often used together with nonhierarchical methods to reduce the number of clusters.

A key question in application of nonhierarchical clustering methods, such as the K-means or K-medoids algorithm, is the choice of the value of K – the number of clusters. This question is general, not related only to the analysis of molecular dynamics trajectories. Interestingly, the solution of this problem by "Clustering by fast search and find of density peaks," was developed by molecular scientists, namely, by Laio and Rodriguez, and became widely used in nonmolecular sciences [29]. This method automatically chooses a suitable number of clusters on the basis of density of points.

The result of a CV-based analysis of a molecular trajectory is a one-, two-, or multidimensional probability distribution or a free energy surface. The result of cluster analysis is a list of clusters with representative structures or

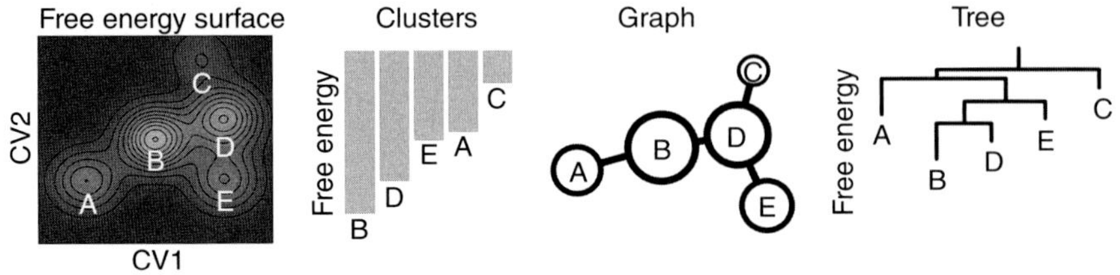

Figure 1.3 Alternative representations of free energy relationships (schematic views).

centroids and with corresponding probabilities or free energies. Alternatively, it is possible to represent clusters in graph-based or tree-based representations. The graph-based representation [30] shows free energy minima as graph nodes. Connection of two nodes by edges usually indicates that a transition between these nodes is kinetically favorable. The tree-based representation [28] shows free energy minima as nodes and transitions as branches. Finally, the Markov chain model is another elegant way to represent free energy surface. This approach is presented in Chapter 4. Different representations of free energy relationships in molecular systems are depicted in Figure 1.3.

1.3 Accuracy of Biomolecular Simulations

The predictive power of molecular simulations depends on their accuracy. The accuracy is influenced by accuracy of simulation methods, molecular mechanics (MM) potentials (also referred to as force fields, mathematical models used to calculate potential energy, and forces based on atomic coordinates) and on completeness of sampling of all relevant states of the studies system. Accuracy of simulation methods has been assured by the development of sophisticated thermostats, barostats, and electrostatics models in the past decades. Application of these models and methods nowadays avoids most simulation artifacts. Nowadays one of the few important method-related artifacts in biomolecular simulations is self-interaction in the periodic boundary condition because many researchers tend to minimize the simulated system to increase the simulation speed.

The second ingredient in biomolecular simulations is the MM force field. Exciting quantum mechanical (QM) or mixed QM/MM simulations are not discussed here. Force fields have been the subject of intensive development focused on their accuracy. Evaluation of the accuracy of molecular simulations is not trivial. For example, force field accuracy can be simply tested by comparing energies calculated by the force field and by an accurate reference method, for example, by some quantum chemistry method. However, this evaluation approach is tricky. Individual bonded and nonbonded force field terms differ significantly in their magnitudes. For example, a small change in a bond angle can be associated with high change of energy. In contrast, formation of non-covalent interactions is usually associated with much lower energy changes. Both these terms can contribute differently to overall accuracy of predictions made by molecular simulations. As a result, a force field that seems to be inaccurate by comparison of energies may be, in fact, pretty accurate in practical application and vice versa.

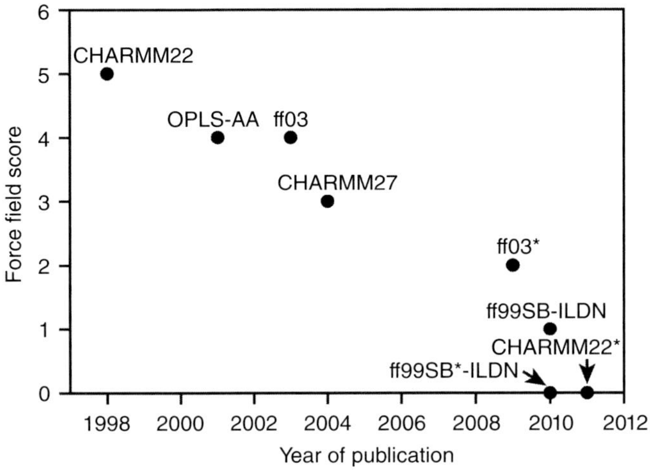

Figure 1.4 Improvement of force fields over time. Each force field was evaluated in three simulation tasks and awarded 0–2 points per task depending on the agreement with experimental data. Low scores indicate good agreement with experiments. Source: Taken from Lindorff-Larsen et al. [31], Creative Commons Attribution License.

The progress in accuracy of MM potential can be illustrated by Figure 1.4 from the work of Lindorff-Larsen et al. [31]. These authors systematically tested MM potentials for proteins developed from 1998 to 2011. These potentials were tested by very long simulations of a folded protein and protein folding process. Each potential was given a score from 0 to 6 depending on agreement of simulations with experimental data (0 for the best agreement). Figure 1.4 shows a steady progress in accuracy, with no major accuracy issues in two force fields published in 2010 and 2011. This progress fits well into the picture of the hype cycle with a slow but steady and systematic improvement in the field in the *Slope of Enlightenment.*

One problematic feature of most MM force fields is the absence of polarizability. Conventional force fields model atoms as charged points. In reality, charge distribution changes dynamically as a response to the environment. Polarizable versions of CHARMM [32] and special AMOEBA force fields [33] were developed.

Main developers of protein force field also develop compatible general force fields for ligands, either under the same title (such as OPLS3 [34]) or under an alternative name (General Amber Force Field, or GAFF [35], for the Amber force field series or CHARMM General Force Field, or CGenFF [36] for the CHARMM force field series). Some force field developers also provide online tools for generation of force field parameters for an uploaded compound in *mol2* or *pdb* format, such as CGenFF web [36] and SwissParm [37] for CHARMM or LigParGen [38] for OPLS-AA. A web-based graphical user interface for CHARMM, known as CHARMM-GUI [39], also provides this functionality, besides other features such as membrane setup for membrane protein simulations.

When comparing protein and general molecule force fields, the situation is not so bright for general molecules. General druglike molecules are much more diverse than 20 amino acid residues. Therefore, at least early force fields for general small molecules contained utterly erroneous terms, for example, wrong hybridization types. Evolution of general force fields corrected most of these errors; nevertheless, development of force fields applicable for all druglike molecules is challenging and these force fields are still inaccurate for many classes of compounds.

Systematic evaluation of force fields by comparison of energies calculated by force fields and by quantum chemistry methods for optimized structures [40] revealed that most problematic molecules are flexible multitorsion molecules or molecules with unusual conjugation of double bonds; however, the relationship between the structure and force field inaccuracy is not clear.

Also, modeling of interactions between a protein and a ligand can be affected by ligand force field inaccuracies or incompleteness. Widely discussed in this context is a halogen bond C—X· · ·A, where X is a halogen (usually other than fluorine) and A is a conventional hydrogen bond acceptor, typically oxygen [41]. It has been shown that this type of bond is common in recognition of druglike molecules [42]. Classical D—H· · ·A hydrogen bond is modeled by most force fields as a combination of electrostatic attraction and van der Waals repulsion between H and A. Since halogens in organic molecules as well as hydrogen bond acceptors are partially negatively charged, interactions between these two groups are rather repulsive. The origin of the halogen bond is in unusual distribution of electrons, referred to as sigma hole, in halogens bound in organic molecules. This phenomenon is usually not modeled by conventional force fields. A new atom type of halogen bond donor atoms has been introduced into the ligand version of optimized potentials for liquid simulations (OPLS) force field and this force field was successfully applied in computational prediction of binding free energies of HIV reverse transcriptase inhibitors [42].

It is possible to improve the accuracy of an individual modeled molecule instead of trying to improve the force field as a whole. Several approaches and tools have been developed for this purpose. For example, it is possible to improve CHARMM force fields using the Force Field Toolkit (ffTK) [43], which is a plugin for a popular visual molecular dynamics (VMD) viewer [44]. Another effort to improve accuracy of simulation of protein–ligand complexes is a repository of ligand parameters. At the website www.ligandbook.org it is possible to find parameters of approximately 3000 molecules in different force fields and for different program packages [45].

1.4 Sampling

The necessity to use femtosecond integration steps together with the fact that each atom in a condensed biomolecular system interacts with another approximately 5000 atoms (considering 2 nm as an interaction cutoff) causes biomolecular simulations that are extremely computationally expensive. The history of biomolecular simulations is tightly connected with availability of computer power. The 1980s were characterized by the introduction of personal computers and a boom in academic supercomputers. The 1990s were characterized by parallelization, i.e. joining of inexpensive computers to larger clusters. Other ideas, such as distributed computing projects using computer power of volunteers' PCs [46], use of GPUs [47], and special purpose computers [48], were introduced later. As a result of the progress in computer power, the first biomolecular simulations studied picosecond timescales, nanosecond simulations became available in the early 1990s, the first microsecond simulations were carried out in the late 1990s, and the milliseconds milestone was reached in around 2010. However, it must be kept in mind that these timescales were typically reached for small molecular systems on cutting-edge hardware and at the time of their publication were far from routine.

Sampling of a biomolecular system can be compared to the situation when a department store manager wants to evaluate the "affinity" of customers to different parts of the department store he manages. It is possible to choose a certain customer and follow his or her route through the department store. It is then possible to calculate probability for individual departments as a ratio of time spent in the department divided by the total time. It is also possible to use Eq. (1.1) to express this probability as free energy (temperature is discussed later). However, this approach, equivalent to the classical molecular dynamics simulation, is inefficient because the customer may stay for a long time in some department and it can take a very long time to sample all departments.

An alternative in the molecular world to running very long simulations is application of enhanced sampling techniques. These techniques were designed to provide equivalent information as several orders of magnitude longer conventional (unenhanced) simulations. There is a group of enhanced sampling techniques that use a bias force or bias potential to accelerate the studied process. Other methods use elevated temperature or other principles. Several hybrid sampling enhancement methods combining multiple principles have been also developed.

Simulations using a bias potential or a bias force, further referred to as biased simulations, include the umbrella sampling method [49], metadynamics [50], steered molecular dynamics [51], local elevation [52], local elevation umbrella sampling [53], adaptively biased molecular dynamics [54], variationally enhanced sampling [55], flying Gaussian method [56], and others. These methods can be divided into two groups depending on whether the bias potential or force is static or dynamic.

The method known as umbrella sampling uses a static bias potential. In the analogy to the department store presented, it is possible to represent it by organizing sales in some unattractive departments and hiking prices in attractive ones. This will make sampling much more efficient. Provided that it is possible to quantify the effect of sales and price elevations, it is possible to calculate the equilibrium probabilities (probabilities under condition of regular prices) from sampling and from price modifications.

Umbrella sampling introduced by Torrie and Valleau in 1977 [49], originally in connection with the Monte Carlo method, represents methods with a static bias potential (some scientists use the term umbrella sampling as a synonym for any simulation with a static bias potential). In the most common design, it is used to enhance sampling along certain CVs (e.g. protein–ligand distance) to predict the corresponding free energy surface. Umbrella sampling is done by running a series of simulations, each with a bias potential $k(s - S_i)^2/2$, where k determines strength of the bias potential, s is the CV, and S_i (for ith simulation) ranges from the initial S_0 and the final state S_N of the simulated process (e.g. bound and unbound state) and is usually uniformly distributed. This potential forces the ligand to sample all states along the binding pathway. Free energy surface can be calculated by, for example, weighted histogram analysis method (WHAM) [57] or by the reweighting formula [58–60]. These methods are explained later; so, briefly, it is possible to calculate unbiased sampling from the knowledge of the biased sampling and the bias potential. An example of umbrella sampling in drug discovery is the study of Bennion et al. [61]. They simulated permeation of drug molecules through the membrane. They used a coordinate perpendicular to the membrane as the CV. This CV was ranging from 0 to 10 nm in 0.1-nm windows (i.e. 100 simulations). They correctly ranked tested compounds as impermeable; low, medium or highly permeable; and in a good quantitative agreement with parallel artificial membrane permeability assays (PAMPA).

Biased simulation with a time-dependent bias potential can be represented by the metadynamics method [50]. In the department store example, it is possible to carry out metadynamics using a device that, at regular intervals, releases some stinky compound. Such a device must be installed onto a customer's shopping basket. If the customer stays for a long time in some department, the device causes the stinky compound to accumulate there. This forces the customer to escape the department and to visit other departments. This makes sampling much more efficient. The free energy surface can be estimated from the amount of the stinky compound, i.e. deep minima require a high amount of the stinky compound.

In the molecular world, that application of metadynamics starts with choice of CVs, typically two. The system is then simulated by conventional simulation for 1 or 2 ps. Then, values of CVs are calculated and recorded as $\mathbf{S}_1$. From this point, a bias potential in the form of a Gaussian hill centered in $\mathbf{S}_1$ is added to the simulated system. The system evolves for another 1 or 2 ps, then another hill is added to $\mathbf{S}_2$, and so forth. The bias potential accumulates in certain free energy minima until this minimum is flooded and the simulation can escape it. This allows for complete sampling of the free energy surface. The free energy surface can be estimated as the negative value of the bias potential [50, 62, 63], because the deeper the free energy minimum, the more hills it needs to flood.

The accuracy of metadynamics (and other biased simulations) is critically dependent on the choice of CVs. Ideally, the CVs must account for all slow degrees of freedom in the simulated system. Existence of some slow degree of freedom not addressed by CVs may cause a significant drop of accuracy. Imagine a simulation of protein–ligand interaction. Naturally, one of the CVs for protein–ligand interaction modeling can be the protein–ligand distance to accelerate binding and unbinding. The second CV should address other slow motions. Imagine the situation that the entrance to the binding site may be occasionally blocked by some amino acid side chain. If the site is blocked, the ligand cannot move inside or outside the binding site. This leads to a huge overestimation or underestimation of the predicted binding free energy.

An ideal solution to this problem would be a second CV that fully addresses side chain motions. It is difficult to design such CVs due to the complexity of the molecular system because there could be multiple problematic side chains or other degrees of freedom. Instead, most researchers rely on sampling. Simulations in timescales of hundreds of nanoseconds or microseconds are usually not long enough to simulate binding and unbinding events, but it is often sufficient to sample such problematic degrees of freedom once binding and unbinding is enhanced.

However, in classical metadynamics, this may cause the problem of hysteresis in the predicted free energy surface due to altering overestimation of the bound and unbound state. This problem can be addressed by well-tempered metadynamics [64]. Well-tempered metadynamics is metadynamics with variable hill heights. The height set by user is scaled by $\exp(-V_{\text{bias}}(s)/k\Delta T)$, where ΔT is the difference between sampling temperature and the temperature of the simulation. Classical metadynamics corresponds to ΔT = infinity and unbiased simulation to $\Delta T = 0$. Flooding of the free energy surface in well-tempered metadynamics slows down until its convergence. The free energy can be calculated as a negative value of the bias potential scaled by $(T + \Delta T)/\Delta T$. The fact that the biasing slows down reduces the hysteresis and increases the accuracy. For this reason, well-tempered metadynamics replaced classical metadynamics in the past decade. Well-tempered metadynamics, together with a funnel method (described later), was used to simulate binding and unbinding and to accurately predict binding free energies for ligands of GPCR, including cannabinoid CB_1 [65], β_2 adrenergic [66], chemokine CXCR3 [67], and vasopressin [68] receptors.

In the previous paragraph I assumed that a single CV cannot address all slow degrees of freedom. However, it is possible to address many slow degrees of freedom by multiple CVs. It has been shown that metadynamics with more than two or three CVs is not efficient [69]. A special variant called bias exchange metadynamics [70] was developed to run metadynamics with multiple CVs. The system is simulated in multiple (N) replicas (usually one per processor CPU), where N is the number of CVs. Metadynamics biases a single CV in each replica (or there could be some unbiased replicas). Occasionally (every few picoseconds) coordinates are exchanged on the basis of an exchange criteria calculated from potential energies and bias potentials in each system. This makes it possible to predict a one-dimensional free energy surface for each CV. Calculation of a multidimensional free energy surface requires a special reweighting procedure [71]. The bias exchange metadynamics has been applied in predicting the binding mode of the compound SSR128129E to fibroblast growth factor receptor [72].

Sampling can be also enhanced by elevated temperature. In the department store example, it is possible to find an analogy between temperature and the music played in the store. It has been shown experimentally that a tempo of music in a supermarket influences the pace of shoppers [73]. It is therefore possible to enhance sampling by playing a fast-paced music. However, by this we would obtain a different free energy surface from the normal music played in the department store. For example, fast moving customers would prefer easy-to-find departments and shelves and would ignore difficult-to-find ones.

Similarly, in a high-temperature molecular simulation, we would obtain a free energy surface different from the normal temperature. Such a free energy surface is usually not interesting. For example, the "native" structure of a protein at a temperature higher than its melting temperature is the unfolded structure.

There is a method that makes it possible to use elevated temperature to enhance sampling and at the same time to obtain normal-temperature free energy surfaces. This method is known as parallel tempering and belongs to the family of replica exchange methods. In the department store analogy, it would be necessary to distribute radios with headphones to multiple customers. Customers would listen to music differing in the tempo. In periodic intervals, their music would be exchanged based on the special criteria. Normal-tempo free energy surface would be obtained by the analysis of trajectories of only those customers who listen to the normal-tempo music.

In a molecular system, it is possible to run parallel tempering by simulation of multiple replicas of the system at different temperatures. These temperatures are chosen so that the lowest is slightly lower than the normal temperature and the highest is high enough to significantly enhance sampling. Replica exchange attempts are evaluated usually every 1 or 2 ps. The potential energy of the ith replica is compared with the potential energy of the $i+1$th replica. If the potential energy of colder replica is lower, the coordinates of replicas are swapped. If not, the Metropolis criterion is calculated as $\exp((E_i - E_{i+1})(1/kT_i - 1/kT_{i+1}))$. If a random number (with a uniform distribution from 0 to 1) is lower than the Metropolis criterion, the coordinates in the replicas are also swapped. If the simulated system adopts an unfavorable (high-energy) structure, it tends to be exchanged for higher temperature replicas and to climb on the temperature ladder. There it can adopt some nice structure with low energy. Once this happens, it would tend to descend on the temperature ladder. Structures sampled at the temperature of interest can be analyzed by Eq. (1.1) to obtain the corresponding free energy surface.

Parallel tempering is a very powerful method for folding of mini-proteins. It is particularly suitable for simulation of small systems because large systems require a huge number of replicas to reach reasonable exchange rates (with a low number of exchanges, the method would behave as a series of independent unbiased simulations). I see the highest potential of parallel tempering in drug design in combination with other sampling enhancement methods. Parallel tempering in combination with metadynamics [74] has been applied to compare wild-type and oncogenic mutants of the epidermal growth factor receptor [75].

An interesting multiple replica method that enhances sampling by cloning and merging replicas is WExplore [76]. This method simulates the system in a constant number of replicas. When two or more replicas sample similar states, they are merged. If a single replica samples some distant state, it is cloned. The free energy method can be obtained from sampling and from cloning and merging history. This method was successfully applied in modeling of the interaction between 1-(1-propanoylpiperidin-4-yl)-3-[4-(trifluoromethoxy)phenyl]urea (TPPU) and its enzyme target epoxide hydrolase [77].

1.5 Binding Free Energy

So far I have presented methods that can be used for general prediction of free energy relationships. Here I present special issues of modeling of protein–ligand

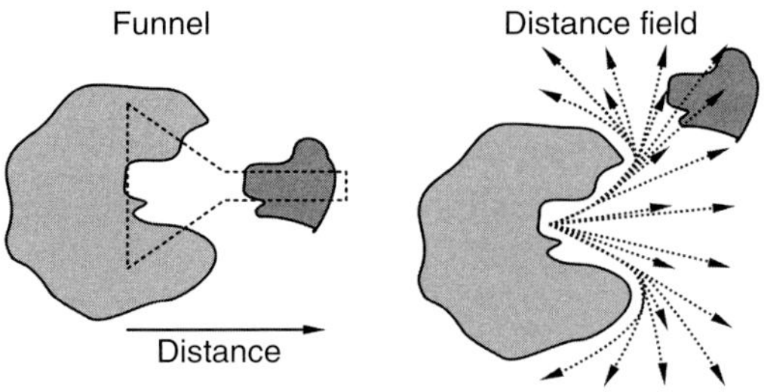

Figure 1.5 Schematic representation of funnel techniques and distance field CV.

interactions and molecular simulation methods especially suited for modeling molecular recognition. An important issue of such simulations is the fact that the entrance into the binding site usually represents only a small part of the overall protein surface. It may seem like a good idea to enhance simulation of protein–ligand binding by biasing the distance between the binding site and the ligand. However, it may happen that the ligand chooses a wrong entrance into the binding site. This significantly slows down the simulation and makes the predicted free energy surface inaccurate. There are two major approaches developed to address this problem. First, is the abovementioned application of a funnel (Figure 1.5), first introduced as funnel metadynamics [78]. The ligand is restrained into a funnel-shaped space outside the binding site by means of an artificial potential. This prevents the ligand from exploring other entrances into the binding site. The result of such a simulation is the free energy difference between the bound state and the state when the ligand resides at the tip of the funnel. A simple correction can be applied on this value to obtain the absolute binding free energy, considering ligand concentration, volume of the system, and the volume of the funnel [78]. The method has been successfully applied in G protein–coupled receptor (GPCR) research [65–68].

An alternative to a funnel is a distance field (Figure 1.5) [79]. Instead of the Euclidean distance between the binding site and the ligand, it is possible to measure the shortest path from the binding site and the ligand without their collisions. At the beginning, a three-dimensional grid is constructed in the simulation box. For each point on the grid (except those inside the protein) a collision-free distance between the binding site and the ligand is calculated. Next, in the simulation it is possible to estimate this distance from grid points close to the ligand position. This approach has been applied together with Hamiltonian replica exchange simulation to study binding of 14-3-3ζ domains with phosphopeptides [80].

The so-called Alchemistic methods can be used to predict binding free energy without simulating the binding process. The term “Alchemistic” indicates that some elements change into other elements, similarly to medieval alchemists attempting to produce gold from inexpensive metals. These methods typically do not provide absolute binding free energies. Instead, they make it possible to predict an outcome of a modification of the ligand, for example, change of hydrogen to halogen, addition of a small group, or other minor modifications. More complex modifications can be studied by combination of multiple Alchemistic simulations.

Alchemistic methods such as free energy perturbation, thermodynamic integration, or Bennett acceptance ratio method use a series of nonphysical processes to study a physical process. For example, it is possible to predict the outcome of a replacement of a hydrogen atom in a ligand by chlorine, i.e. the difference between binding free energy of the ligand L–Cl and the ligand L–H. First, the complex protein – L–H is simulated and its force field parameters are gradually changed (linearly or nonlinearly) into parameters of L–Cl; that is, the increase in mass from 1 to 35.45, the increase in bond length from ~1 to ~1.8 Å, etc. The response of energy of the system is monitored. This response makes it possible to predict the free energy difference of a nonphysical (experimentally unfeasible) process of changing of H to Cl on a protein-bound ligand. In addition, it is possible to do the same calculation for an unbound ligand and to construct a thermodynamic cycle comprising (i) binding of L–H to protein, (ii) change of bound L–H to L–Cl, (iii) unbinding of L–Cl, and (iv) change of unbound L–Cl to L–H. Despite the fact that two of these processes are nonphysical, the overall free energy change of the thermodynamic cycle is zero. It is therefore possible to predict $\Delta\Delta G$ (the difference of binding ΔG of L–Cl versus L–H). This can give an answer to whether the change of H to Cl strengthens or weakens the binding to the protein. A good example of application of Alchemistic simulation is the campaign leading from a weakly binding docking hit to a picomolar inhibitor of HIV integrase by Jorgensen's group [81–84].

Finally, several methods have been developed to predict binding free energies from molecular simulations without simulating the binding process. These methods assume that the affinity is determined by the strength of non-covalent intermolecular interactions. The ligand is simulated as a complex in the target and, in parallel, in a solvent. Non-covalent interactions are monitored in both simulations and they are used to predict binding free energy and the effect of ligand desolvation. Examples are linear interaction energy [85] and methods combining MM with implicit solvent models (molecular mechanics/Poisson–Boltzmann surface area (MM/PBSA) and molecular mechanics/generalized born surface area (MM/GBSA)) [86]. Wright et al. used the MM/PBSA and MM/GBSA method to predict binding free energies of nine HIV-protease inhibitors approved for HIV treatment [87]. This study is an example of replications in simulations. The authors used short simulations (4 ns) done in 50 independent replicates for each molecule to obtain a robust model with a good agreement with experiment.

1.6 Convergence of Free Energy Estimates

Experimental researchers use replication to assess and improve accuracy of their predictions. In the spirit of the central limit theorem, measurements done in multiple replicates can be averaged to estimate the mean value. Standard deviation or standard error of the mean can be used to assess the accuracy. Measurements done in replicates are also used to statistically test research hypotheses.

In principle, replications can also be used in biomolecular simulations; however, most researchers prefer prolonging their simulations rather than replicating them. It is possible to use experiment, such as nuclear magnetic resonance (NMR) measurement, to determine a dissociation constant of a protein–ligand complex. By a properly designed NMR experiment it is possible to determine concentrations of the free ligand, free protein, and the protein–ligand complex (or at least ratios of their concentrations). In the other words, it is possible to determine the number of molecules in different states (free protein, free ligand, and free complex) in the studied system at a certain moment.

Instead, biomolecular simulations study a single biomolecule as a sample of the whole biomolecular system. They calculate how long the single studied system spends in different forms. A dissociation constant of the protein–ligand complex can be calculated as the ratio of times spent in the ligand-bound and the unbound form. Both concepts, concentration and time ratios, can be generalized in the way that both quantities are proportional to probabilities of states, i.e. dissociation constant can be determined as the ratio of equilibrium probabilities of different forms of the studied systems.

The main reason why replication is rarely used in biomolecular simulation is the fact that it is difficult to generate independent starting conditions. Basic molecular dynamics simulation is a deterministic method. Running two simulations from the same starting coordinates with the same starting velocity vectors should give identical trajectories. Random initialization by different starting velocities usually does not provide the satisfactory level of independence. The second reason is that many biologically interesting quantities, such as dissociation constants, require sampling of multiple transitions between the relevant states of the system.

Nevertheless, errors of some quantities of the molecular system can be calculated by a "standard" way used by experimental scientists who average results of independent experiments. These quantities include temperature, pressure, membrane surface tension, number of non-covalent interactions, experiment-related properties (e.g. fluorescence resonance energy transfer (FRET), pair and radial distribution functions or NMR quantities), molecular surface, forces acting on selected molecular degree of freedom, and others. Calculation of these properties requires that the system exist only in one form whose property we want to calculate or the transitions between forms are rapid enough.

Most interesting from the point of view of drug design is prediction of thermodynamic and kinetic quantities, especially association/dissociation constants and binding/unbinding rates of protein–ligand complexes. Calculation of these quantities requires sampling of multiple transitions between the forms of the molecular system. The equilibrium constant of the transition from form A to B can be predicted as the time spent in form B divided by the time spent in form A. A 1-μs simulation with a single transition from A to B at ~0.5 μs would give free energy difference estimate around 0 kJ mol^{-1} (i.e. $-kT\log(0.5/0.5)$). However, it is possible that the system would have stayed in state B for another 100 μs, so the real free energy difference is approximately −13 kJ mol^{-1} (i.e. $-kT\log(100.5/0.5)$). On the other hand, a simulation with many A to B and B to A transitions provides good confidence that the calculated binding free energy is accurate, at least in terms of sampling.

This phenomenon can be addressed by a block analysis [88–91]. Simulation trajectories are separated into M equivalently sized blocks with $n = 1$ to N, where

$M = N/n$, N is the number of samples in the trajectory, and n is the number of samples in a block. The calculated value, for example, the population of the state B P_B, is averaged in each ith block yielding $\langle P_B \rangle_i$. Next, standard deviation and standard error (block standard error, BSE) is estimated for each value of n from $\langle P_B \rangle_i$ as

$$\sigma(P, n) = \sqrt{\frac{\sum (\langle P_B \rangle_i - \langle\langle P_B \rangle\rangle)^2}{M-1}} \tag{1.3}$$

$$\mathrm{BSE}(P, n) = \frac{\sigma(P, n)}{\sqrt{M}} \tag{1.4}$$

where $\langle\langle\ \rangle\rangle$ is average across the block size n. This procedure can be demonstrated on sampling of a model one-dimensional energy profile with two minima at CV equal to approximately 3 (minimum A) and 7 (minimum B). These minima have the same depth, so the free energy difference is 0 and equilibrium constant is 1. It was sampled by the Monte Carlo method with CV profiles depicted in Figure 1.6. The top profile shows sampling at low temperature with few A to B and B to A transitions. A block analysis with $n = 1$–100 gives a divergent estimate of BSE(P, n). The value for $n = 1$ corresponds to classical standard error of the mean calculated for independent samples in many fields of experimental sciences. This value is strongly underestimated due to autocorrelation of values of the CV in the trajectory. If the system is in state A, it is highly probable that it will be in state A in the next step or 10 steps later. The value of P_B was calculated as 0.503 (equilibrium constant 1.01). The block analysis shows that the value of BSE(P, n) rises for $n = 1$–100 and is not convergent (extending n does not help; data not shown). It would be therefore necessary to prolong the simulation in order to obtain a convergent estimate of standard error. The situation is different in the simulation at a higher temperature depicted in the bottom profile. The P_B was calculated as 0.453 and the number of A to B and B to A transitions was higher. The result of the higher number of transitions is a convergent profile of BSE. Highest BSE value (0.08) can be used as a standard error estimate, i.e. P_B is equal to 0.45 ± 0.08 (mean $\pm$ BSE).

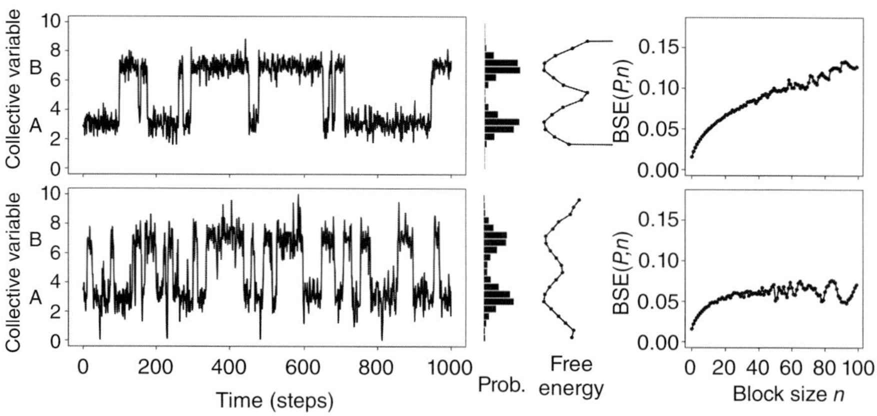

Figure 1.6 Demonstration of block error analysis on sampling of a model energy profile.

A similar analysis can be applied on biased simulations. The easiest free energy estimation can be done from metadynamics simulations. In classical metadynamics [50, 69], it is possible to use a negative value of the bias potential as an estimate of the free energy surface [62, 63]. In well-tempered metadynamics, the free energy can be predicted as a negative value of the bias potential scaled by a constant factor $((T + \Delta T)/\Delta T)$. However, this approach does not provide an estimate of its accuracy. Simple averaging of free energy differences $\Delta G_{\mathrm{A}\to\mathrm{B}}$ along the simulation suffers the problem of autocorrelation in simulation trajectories. It is possible to plot the profile $\Delta G_{\mathrm{A}\to\mathrm{B}}$ along a metadynamics simulation with a nice convergence, but the converged $\Delta G_{\mathrm{A}\to\mathrm{B}}$ can be completely wrong due to a low number of A to B and B to A transitions.

The problem can be addressed by block analysis also in biased simulations [92]. As an alternative to calculating free energy surface from the bias potential, it is possible to calculate it from the combination of the bias potential and sampling. Equilibrium (unbiased) probabilities can be predicted from biased sampling by reweighting formula [58–60]:

$$P(\mathbf{S}) = \frac{\sum \delta(\mathbf{s}(t) - \mathbf{S}) \exp(+V_{\mathrm{bias}}(t)/kT)}{\sum \exp(+V_{\mathrm{bias}}(t)/kT)} \tag{1.5}$$

where $\mathbf{S}$ is a multidimensional vector of CVs and $\mathbf{s}(t)$ is the vector of CVs sampled at time t. In other words, equilibrium probabilities from biased simulations are calculated in the same way as from unbiased except that they are scaled by the factor $\exp(+V_{\mathrm{bias}}(t)/kT)$. This is a generalization of Eq. (1.1), where $\exp(+V_{\mathrm{bias}}(t)/kT) = 1$ in the absence of the biased potential. Similarly, the bias potential in a non-well-tempered metadynamics is constructed to make sampling of all values of $\mathbf{S}$ with the same probability, i.e. $\delta(\mathbf{s}(t) - \mathbf{S})$ is constant. This is true only if $P(\mathbf{S}) = \exp(-F(\mathbf{S})/kT)$, i.e. $F = -V_{\mathrm{bias}}$. This idea can be extended for well-tempered metadynamics. Prediction of $P(\mathbf{S})$ using reweighting formula makes it possible to analyze the data by block analysis to predict BSE.

The problem of reweighting formula is that it should be used together with a static (time-independent) bias potential. The bias potential of metadynamics is time-dependent. With caution it is possible to use reweighting formula and considering the metadynamics bias potential as quasi-static. Alternatively, it is possible to apply corrections developed by Tiwary and Parrinello [93].

Another possibility to predict the free energy surface is application of WHAM. It should be noted that some researchers use the term umbrella sampling for any biased simulation with a static bias potential. The same researchers would call the reweighting formula in Eq. (1.5) as WHAM. However, most scientists use the term umbrella sampling for biased simulations carried out in multiple windows, where different bias potentials are used in each window and all windows cover the whole range of the CV [57]. The pair of WHAM equations

$$P(\mathbf{S}) = \frac{\sum \delta(\mathbf{s}(t) - \mathbf{S})}{\sum \exp([F_i - V_{\mathrm{bias}}(t)]/kT)} \tag{1.6}$$

$$F_i = -kT \log \sum [P(\mathbf{S}) \exp(-V_{\mathrm{bias}}(\mathbf{S})/kT)] \tag{1.7}$$

is solved iteratively to self-consistency to obtain $P(\mathbf{S})$ and F_i. It can be intuitively explained that the free energy is calculated in small fragments using reweighting formula for each window. Simultaneously, free energy shifts F_i of these fragments are calculated. The whole free energy surface is reconstructed by merging fragments of the free energy surface shifted by F_i. Block analysis has also been applied in WHAM [94].

Prediction of kinetics of drug binding and unbinding has become attractive for drug design [95]. Partially, this is because it is easier to sample a single drug binding or single unbinding event compared to sampling of numerous binding and unbinding events, so researchers make a virtue of necessity. Beside this, numerous experimental results show that binding or unbinding kinetics can be equivalently or even more useful in drug design compared to thermodynamics. Prediction of kinetics from unbiased and biased simulations and assessment of the accuracy of these predictions is not as developed as for thermodynamics, but there are several examples of extraction of kinetic information from unbiased simulations [96] or metadynamics [97, 98]. The Markov chain model made from biomolecular simulations is presented in Chapter 4.

1.7 Future Outlook

The question is: What are the limits in the predictive power of biomolecular simulations? In 1998, Xie and coworkers measured the rate of enzymatic reaction at the single-molecule level using a special fluorescence microscopy technique [99]. They found significant heterogeneity in kinetic parameters in individual enzyme molecules (standard deviation of 70% for k_{cat}). These heterogeneities were explained by conformational heterogeneities. A similar heterogeneity was observed in signaling by the β_2 adrenergic receptor [100]. In other words, individual enzyme or receptor molecules are highly heterogeneous in their ligand-binding or catalytic properties not only *in vivo* but also *in vitro*. The experimentally measured kinetic or thermodynamic parameters represent an averaged value across all molecules in the system. Biomolecular simulations study a single molecule. It is therefore natural that predicted parameters of biomolecular simulation may differ from experimental results due to the heterogeneity in target molecules. This problem can be, in principle, solved by replication of simulations or by enhancement of sampling of degrees of freedom associated with such heterogeneity, but none of these approaches is simple.

At the beginning of this chapter I introduced three designs of molecular simulations: evaluative, refinement, and equilibrium. The examples of studies presented later in this chapter follow almost always the equilibrium design. This can be explained by the fact that biomolecular simulations in drug design are mostly the domain of physical chemists. A typical physical chemist approaches the problem from the bottom-up perspective. This starts with a precise development and tuning of simulation methods, force fields, and sampling enhancement tools, walking stepwise from simple systems to complicated ones. Other approaches in computational drug design such as protein–ligand docking or pharmacophore modeling are the domain of chemoinformaticians. Chemoinformaticians are more open to heuristic approaches. They typically train a model on a training set and validate it on a validation set of data. If the model helps distinguish between good and bad ligands with statistical significance, it is acceptable for drug design, no matter how solid is its physical basis. I believe that more and more researchers will use biomolecular simulations in such chemoinformatics spirit. Instead of tuning methods on low number of systems, they will test practical impacts on large numbers of systems. This is the area where evaluative and refinement design of molecular simulations can be used.

Predictive power of biomolecular simulations is determined by availability of computer power, partially because longer computational times provide better sampling, partially because long computational times make it possible to identify and correct other limiting factors of biomolecular simulations, such as force field inaccuracies. In the area of DNA sequencing, there has been an enormous jump in performance due to introduction of parallel sequencing machines. The question is whether we can expect a similar jump in biomolecular simulations or whether we can expect evolution rather than revolution. Two emerging technologies have a certain potential to cause such a jump in performance of biomolecular simulations; these technologies are machine learning and quantum computing.

References

1 Metropolis, N., Rosenbluth, A.W., Rosenbluth, M.N. et al. (1953). Equation of state calculations by fast computing machines. *J. Chem. Phys.* 21: 1087–1092.

2 Alder, B.J. and Wainwright, T.E. (1957). Phase transition for a hard sphere system. *J. Chem. Phys.* 27: 1208–1209.

3 Rahman, A. (1964). Correlations in the motion of atoms in liquid argon. *Phys. Rev.* 136: A405–A410.

4 Rahman, A. and Stillinger, F.H. (1971). Molecular dynamics study of liquid water. *J. Chem. Phys.* 55: 3336–3359.

5 McCammon, J.A., Gelin, B.R., and Karplus, M. (1977). Dynamics of folded proteins. *Nature* 267: 585–590.

6 Karplus, M. (2013). Development of multiscale models for complex chemical systems from $H+H_2$ to biomolecules. Nobel Prize lecture.

7 www.gartner.com.

8 Cavalli, A., Bottegoni, G., Raco, C. et al. (2014). A computational study of the binding of propidium to the peripheral anionic site of human acetylcholinesterase. *J. Med. Chem.* 47 (16): 3991–3999.

9 Decherchi, S., Berteotti, A., Bottegoni, G. et al. (2015). The ligand binding mechanism to purine nucleoside phosphorylase elucidated via molecular dynamics and machine learning. *Nat. Commun.* 6: 6155.

10 Pan, A.C., Xu, H., Palpant, T., and Shaw, D.E. (2017). Quantitative characterization of the binding and unbinding of millimolar drug fragments with molecular dynamics simulations. *J. Chem. Theory Comput.* 13 (7): 3372–3377.

11 Erlanson, D.A. and Jahnke, W. (eds.) (2016). *Fragment-Based Drug Discovery: Lessons and Outlook*. Wiley VCH. ISBN: 978-3-527-33775-0.

12 Iannuzzi, M., Laio, A., and Parrinello, M. (2003). Efficient exploration of reactive potential energy surfaces using Car-Parrinello molecular dynamics. *Phys. Rev. Lett.* 90: 238302.

13 Pietrucci, F. and Laio, A. (2009). A collective variable for the efficient exploration of protein beta-sheet structures: application to SH3 and GB1. *J. Chem. Theory Comput.* 5 (9): 2197–2201.

14 Amadei, A., Linssen, A.B., and Berendsen, H.J. (1993). Essential dynamics of proteins. *Proteins Struct. Funct. Bioinf.* 17 (4): 412–425.

15 Spiwok, V., Lipovová, P., and Králová, B. (2007). Metadynamics in essential coordinates: free energy simulation of conformational changes. *J. Phys. Chem. B* 111 (12): 3073–3076.

16 Das, P., Moll, M., Stamati, H. et al. (2006). Low-dimensional, free-energy landscapes of protein-folding reactions by nonlinear dimensionality reduction. *Proc. Natl. Acad. Sci. U.S.A.* 103 (26): 9885–9890.

17 Spiwok, V. and Králová, B. (2011). Metadynamics in the conformational space nonlinearly dimensionally reduced by isomap. *J. Chem. Phys.* 135 (22): 224504.

18 Ceriotti, M., Tribello, G.A., and Parrinello, M. (2011). Simplifying the representation of complex free-energy landscapes using sketch-map. *Proc. Natl. Acad. Sci. U.S.A.* 108 (32): 13023–13028.

19 Abraham, M.J., Murtola, T., Schulz, R. et al. (2015). GROMACS: high performance molecular simulations through multi-level parallelism from laptops to supercomputers. *SoftwareX* 1–2: 19–25.

20 Christen, M., Hünenberger, P.H., Bakowies, D. et al. (2005). The GROMOS software for biomolecular simulation: GROMOS05. *J. Comput. Chem.* 26: 1719–1751.

21 Salomon-Ferrer, R., Case, D.A., and Walker, R.C. (2013). An overview of the Amber biomolecular simulation package. *WIREs Comput. Mol. Sci.* 3: 198–210.

22 Skjærven, L., Yao, X.Q., Scarabelli, G., and Grant, B.J. (2014). Integrating protein structural dynamics and evolutionary analysis with Bio3D. *BMC Bioinf.* 15: 399.

23 Michaud-Agrawal, N., Denning, E.J., Woolf, T.B., and Beckstein, O. (2011). MDAnalysis: a toolkit for the analysis of molecular dynamics simulations. *J. Comput. Chem.* 32: 2319–2327.

24 McGibbon, R.T., Beauchamp, K.A., Harrigan, M.P. et al. A modern open library for the analysis of molecular dynamics trajectories. *Biophys. J.* 109 (8): 1528–1532.

25 Lloyd, S.P. (1982). Least square quantization in PCM. *IEEE Trans. Inf. Theory* 28 (2): 129–137.

26 Kaufman, L. and Rousseeuw, P.J. (1987). Clustering by means of medoids. In: *Statistical Data Analysis Based on the L1-Norm and Related Methods* (ed. Y. Dodge), 405–416. North-Holland: Springer.

27 Daura, X., Gademann, K., Jaun, B. et al. (1999). Peptide folding: when simulation meets experiment. *Angew. Chem. Int. Ed.* 38: 236–240.

28 Wales, D. (2004). *Energy Landscapes: Applications to Clusters, Biomolecules and Glasses.* Cambridge University Press.
29 Rodriguez, A. and Laio, A. (2014). Clustering by fast search and find of density peaks. *Science* 344 (6191): 1492–1496.
30 Gfeller, D., De Los Rios, P., Caflisch, A., and Rao, F. (2007). Complex network analysis of free-energy landscapes. *Proc. Natl. Acad. Sci. U.S.A.* 104 (6): 1817–1822.
31 Lindorff-Larsen, K., Maragakis, P., Piana, S. et al. (2012). Systematic validation of protein force fields against experimental data. *PLoS One* 7 (2): e32131.
32 Vanommeslaeghe, K. and MacKerell, A.D. Jr., (2015). CHARMM additive and polarizable force fields for biophysics and computer-aided drug design. *Biochim. Biophys. Acta* 1850 (5): 861–871.
33 Shi, Y., Xia, Z., Zhang, J.H. et al. (2013). Polarizable atomic multipole-based AMOEBA force field for proteins. *J. Chem. Theory Comput.* 9 (9): 4046–4063.
34 Harder, E., Damm, W., Maple, J. et al. (2015). OPLS3: a force field providing broad coverage of drug-like small molecules and proteins. *J. Chem. Theory Comput.* 12 (1): 281–296.
35 Wang, J., Wolf, R.M., Caldwell, J.W. et al. (2004). Development and testing of a general AMBER force field. *J. Comput. Chem.* 25: 1157–1174.
36 Vanommeslaeghe, K., Hatcher, E., Acharya, C. et al. (2010). CHARMM general force field: a force field for drug-like molecules compatible with the CHARMM all-atom additive biological force field. *J. Comput. Chem.* 31: 671–690.
37 Zoete, V., Cuendet, M.A., Grosdidier, A., and Michielin, O. (2011). SwissParam, a fast force field generation tool for small organic molecules. *J. Comput. Chem.* 32 (11): 2359–2368.
38 Dodda, L.S., Cabeza de Vaca, I., Tirado-Rives, J., and Jorgensen, W.L. (2017). LigParGen web server: an automatic OPLS-AA parameter generator for organic ligands. *Nucleic Acids Res.* 45 (W1): W331–W336.
39 Jo, S., Kim, T., Iyer, V.G., and Im, W. (2008). CHARMM-GUI: a web-based graphical user interface for CHARMM. *J. Comput. Chem.* 29 (11): 1859–1865.
40 Kanal, I.Y., Keith, J.A., and Hutchison, G.R. (2018). A sobering assessment of small-molecule force field methods for low energy conformer predictions. *Int. J. Quantum Chem.* 118: e25512.
41 Hobza, P. and Muller-Dethlefs, K. (2009). *Non-covalent Interactions: Theory and Experiment*. RSC Publishing. ISBN: 978-1-84755-853-4.
42 Jorgensen, W.L. and Schyman, P. (2012). Treatment of halogen bonding in the OPLS-AA force field: application to potent anti-HIV agents. *J. Chem. Theory Comput.* 8 (10): 3895–3901.
43 Mayne, C.G., Saam, J., Schulten, K. et al. (2013). Rapid parameterization of small molecules using the force field toolkit. *J. Comput. Chem.* 34: 2757–2770.

44 Humphrey, W., Dalke, A., and Schulten, K. (1996). VMD – visual molecular dynamics. *J. Mol. Graphics* 14: 33–38.

45 Domanski, J., Beckstein, O., and Iorga, B.I. (2017). Ligandbook: an online repository for small and drug-like molecule force field parameters. *Bioinformatics* 33 (11): 1747–1749.

46 Shirts, M. and Pande, V.S. (2000). Screen savers of the world unite. *Science* 290 (5498): 1903–1904.

47 Kutzner, C., Páll, S., Fechner, M. et al. (2015). Best bang for your buck: GPU nodes for GROMACS biomolecular simulations. *J. Comput. Chem.* 36 (26): 1990–2008.

48 Shaw, D.E., Deneroff, M.M., Dror, R.O. et al. (2008). Anton, a special-purpose machine for molecular dynamics simulation. *Commun. ACM* 51 (7): 91–97.

49 Torrie, G.M. and Valleau, J.P. (1977). Nonphysical sampling distributions in Monte Carlo free-energy estimation – umbrella sampling. *J. Comput. Phys.* 23: 187–199.

50 Laio, A. and Parrinello, M. (2002). Escaping free-energy minima. *Proc. Natl. Acad. Sci. U.S.A.* 99 (20): 12562–12566.

51 Colizzi, F., Perozzo, R., Scapozza, L. et al. (2010). Single-molecule pulling simulations can discern active from inactive enzyme inhibitors. *J. Am. Chem. Soc.* 132 (21): 7361–7371.

52 Huber, T., Torda, A.E., and van Gunsteren, W.F. (1994). Local elevation: a method for improving the searching properties of molecular dynamics simulation. *J. Comput.-Aided Mol. Des.* 8: 695–708.

53 Hansen, H.S. and Hünenberger, P.H. (2010). Using the local elevation method to construct optimized umbrella sampling potentials: calculation of the relative free energies and interconversion barriers of glucopyranose ring conformers in water. *J. Comput. Chem.* 31: 1–23.

54 Babin, V., Roland, C., and Sagui, C. (2008). Adaptively biased molecular dynamics for free energy calculations. *J. Chem. Phys.* 128: 134101.

55 Valsson, O. and Parrinello, M. (2014). Variational approach to enhanced sampling and free energy calculations. *Phys. Rev. Lett.* 113 (9): 090601.

56 Šućur, Z. and Spiwok, V. (2016). Sampling enhancement and free energy prediction by flying Gaussian method. *J. Chem. Theory Comput.* 12 (9): 4644–4650.

57 Kumar, S., Bouzida, D., Swendsen, R.H. et al. (1992). The weighted histogram analysis method for free-energy calculations on biomolecules. I. The method. *J. Comput. Chem.* 13: 1011–1021.

58 Dickson, B.M. (2011). Approaching a parameter-free metadynamics. *Phys. Rev. E: Stat. Nonlinear Soft Matter Phys.* 84: 037701.

59 Tribello, G.A., Ceriotti, M., and Parrinello, M. (2012). Using sketch-map coordinates to analyze and bias molecular dynamics simulations. *Proc. Natl. Acad. Sci. U.S.A.* 109: 5196–5201.

60 Bonomi, M., Barducci, A., and Parrinello, M. (2009). Reconstructing the equilibrium Boltzmann distribution from well-tempered metadynamics. *J. Comput. Chem.* 30: 1615–1621.

61 Bennion, B.J., Be, N.A., McNerney, M.W. et al. (2017). Predicting a drug's membrane permeability: a computational model validated with in vitro permeability assay data. *J. Phys. Chem. B* 121 (20): 5228–5237.
62 Bussi, G., Laio, A., and Parrinello, M. (2006). Equilibrium free energies from nonequilibrium metadynamics. *Phys. Rev. Lett.* 96: 090601.
63 Hošek, P. and Spiwok, V. (2016). Metadyn view: fast web-based viewer of free energy surfaces calculated by metadynamics. *Comput. Phys. Commun.* 198: 222–229.
64 Barducci, A., Bussi, G., and Parrinello, M. (2008). Well-tempered metadynamics: a smoothly converging and tunable free-energy method. *Phys. Rev. Lett.* 100: 020603.
65 Saleh, N., Hucke, O., Kramer, G. et al. (2018). Multiple binding sites contribute to the mechanism of mixed agonistic and positive allosteric modulators of the cannabinoid CB_1 receptor. *Angew. Chem. Int. Ed.* doi: 10.1002/anie.201708764.
66 Saleh, N., Ibrahim, P., and Clark, T. (2017). Differences between G-protein-stabilized agonist-GPCR complexes and their nanobody-stabilized equivalents. *Angew. Chem. Int. Ed.* 56 (31): 9008–9012.
67 Milanos, L., Saleh, N., Kling, R.C. et al. (2016). Identification of two distinct sites for antagonist and biased agonist binding to the human chemokine receptor CXCR3. *Angew. Chem. Int. Ed. Engl.* 55 (49): 15277–15281.
68 Saleh, N., Saladino, G., Gervasio, F.L. et al. (2016). A three-site mechanism for agonist/antagonist selective binding to vasopressin receptors. *Angew. Chem. Int. Ed. Engl.* 55 (28): 8008–8012.
69 Laio, A., Rodriguez-Fortea, A., Gervasio, F.L. et al. (2005). Assessing the accuracy of metadynamics. *J. Phys. Chem. B* 109 (14): 6714–6721.
70 Piana, S. and Laio, A. (2007). A bias-exchange approach to protein folding. *J. Phys. Chem. B* 111: 4553–4559.
71 Marinelli, F., Pietrucci, F., Laio, A., and Piana, S. (2009). A kinetic model of Trp-cage folding from multiple biased molecular dynamics simulations. *PLoS Comput. Biol.* 5 (8): e1000452.
72 Herbert, C., Schieborr, U., Saxena, K. et al. (2013). Molecular mechanism of SSR128129E, an extracellularly acting, small-molecule, allosteric inhibitor of FGF receptor signaling. *Cancer Cell* 23 (4): 489–501.
73 Milliman, R.E. (1982). Using background music to affect the behavior of supermarket shoppers. *J. Mark.* 46 (3): 86–91.
74 Bussi, G., Gervasio, F.L., Laio, A., and Parrinello, M. (2006). Free-energy landscape for β hairpin folding from combined parallel tempering and metadynamics. *J. Am. Chem. Soc.* 128 (41): 13435–13441.
75 Sutto, L. and Gervasio, F.L. (2013). Effects of oncogenic mutations on the conformational free-energy landscape of EGFR kinase. *Proc. Natl. Acad. Sci. U.S.A.* 110 (26): 10616–10621.
76 Dickson, A. and Brooks, C.L. (2014). WExplore: hierarchical exploration of high-dimensional spaces using the weighted ensemble algorithm. *J. Phys. Chem. B* 118 (13): 3532–3542.

77 Lotz, S.D. and Dickson, A. (2018). Unbiased molecular dynamics of 11 min timescale drug unbinding reveals transition state stabilizing interactions. *J. Am. Chem. Soc.* 140 (2): 618–628.

78 Limongelli, V., Bonomi, M., and Parrinello, M. (2013). Funnel metadynamics as accurate binding free-energy method. *Proc. Natl. Acad. Sci. U.S.A.* 110 (16): 6358–6363.

79 de Ruiter, A. and Oostenbrink, C. (2013). Protein-ligand binding from distancefield distances and Hamiltonian replica exchange simulations. *J. Chem. Theory Comput.* 9 (2): 883–892.

80 Nagy, G., Oostenbrink, C., and Hritz, J. (2017). Exploring the binding pathways of the 14-3-3ζ protein: structural and free-energy profiles revealed by Hamiltonian replica exchange molecular dynamics with distance field distance restraints. *PLoS One* 12 (7): e0180633.

81 Bollini, M., Domaoal, R.A., Thakur, V.V. et al. (2011). Computationally-guided optimization of a docking hit to yield catechol diethers as potent anti HIV agents. *J. Med. Chem.* 54 (24): 8582–8591.

82 Lee, W.G., Gallardo-Macias, R., Frey, K.M. et al. (2013). Picomolar inhibitors of HIV reverse transcriptase featuring bicyclic replacement of a cyanovinylphenyl group. *J. Am. Chem. Soc.* 135 (44): 16705–16713.

83 Frey, K.M., Puleo, D.E., Spasov, K.A. et al. (2015). Structure-based evaluation of non-nucleoside inhibitors with improved potency and solubility that target HIV reverse transcriptase variants. *J. Med. Chem.* 58 (6): 2737–2745.

84 Kudalkar, S.N., Beloor, J., Quijano, E. et al. (2018). From in silico hit to long-acting late-stage preclinical candidate to combat HIV-1 infection. *Proc. Natl. Acad. Sci. U.S.A.* 115 (4): E802–E811.

85 Gutiérrez-de-Terán, H. and Åqvist, J. Linear interaction energy: method and applications in drug design. *Methods Mol. Biol.* 819: 305–323.

86 Genheden, S. and Ryde, U. (2015). The MM/PBSA and MM/GBSA methods to estimate ligand-binding affinities. *Expert Opin. Drug Discovery* 10 (5): 449–461.

87 Wright, D.W., Hall, B.A., Kenway, O.A. et al. (2014). Computing clinically relevant binding free energies of HIV-1 protease inhibitors. *J. Chem. Theory Comput.* 10 (3): 1228–1241.

88 Flyvbjerg, H. and Petersen, H.G. (1998). Error estimates on averages of correlated data. *J. Chem. Phys.* 91: 461–466.

89 Romo, T.D. and Grossfield, A. (2011). Block covariance overlap method and convergence in molecular dynamics simulation. *J. Chem. Theory Comput.* 7 (8): 2464–2472.

90 Grossfield, A. and Zuckerman, D.M. (2009). Quantifying uncertainty and sampling quality in biomolecular simulations. *Annu. Rep. Comput. Chem.* 5: 23–48.

91 Klimovich, P.V., Shirts, M.R., and Mobley, D.L. (2015). Guidelines for the analysis of free energy calculations. *J. Comput.-Aided Mol. Des.* 29 (5): 397–411.

92 https://plumed.github.io/doc-v2.4/user-doc/html/trieste-2.html.

93 Tiwary, P. and Parrinello, M. (2015). A time-independent free energy estimator for metadynamics. *J. Phys. Chem. B* 119 (3): 736–742.

94 Zhu, F. and Hummer, G. (2012). Convergence and error estimation in free energy calculations using the weighted histogram analysis method. *J. Comput. Chem.* 33 (4): 453–465.

95 Erlanson, D.A. and Jahnke, W. (2016). *Fragment-Based Drug Discovery: Lessons and Outlook*. Wiley-VCH 9783527337750.

96 Zhang, Y. and McCammon, J.A. (2003). Studying the affinity and kinetics of molecular association with molecular-dynamics simulation. *J. Chem. Phys.* 118: 1821.

97 Tiwary, P. and Parrinello, M. (2013). From metadynamics to dynamics. *Phys. Rev. Lett.* 111: 230602.

98 Tiwary, P., Limongelli, V., Salvalaglio, M., and Parrinello, M. (2015). Kinetics of protein-ligand unbinding: predicting pathways, rates, and rate-limiting steps. *Proc. Natl. Acad. Sci. U.S.A.* 112 (5): E386–E391.

99 Lu, H.P., Xun, L., and Xie, X.S. (1998). Single-molecule enzymatic dynamics. *Science* 282 (5395): 1877–1882.

100 Peleg, G., Ghanouni, P., Kobilka, B.K., and Zare, R.N. (2001). Single-molecule spectroscopy of the β_2 adrenergic receptor: observation of conformational substates in a membrane protein. *Proc. Natl. Acad. Sci. U.S.A.* 98 (15): 8469–8474.

2

Molecular Dynamics–Based Approaches Describing Protein Binding

Andrea Spitaleri and Walter Rocchia

CONCEPT Lab, Istituto Italiano di Tecnologia, via Morego, 30, 16163 Genoa, Italy

2.1 Introduction

Biomolecular recognition is at the heart of all biological processes that take place in living organisms. Understanding how a ligand binds to a biological receptor, how proteins interact with each other, how lipids and proteins aggregate in the cell membrane, and how these events trigger or block a wide range of biochemical reactions is of paramount importance, not only for the field of biophysics but also for other disciplines such as rational drug design.

The first breakthrough in the theory of biomolecular recognition was provided by Fischer, who as early as 1894, proposed his popular lock-and-key model for ligand binding [1]. According to this model, which was originally developed in the context of enzyme catalysis, ligands involved in biological reactions fit perfectly into their targets like a key into a lock. In this process, no changes in conformations are allowed. Later, however, the lock-and-key model based on rigid body interactions between ligands and proteins was severely challenged by the conformational plasticity of macromolecules, as revealed by X-ray crystallography, nuclear magnetic resonance (NMR) spectroscopy, and single-molecule fluorescence detection [2].

In recent years, more dynamic models of molecular recognition have superseded Emil Fischer's rigid lock-and-key binding paradigm. The idea of induced fit, introduced by Koshland in 1958, aimed at overcoming the limitations of the lock-and-key model, relies on the formation of an initial loose ligand–receptor complex that induces a conformational change in the protein, resulting in a series of rearrangements that lead to a complex with tighter binding.

In 1999, Nussinov and coworkers coined the term conformational selection, also known as population shift, selected fit, and population selection, which is based on the idea that all conformations are present when the ligand is not bound to the receptor and, then, the ligand acts to selectively stabilize specific receptor conformations, causing a shift in the populations observed in the unbound ensemble toward this specific conformational state [2, 3].

Although both theories, induced-fit and conformational selection, appear to be antagonistic, they are not necessarily mutually exclusive. Nowadays it is

Biomolecular Simulations in Structure-Based Drug Discovery,
First Edition. Edited by Francesco L. Gervasio and Vojtech Spiwok.

increasingly seen that these two major mechanisms contribute to molecular recognition, which requires a dynamic description over timescales spanning several orders of magnitude to be fully understood. Moreover, water molecules often play a crucial role in molecular association. In fact, experimental and theoretical studies have pointed out the capital importance of both entropic and enthalpic contributions of water networks to the free energy of binding [4]. All these findings not only served to advance the field toward a better understanding of receptor–ligand (e.g. protein, peptide, nuclei acid, and ligand) binding but also introduced an extra degree of complexity to the description of biomolecular recognition processes. To comprehend how biomolecular recognition occurs, we first need to understand the role of all different partners involved in this association process.

2.1.1 Protein Binding: Molecular Dynamics Versus Docking

Molecular docking programs are widely used to predict receptor–ligand complexes and are extensively and routinely used in computer-aided drug discovery, mostly in the framework of virtual screening (VS) campaigns [5]. Molecular docking, in practice, has two essential requirements: structural data, for candidate ligands and the protein target of interest, and a procedure to estimate protein–ligand interaction poses and strengths. Various molecular docking algorithms are available to predict receptor–ligand, e.g. protein, peptide, nucleic acid, and ligand, poses and to rank them based on scoring functions implemented in each specific docking approach.

In spite of the large usage in structure-based drug discovery (SBDD), docking retains crucial liabilities. The high computational speed in determining the putative poses comes at the cost of accuracy, especially when target rearrangements are required upon ligand binding. Sampling the conformational space remains challenging. All docking algorithms exhibit sensitivity to initial conditions; therefore, some variation in the poses and scores is expected when starting from different input conformations of a particular ligand. The vast majority of molecular docking applications considers the ligand conformational flexibility either during docking calculations or using libraries of conformers. Conversely, the flexibility of the receptor is usually neglected as the number of degrees of freedom, which should be considered in the calculations, is extremely computationally demanding. Moreover, docking algorithms lack explicit water treatments, which are crucial for reproducing specific drug–target complexes. All these approximations constitute a major drawback in docking calculations, limiting the success rate of the drug discovery campaign.

These limitations can be addressed by molecular dynamics (MD) simulations. Thanks to the rapid development of faster architectures (graphics processing unit (GPU)-based clusters) and better algorithms for high-level computations, classical MD simulations nowadays allow the implementation of SBDD strategies that account for the structural flexibility of drug–target systems at a fully atomistic description. Therefore, MD simulations can be exploited to characterize the protein–ligand binding process at a fully dynamic level. As a result, classical MD is no longer considered prohibitive for effective drug design. Instead, it is pushing the frontiers of computationally driven drug discovery in both academia and industry [6].

2.1.2 Molecular Dynamics – The Current State of the Art

MD is a physical method for studying the interaction and motion of atoms and molecules according to Newton's physics. A force field is used to estimate the forces between interacting atoms and calculate the overall energy of the system. MD simulations are used today to study nearly every type of macromolecule, including proteins, nucleic acids, and carbohydrates, of biological or medicinal interest. Simulations span wide spatial and temporal ranges and resolutions. To be explicit, all-atom MD, thousands to millions of individual atoms representing, for instance, all the atoms of a protein and surrounding water molecules, move in a series of short (e.g. 2 fs), discrete time steps. At each step, the forces on each atom, determined from the "force field," e.g. a collection of physics-based parameters that represent both bonded and nonbonded (e.g. van der Waals) interatomic forces, are computed and the atomic position and velocity are updated according to Newton's laws of motion [7]. Recently, computational speed has grown rapidly with the introduction of parallel computing based on GPU computing techniques and specialized computing machines, e.g. high-performance supercomputers [8], designing of specialized hardware [9], and development of intelligent MD protocols [10]. As a consequence, nowadays simulations lasting up to a few milliseconds are now possible. Assuming that the ergodic hypothesis holds, an infinitely long MD trajectory should be able to sample the entire conformational space. However, except for a few cases published recently [11, 12], long MD simulations are often relegated to some local minimum in the rugged and complex free energy surface (FES) that is accessible to the system. At room temperature, the probability of crossing high-energy barriers is often too small to be observed during a finite MD simulation. Most likely, even with several hundred nanoseconds of simulations, the system might be confined to limited regions of the conformational space. Therefore, besides the increasing computational power, various modeling techniques have been developed to accelerate simulations and/or to sample rare events during ligand binding processes. A solution usually applied to overcome the limited sampling efficiency of MD simulations at room temperature consists in raising the simulation temperature. The additional kinetic energy available at higher temperature allows the crossing of high-energy barriers, thus ensuring a wider sampling of the conformational space. This methodology is the basis of two computational approaches, simulated annealing (SA) and parallel tempering (also named replica-exchange molecular dynamics, REMD). SA consists in heating up the system in order to jump out from the initial local minimum to explore other minima [13]. The heating step is followed by a gradual cooling, which allows the system to slowly settle down to a lower energy minimum. REMD [14] is based on the run of multicopy MD simulations randomly initialized, at different temperatures. The conformations are then exchanged at different temperatures following the Metropolis criterion. The strength and robustness of this method allows sampling of both low- and high-energy configurations [15].

The well-recognized limitations of sampling in atomistic dynamics have led to many innovative alternatives to enhance the coverage of the thermally accessible conformational space and to capture *rare events* (events that might happen on a long timescale) [16]. One trick that is commonly used to address this problem is to add a potential bias in order to force the *rare event* to occur. In this context, several techniques, including the local elevation method [17], taboo search [18], the Wang–Landau method [19], adaptive force bias [20] conformational flooding [21], umbrella sampling [22], weighted histogram techniques [23], transition state theory and path sampling [24], steered MD [25], free energy–guided sampling [26], and metadynamics [27], have been developed to address the sampling problem. In this context, metadynamics has emerged as a powerful coarse-grained non-Markovian MD approach for the acceleration of rare events and the efficient and rapid computation of multidimensional FES as a function of a restricted number of degrees of freedom, named collective variables (CVs). Different from other sampling methods, in which the calculation of FES requires an additional step (such as the weighted histogram analysis method (WHAM) [19]), metadynamics directly provides a good estimate of the free energy of the system projected into the CVs. On one hand, the main advantage of metadynamics along with other methods based on CVs is that the sampling is enhanced toward the specific event of interest by biasing the MD simulations along chosen CVs, which are functions of the atomic coordinates. In this way, free energy barriers can be efficiently crossed. On the other hand, it can be difficult to choose the right set of CVs *a priori*, particularly when the system's reaction mechanism is not yet known. In the latter case, tempering methods, such as REMD, accelerated molecular dynamics (aMD) [28], and potential scaled [29] seem to be more appropriate since they act by heating all degrees of freedom of the system or by modifying the Hamiltonian [30].

Finally, MD simulations are capable of providing quantitative binding estimates with high accuracy, i.e. free energy of binding, but it has been typically regarded as too computationally expensive. A variety of free energy simulation methods, such as free energy perturbation (FEP), thermodynamic integration (TI), and λ dynamics, employ an analysis of atomistic MD simulations to determine the free energy difference between two related ligands via either a chemical or alchemical path. In recent years, FEP calculations have benefited from improved force fields, new sampling algorithms, and the emergence of low-cost parallel computing, which have resulted in the level of accuracy and turnaround time needed to impact lead optimization efforts.

The examples in the following sections indicate some of the current capabilities of MD simulations in three different fields: protein–protein, protein–peptide, and protein–ligand binding and the insights they can provide.

2.2 Protein–Protein Binding

Protein–protein interactions (PPIs) are ubiquitous in biological systems as they are the building blocks for the complex molecular machinery that drives critical functions within the cell. Protein–protein association signals transduction and inhibition, immune response, and protein recruitment to membrane scaffolds. Mutations and perturbations within the proteins that result in aberrant interactions often lead to a number of diseases, including cancer. It has been estimated that the human interactome involves between 130 000 and 600 000 PPIs [31–33]. Although the Protein Data Bank (PDB) contains more than 100 000 entries, structural data on protein–protein complexes remain scarce [34]. In this context, MD simulations along with docking aim at offering opportunities to fill this gap and provide valuable information on the structure of PPIs as well as their specific interactions, and furthermore support the design of novel PPI modulators or inhibitors. PPIs are often characterized by the cooperative formation of many weak interatomic interactions over an extended binding surface rather than by few strong interactions, as in the case of protein–ligand binding. In comparison to classical small-molecule targets, the binding site of protein–protein interfaces is relatively large, with an area of about 1500–3000 $Å^2$ [35]. Nevertheless, slight conformational changes can be temporarily induced to enable PPIs, which mostly involve side-chain motions or small fluctuations of loops [36].

Despite the availability of several experimental data on the PPIs, such as NMR, X-ray structure, and isothermal titration calorimetry (ITC) thermodynamic data, the description of the protein–protein system is limited to static snapshots, which are not sufficient to describe coherently the molecular processes occurring upon binding. In this scenario, atomistic simulations could greatly contribute to linking binding mechanism and function. MD simulation could adequately represent the complex physics and thermodynamics behind the protein–ligand complex formation in order to investigate the underlying mechanisms of complex formation and modulation or inhibition. The mechanism of regulated recognition involves conformational changes in the binding partners, which are prevented or induced by external signals. This implies that protein–protein binding motifs can be labile and intrinsically flexible with binding sites that are shallow and rather featureless compared to "druggable" sites [36]. In comparison to classical ligand–protein approaches, this step is far from trivial as high-resolution structural information is needed as a prerequisite. These data can be obtained from experiments (NMR, X-ray crystallography, etc.), PPI databases, or modeling approaches such as homology modeling or protein–protein docking. As an alternative to simulating already existing PPIs, it has been shown that the initial formation of PPIs during MD simulations starting from two separate proteins is possible [37]. Nevertheless, this approach still is computationally demanding and requires experimental validation. This section highlights the benefits of applying MD to PPI investigations.

With MD simulations, protein interactions, flexibility, and sampling of the conformational space can be characterized, which is of great value for the investigation of the interacting surfaces of protein–protein complexes. MD simulations can also be used to identify binding hot spots and transient pockets that play a role in protein–protein binding but could not be captured by crystallography [38, 39]. Due to the dynamic nature of protein–protein interfaces, transient pockets and buried binding hot spots can emerge on the protein surfaces; and binding of small molecules to these transient areas have been reported [40].

Combining MD and principal component analysis (PCA), Dixit and Verkhivker identified conserved functional regions of Hsp90, an important chaperone protein [41]. These findings help in the design of inhibitors targeting these protein–protein interfaces and furthermore provide mechanistic insights into the allosteric regulation of the complex dynamics and motions of the chaperone machinery.

Bastianelli et al. exploited MD simulations to refine the protein–protein docking results and to perform free energy calculations [42]. In this work, the authors were able to predict the interactions between two proteins, i.e. PcFK1, a small protein extracted from spider venom, and PfSUB1, a subtilisin-like serine protease. The results were then confirmed by experimental validation.

Although the interaction area between proteins is relatively large, it has been shown that only a small subset of key residues contribute the most to high-affinity binding in PPIs [43]. These protein domains with key interacting residues are termed hot spots, and hot spot detection is an efficient strategy to identify potential druggable areas of PPIs [43]. This can be done by computational alanine scanning (CAS), by which a single residue is mutated in alanine in order to determine its energetic contribution to identifying hot spots. Using the CAS method, Sousa et al. provided an evaluation of the energetic contribution of the amino acid residues at the subunit interface and also potential starting points for HIV inhibitors that block the dimerization of the protein [44].

2.3 Protein–Peptide Binding

In many PPIs, it has been shown that the interactions are dominated by short linear recognition motifs that are part of the larger protein [45]. Protein–peptide recognition has a crucial role in various fundamental aspects of cellular homeostasis, such as signal transduction, protein trafficking, and immune response. Peptides often serve as signaling vehicles as, for example, in the case of hormones and neurotransmitters. Peptides in organisms often occur as part of the breakdown of proteins. One important example is the amyloid beta (Abeta) peptide which results from the hydrolysis of the APP protein on the surface of neurons. Plaques formed by the aggregation of Abeta peptides are recognized as one of the hallmarks of Alzheimer's disease [46]. Antimicrobial peptides are an important component of the innate immune system, and peptides presented and recognized on the surface of antigen-presenting cells by the fragmentation of foreign proteins are recognized as one of the main processes of the adaptive immune response [47].

Protein–peptide complexes often display interaction patterns different from the ones often encountered in protein–protein complexes. For example, protein–peptide interactions tend to be more hydrogen bond–dependent than PPIs, usually involving the peptide backbone. Finally, the protein–peptide binding process frequently involves significant conformational rearrangements of peptide chains (e.g. from unfolded to folded structure).

The efficient treatment of the large-scale changes remains one of the major challenges for molecular docking and dynamics simulations [48]. The inefficiency in classic MD sampling is a result of the highly rugged nature of the folding energy landscape of the peptide chains. The prediction of the structures of protein–peptide complexes and the quantitative prediction of their binding affinities are far from a solved problem. In the following, we report the few examples in literature exploiting MD simulations of protein–peptide systems. In 2009, Pietrucci et al. used the bias-exchange metadynamics technique to successfully describe the binding mechanism of a small peptide to the HIV-1 protease [49]. Even though the authors accurately computed the free energy associated with ligand binding and unbinding as a function of 7 CVs, almost 2 μs of simulation were required to converge the free energy. Despite this, they managed to characterize the kinetics of the binding/unbinding process using a discrete-states kinetic model, including the relevant metastable states along the recognition pathway.

Tortorici et al. performed MD simulations in explicit solvent as well as in temperature replica exchange MD simulations to explore the peptide conformations derived from the "SNAG" domain of SNAIL1 in solution [50]. True to the experimental data, MD simulations revealed that the peptides exist as random structures, showing transient secondary structures with an α-helical propensity of only about 7%. The authors used this evidence to suggest that the peptides do indeed exist as disordered conformers in solution and only form a secondary structure upon binding, in agreement with experimental data (CD spectra) and with the analysis of seven crystal structures of SNAIL-derived peptides bound to LSD1/CoREST in which the peptides were bound in a helical structure.

Efficiently predicting binding affinities of protein–peptide binding affinities from structural models would be an extremely important step in our overall understanding of the regulation of biological systems and critical for drug design and development. MD simulations together with binding free energy calculation under the molecular mechanics-generalized born surface area (MM-GBSA) approximation was used to study the binding affinities of a series of short peptides against the catalytic site of protein kinase A [51]. To reduce computational cost, the entropy term was not computed. This approach was justified by the observation that other studies on similar systems had concluded that the entropy term did not contribute significantly to the binding affinity and/or would not vary appreciably across the peptide library. The authors reported high correlation between experimental free energy values and computational values ($r^2 = 0.98$). The authors also concluded that the most influential term in the binding free energy is the electrostatic term, again showing that the success of this model is intimately tied to the electrostatic interactions present.

Another important field in protein–peptide binding is the epigenetic recognition by histone tails. Epigenetic regulation of gene transcription relies on an array of recurring structural domains that have evolved to recognize posttranslational modifications on histones. Misinterpretation of epigenetic mechanisms is often associated with disease conditions. Epigenetic readers are therefore rapidly becoming a privileged target for drug discovery study. Plant homeodomain (PHD) fingers represent one of the largest families of epigenetic readers capable of decoding posttranslationally modified or unmodified histone H3 tails. Spiliotopoulos et al. presented a computational case study describing the dynamics and the thermodynamics at the basis of unmodified histone H3 (H3K4me0) recognition by the first PHD finger of the transcriptional activator Autoimmune Regulator (AIRE-PHD1) [52]. MD simulations together with PCA showed that histone peptide binding strongly affects AIRE-PHD1 internal correlation motions. In particular, analysis of the essential domain motions highlighted the presence of a "flapping" movement in free AIRE-PHD1, which might be relevant for domain function. Importantly, this intrinsic domain "breathing" was blocked upon H3K4me0 binding. Moreover, calculations of the free energy of binding obtained through molecular mechanics/Poisson–Boltzmann surface area (MM/PBSA) methodology allowed us to dissect the energetic terms associated with native and alanine mutants of AIRE-PHD1/H3K4me0 complexes. The calculated binding free energies were in good qualitative agreement with experiments. Prompted by these encouraging results, the authors extended MM/PBSA calculations to other PHD fingers recognizing the unmodified status of histone H3K4 in order to investigate whether this subclass of histone readers shared similar thermodynamic behavior. Indeed, comparison of calculations with experimental binding data confirmed that MM/PBSA can be considered as a valuable tool to rapidly analyze the energetic determinants dictating histone decoding by this class of epigenetic effectors.

Unlike endpoint methods, such as MM/PBSA/GBSA, pathway free energy methods compute the difference in binding free energy between the unbound and bound states of the ligand–protein complex by linking them through a suitable thermodynamic path. Oftentimes the path is parameterized by a progress parameter λ ranging from 0 to 1. A λ-dependent potential energy function $U(x_A, x_B, \zeta_B, r_s; \lambda)$ is designed in such a way that at $\lambda = 0$ it represents the unbound state of the complex (when protein and ligand are not interacting), and at $\lambda = 1$ it represents the bound state of the complex. The reversible work along the path yields the binding free energy of the complex [53]. As the decoupling transformations are too extensive and the corresponding free energies are too large to routinely obtain converged binding free energy estimates, there has been only one report of the application of the double decoupling method to the calculation of absolute binding free energies of peptides [54].

2.4 Protein–Ligand Binding

MD simulations have been successfully and widely utilized in drug design and development [30]. Usually, protein–ligand binding is studied by docking simulations. If we are interested in determining a plausible pose, the receptor flexibility during the docking is modeled using an ensemble of structures, obtained through either experimental techniques (X-ray crystallography or NMR spectroscopy) or by exploiting sampling engines such as Monte Carlo (MC) or MD. The latter one has the advantage of taking into account the global flexibility of the protein, and the only drawback is the narrowness of sampling. First, a long MD simulation of the protein in the unbound state is performed to exhaustively sample the conformational space. Then, MD snapshots are selected either at regular time intervals or after cluster analysis to perform docking simulations.

Moving beyond protocols that use MD to incorporate target flexibility into standard docking calculations, it is now possible to run MD simulations for long enough to explore the free energy landscape and kinetic profile associated with the overall drug-binding process (i.e. from the drug fully solvated in water to the drug–target bound state) [55]. The major advantage of MD simulation with respect to docking simulation in studying protein–ligand is twofold: (i) investigating possible allosteric binding mode; (ii) structural characterization of the binding path and *a posteriori* calculations of the binding free energy minimum path. Even though the MD framework is certainly appealing and will most likely be employed more in the future, at present, some efficiency limitations prevent its widespread use in SBDD. Long *plain* MD simulations have already described the spontaneous binding of small organic molecules to biological targets of pharmacological interest, demonstrating both the adequacy of the present force fields and the feasibility, at least in some cases, of observing the rare event.

A seminal example of the useful insights offered by MD simulations in structure-based drug design was reported in 2011, when Buch and Giorgino performed on GPU-based infrastructures 495 MD simulations of 100 ns each to simulate the complete binding process of the inhibitor benzamidine to β-trypsin [56]. The approach allowed the identification of the lowest energy binding mode of a ligand to a receptor. Monitoring of the binding process at an atomic resolution can potentially assist the development of drugs able to control and modulate the ligand–receptor recognition process. In this context, Shaw and collaborators formulated a mechanism for the flipping of a conserved motif of Abl tyrosine kinases combining microsecond MD simulations with crystallographic and kinetic experiments. Importantly, the conformation of this motif was crucial to discriminate between active/inactive kinase conformations. Their results led to the identification of a class of potent inhibitors of both Src and Abl that recognize the inactive kinase conformations [12].

The G protein–coupled receptor β_2-adrenergic (β_2-AR) is an important target for hypertension and several heart diseases. Shaw and coworkers have used extensive plain MD to study how this receptor interacts and binds with the inverse agonist alprenolol [11]. The binding paths followed a two-step molecular mechanism. First, alprenolol reorients so that its hydrophobic group binds to the extracellular vestibule surface. Then, it penetrates into the binding pocket through the gate closing–opening mechanism between Tyr308 and Phe193 to form the salt bridges between the ammonium moiety and the carboxylate of Asp113.

More recently, Decherchi et al. ran extensive unbiased MD simulations (about 1 μs each) to investigate the tight binding event of a transition state analog (DADMe-immucilin H) into the pharmaceutical target, purine nucleoside phosphorylase [57]. Despite observing spontaneous binding through different routes, the authors used the pathways obtained by MD simulations to determine the free energy profiles associated with the diverse binding mechanisms.

The aforementioned examples require a huge amount of computational resources, far more than the average time allocated to a real-world drug discovery project. A valid alternative is to accelerate the sampling while preserving a statistical mechanics distribution of states through so-called enhanced sampling methods. Researchers have recently demonstrated the power of these methods for studying protein–ligand binding and estimating the associated free energy and kinetics [38, 58].

Among all these, the metadynamics-based method, which allows fast exploration of the underlying free energy landscape of rare events, has been applied to a number of ligand–target complexes, demonstrating its ability to characterize binding and unbinding paths, to treat conformation flexibility, and to compute free energy profiles. Since Gervasio et al. first applied metadynamics to ligand–target complexes, several other informative studies have been reported [58, 59].

2.5 Future Directions

Nowadays, computational power has increased with respect to the past to such an extent that most MD simulations are run on personal powerful workstations. The increase in computational power implies a different future scenario. Simulations will become much larger and will reach longer timescales. This reach for ever larger and longer simulations will also be increasingly aided by improved algorithmic methods to increase the sampling of conformational space. The unbiased assessment of protein–protein and protein–nucleic acid interactions, in particular, will present new drug discovery opportunities. MD simulations can address two aspects of this problem. First, how and where do the partners interact? And as they interact, are novel ligand binding sites created, at the interface or at allosteric sites? MD simulations seem to be good at identifying low-energy protein conformations that harbor cryptic drug-binding sites. Second, how can small-molecule ligands modulate those interactions? How can we optimize their binding and druglike properties? Simulations of macromolecular assembly will also extend to include very large complexes, such as the nuclear pore complex or even entire organelles.

Although we have been able to simulate on a millisecond timescale, there are still some parameters including torsions in amino acids, carbohydrates, and single-stranded nucleic acids for which the force fields need further improvement. The existing force fields are not efficient for studying the crowded environment inside the cells, since these interactions involve multiple factors in real time. Therefore, the improved force fields may provide the opportunities for their wider applications on the complex biosystems in diverse cellular conditions.

2.5.1 Modeling of Cation-p Interactions

These interactions, driven by electric quadrupole moments and polarization effects, are now recognized to be quite important to both protein structure (e.g. arginine–tryptophan ladders) and protein function, for instance, in a wide variety of protein–ligand interactions. Polarizable models should enable us to more accurately describe (inter)molecular interactions, and, indeed, several polarizable force fields have begun to demonstrate their value. The AMOEBA force field, for instance, includes some cross-terms and, significantly, polarizable atomic multipoles (up to quadrupoles) replace fixed partial charges [6].

2.6 Grand Challenges

There are several grand challenges that will help make MD simulations a standard "tool" to study and describe life sciences. First of all, free energy calculations should be more reliable and rapid. The calculation of free energy should answer to different questions for both macromolecule–ligand and macromolecule–macromolecule interactions. The importance of this computationally demanding task is crucial to answer whether either a ligand or a macromolecule will interact or not. Moreover, the magnitude of the interactions needs to be accurate (less than 1 kcal mol^{-1}). Finally, the free energy is of vital importance in drug discovery. Improvement of free energy calculations has to go through both the exhaustive sampling of configurational space and better force field. The abovementioned force field improvements should help significantly. Converged calculations will enable rigorous determination of both force field accuracy and the need for specific force field improvements.

References

1 Changeux, J.-P. and Edelstein, S. (2011). Conformational selection or induced fit? 50 years of debate resolved. *F1000 Biol. Rep.* 3: 19.

2 Boehr, D.D., Nussinov, R., and Wright, P.E. (2009). The role of dynamic conformational ensembles in biomolecular recognition. *Nat. Chem. Biol.* 5: 789–796.

3 Ma, B., Kumar, S., Tsai, C.J., and Nussinov, R. (1999). Folding funnels and binding mechanisms. *Protein Eng.* 12 (9): 713–720.

4 Breiten, B., Lockett, M.R., Sherman, W. et al. (2013). Water networks contribute to enthalpy/entropy compensation in protein-ligand binding. *J. Am. Chem. Soc.* 135: 15579–15584.

5 Jorgensen, W.L. (2004). The many roles of computation in drug discovery. *Science* 303: 1813–1818.

6 Borhani, D.W. and Shaw, D.E. (2012). The future of molecular dynamics simulations in drug discovery. *J. Comput.-Aided Mol. Des.* 26: 15–26.

7 Karplus, M. and Kuriyan, J. (2005). Molecular dynamics and protein function. *Proc. Natl. Acad. Sci. U.S.A.* 102: 6679–6685.

8 Shaw, D.E., Chao, J.C., Eastwood, M.P. et al. (2008). Anton, a special-purpose machine for molecular dynamics simulation. *Commun. ACM* 51: 91.

9 Harvey, M.J., Giupponi, G., and Fabritiis, G.D. (2009). ACEMD: accelerating biomolecular dynamics in the microsecond time scale. *J. Chem. Theory Comput.* 5: 1632–1639.

10 Pronk, S., Pall, S., Schulz, R. et al. (2013). GROMACS 4.5: a high-throughput and highly parallel open source molecular simulation toolkit. *Bioinformatics* 29: 845–854.

11 Dror, R.O., Pan, A.C., Arlow, D.H. et al. (2011). Pathway and mechanism of drug binding to G-protein-coupled receptors. *Proc. Natl. Acad. Sci. U.S.A.* 108: 13118–13123.

12 Shan, Y., Kim, E.T., Eastwood, M.P. et al. (2011). How does a drug molecule find its target binding site? *J. Am. Chem. Soc.* 133: 9181–9183.

13 Kirkpatrick, S., Gelatt, C., D. Jr., and Vecchi, M.P. (1983). Optimization by simulated annealing. *Science* 220: 671–680.

14 Sugita, Y. and Okamoto, Y. (1999). Replica-exchange molecular dynamics method for protein folding. *Chem. Phys. Lett.* 314: 141–151.

15 Ostermeir, K. and Zacharias, M. (2013). Advanced replica-exchange sampling to study the flexibility and plasticity of peptides and proteins. *Biochim. Biophys. Acta* 1834: 847–853.

16 Mitsutake, A. and Mori, Y. (2013). *Enhanced Sampling Algorithms*. Springer.

17 Huber, T., Torda, A.E., and van Gunsteren, W.F. (1994). Local elevation: a method for improving the searching properties of molecular dynamics simulation. *J. Comput.- Aided Mol. Des.* 8: 695–708.

18 Cvijovicacute, D. and Klinowski, J. (1995). Taboo search: an approach to the multiple minima problem. *Science* 267: 664–666.

19 Wang, F. and Landau, D.P. (2001). Efficient, multiple-range random walk algorithm to calculate the density of states. *Phys. Rev. Lett.* 86: 2050–2053.

20 Darve, E., Rodriguez-Gomez, D., and Pohorille, A. (2008). Adaptive biasing force method for scalar and vector free energy calculations. *J. Chem. Phys.* 128: 144120.

21 Grubmuller, H. (1995). Predicting slow structural transitions in macromolecular systems: conformational flooding. *Phys. Rev. E Stat. Phys. Plasmas Fluids Relat. Interdisciplin. Top.* 52: 2893–2906.

22 Patey, G.N. and Valleau, J.P. (1975). Monte-Carlo method for obtaining interionic potential of mean force in ionic solution. *J. Chem. Phys.* 63: 2334–2339.

23 Ferrenberg, A.M. and Swendsen, R.H. (1988). New Monte Carlo technique for studying phase transitions. *Phys. Rev. Lett.* 61: 2635–2638.

24 Bolhuis, P.G., Chandler, D., Dellago, C. et al. (2002). Transition path sampling: throwing ropes over rough mountain passes, in the dark. *Annu. Rev. Phys. Chem.* 53: 291–318.

25 Grubmüller, H., Heymann, B., and Tavan, P. (1996). Ligand binding: molecular mechanics calculation of the streptavidin-biotin rupture force. *Science* 271: 997–999.

26 Zhou, T. and Caflisch, A. (2012). Free energy guided sampling. *J. Chem. Theory Comput.* 8: 2134–2140.

27 Laio, A. and Parrinello, M. (2002). Escaping free-energy minima. *Proc. Natl. Acad. Sci. U.S.A.* 99: 12562–12566.

28 Hamelberg, D., Mongan, J., and McCammon, J.A. (2004). Accelerated molecular dynamics: a promising and efficient simulation method for biomolecules. *J. Chem. Phys.* 120: 11919–11929.

29 Mark, A.E., van Gunsteren, W.F., and Berendsen, H.J.C. (1991). Calculation of relative free energy via indirect pathways. *J. Chem. Phys.* 94: 3808–3816.

30 De Vivo, M., Masetti, M., Bottegoni, G., and Cavalli, A. (2016). Role of molecular dynamics and related methods in drug discovery. *J. Med. Chem.* 59: 4035–4061.
31 Stumpf, M.P.H. et al. (2008). Estimating the size of the human interactome. *Proc. Natl. Acad. Sci. U.S.A.* 105: 6959–6964.
32 Venkatesan, K. et al. (2009). An empirical framework for binary interactome mapping. *Nat. Methods* 6: 83–90.
33 Cunningham, A.D. et al. (2017). Peptides and peptidomimetics as regulators of protein–protein interactions. *Curr. Opin. Struct. Biol.* 44: 59–66.
34 Szilagyi, A. and Zhang, Y. (2014). Template-based structure modeling of protein–protein interactions. *Curr. Opin. Struct. Biol.* 24: 10–23.
35 Wells, J.A. and McClendon, C.L. (2007). Reaching for high-hanging fruit in drug discovery at protein–protein interfaces. *Nature* 450: 1001–1009.
36 Arkin, M.R. and Wells, J.A. (2004). Small-molecule inhibitors of protein–protein interactions: progressing towards the dream. *Nat. Rev. Drug. Discov.* 3: 301–317.
37 Shan, Y., Gnanasambandan, K., Ungureanu, D. et al. (2014). Molecular basis for pseudokinase-dependent autoinhibition of JAK2 tyrosine kinase. *Nat. Struct. Mol. Biol.* 21: 579–584.
38 Eyrisch, S. and Helms, V. (2007). Transient pockets on protein surfaces involved in protein–protein interaction. *J. Med. Chem.* 50: 3457–3464.
39 Eyrisch, S., Medina-Franco, J.L., and Helms, V. (2012). Transient pockets on XIAP-BIR2: toward the characterization of putative binding sites of small-molecule XIAP inhibitors. *J. Mol. Model.* 18: 2031–2042.
40 Arkin, M.R., Randal, M., DeLano, W.L. et al. (2003). Binding of small molecules to an adaptive protein–protein interface. *Proc. Natl. Acad. Sci. U.S.A.* 100: 1603–1608.
41 Dixit, A. and Verkhivker, G.M. (2012). Probing molecular mechanisms of the Hsp90 chaperone: biophysical modeling identifies key regulators of functional dynamics. *PLoS One* 7: e37605.
42 Bastianelli, G., Bouillon, A., Nguyen, C. et al. (2011). Computational reverse-engineering of a spider-venom derived peptide active against Plasmodium falciparum SUB1. *PLoS One* 6: e21812.
43 Moreira, I.S., Fernandes, P.A., and Ramos, M.J. (2007). Hot spots-a review of the protein–protein interface determinant amino-acid residues. *Proteins* 68: 803–812.
44 Sousa, S.F., Tamames, B., Fernandes, P.A., and Ramos, M.J. (2011). Detailed atomistic analysis of the HIV-1 protease interface. *J. Phys. Chem. B* 115: 7045–7057.
45 London, N., Movshovitz-Attias, D., and Schueler-Furman, O. (2010). The structural basis of peptide-protein binding strategies. *Structure* 18 (2): 188–199.
46 Hardy, J. and Allsop, D. (1991). Amyloid deposition as the central event in the aetiology of Alzheimer's disease. *Trends Pharmacol. Sci.* 12: 383–388.
47 Pino-Angeles, A., Leveritt, J.M. III, and Lazaridis, T. (2016). Pore structure and synergy in antimicrobial peptides of the magainin family. *PLoS Comput. Biol.* 12 (1).

48 Antunes, D.A., Devaurs, D., and Kavraki, L.E. (2015). Understanding the challenges of protein flexibility in drug design. *Expert Opin. Drug Discovery* 10: 1301–1313.

49 Pietrucci, F., Marinelli, F., Carloni, P., and Laio, A. (2009). Substrate binding mechanism of HIV-1 protease from explicit-solvent atomistic simulations. *J. Am. Chem. Soc.* 131: 11811–11818.

50 Tortorici, M., Borrello, M.T., Tardugno, M. et al. (2013). Protein recognition by short peptide reversible inhibitors of the chromatin-modifying LSD1/CoREST lysine demethylase. *ACS Chem. Biol.* 8 (8): 1677–1682.

51 Mena-Ulecia, K., Vergara-Jaque, A., Poblete, H. et al. (2014). Study of the affinity between the protein kinase PKA and peptide substrates derived from kemptide using molecular dynamics simulations and MM/GBSA. *PLoS One* 9 (10): e109639.

52 Spiliotopoulos, D., Spitaleri, A., and Musco, G. Exploring PHD fingers and H3K4me0 interactions with molecular dynamics simulations and binding free energy calculations: AIRE-PHD1, a comparative study. *PLoS One* 7 (10): e46902.

53 de Ruiter, A. and Oostenbrink, C. (2011). Free energy calculations of protein–ligand interactions. *Curr. Opin. Chem. Biol.* 15 (4): 547–552.

54 Gumbart, J.C., Roux, B., and Chipot, C. (2012). Standard binding free energies from computer simulations: what is the best strategy? *J. Chem. Theory Comput.* 9 (1): 794–802.

55 Klepeis, J.L., Lindorff-Larsen, K., Dror, R.O., and Shaw, D.E. (2009). Long-timescale molecular dynamics simulations of protein structure and function. *Curr. Opin. Struct. Biol.* 19: 120–127.

56 Buch, I. and Giorgino, T. (2011). Complete reconstruction of an enzyme-inhibitor binding process by molecular dynamics simulations. *Proc. Natl. Acad. Sci. U.S.A.* 108 (25): 10184–10189.

57 Decherchi, S., Berteotti, A., Bottegoni, G. et al. (2015). The ligand binding mechanism to purine nucleoside phosphorylase elucidated via molecular dynamics and machine learning. *Nat. Commun.* 6: 6155.

58 Cavalli, A., Spitaleri, A., Saladino, G., and Gervasio, F.L. (2015). Investigating drug–target association and dissociation mechanisms using metadynamics-based algorithms. *Acc. Chem. Res.* 48 (2): 277–285.

59 Gervasio, F.L., Laio, A., and Parrinello, M. (2005). Flexible docking in solution using metadynamics. *J. Am. Chem. Soc.* 127: 2600–2607.

Part II

Advanced Algorithms

3

Modeling Ligand–Target Binding with Enhanced Sampling Simulations

Federico Comitani[1] and Francesco L. Gervasio[1,2]

[1] *University College London, Department of Chemistry, 20 Gordon Street, London WC1H 0AJ, UK*
[2] *University College London, Institute of Structural and Molecular Biology, Department of Chemistry, 20 Gordon street, London WC1H 0AJ, UK*

List of Abbreviations

aMD	accelerated molecular dynamics
CADD	computer-aided drug discovery
CV	collective variable
FEP	free energy perturbation
MetaD	metadynamics
MD	molecular dynamics
MSM	Markov state models
PT	parallel tempering
RE	replica exchange
REMD	replica-exchange molecular dynamics
ST	simulated tempering
TI	thermodynamic integration

3.1 Introduction

A detailed and quantitative knowledge of the mechanism of drugs binding to their pharmaceutical targets can greatly simplify the task of optimizing lead compounds and designing novel biologically active ones. As discussed in the previous chapters, in computer-aided drug discovery (CADD), atomistic molecular dynamics (MD) simulations combined with accurate protein and ligand force fields and advanced analysis tools such as Markov state models, can provide atomic resolution insights in both the binding mechanisms and the associated free energies and kinetics. One major challenge faced by MD is the limited timescales accessible even on supercomputers; MD can typically simulate a few to tens of microseconds, while most events of interest in pharmaceutical chemistry, such as target conformational changes and ligand unbinding, require much longer times. This *timescale problem* (reflected in the computational

Biomolecular Simulations in Structure-Based Drug Discovery,
First Edition. Edited by Francesco L. Gervasio and Vojtech Spiwok.

cost of running sufficiently long MD simulations) has impacted the usefulness of plain MD simulations in drug discovery, encouraging the development of a number of smart approaches, including *enhanced sampling algorithms*. This chapter summarizes some of the most successful methods belonging to this group, specifically applied to ligand–target binding.

3.2 The Limits of Molecular Dynamics

Since MD was first applied to the study of a protein, the bovine pancreatic trypsin inhibitor, in 1977 by Karplus and McCammon [1–6], its potential in drug discovery became clear. The possibility of observing the evolution of structural properties of the drug targets and the interactions of the ligand–target complexes at the atomic level has a significant potential in helping the design and optimization of biologically active ligands. The case of raltegravir is a typical MD success story. A potent HIV integrase inhibitor, raltegravir, was developed after the discovery, through MD-based methods, of a transient binding area which was not apparent in X-ray crystallographic structures [7, 8]. However, in spite of their potential, MD techniques are still held back by the timescale issue, as well as force field limitations and the size of the systems that can be simulated.

Let us start with the latter, the systems' size. Thanks to efficient parallel MD codes [9, 10] up to a few million atoms can now be efficiently simulated in a single system, allowing MD to model most drug targets. However, to obtain accurate predictions, the proteins and their crowded environment need often to be fully accounted for. When working with targets such as membrane proteins or multi-domain structures, the simplified representations of the system generally used may lead to inaccuracies. For instance, while homogeneous bilayers are commonly employed, experiments have shown that the heterogeneous composition of the membrane plays a fundamental role in maintaining transmembrane proteins structurally intact. Cholesterol has been observed regulating the gating process of ion channels [11–13], while the formation of rafts can directly affect their function [14]. Thus, as our knowledge of environmental effects on the activity of biological systems increases, the need for larger and more complex models also grows. A number of approaches are currently being developed to handle such large models, ranging from multi-scale models [15, 16] and specialized hardware such as Anton 2 [17] to exascale supercomputer architectures.

The accuracy of the ligand force fields and their parameterization still remains a concern. While there has been substantial progress in the accuracy of protein force fields during the years [18–20], ligand parameters still lag behind, and in spite of recent progress [21] time-consuming re-parameterization of each ligand by fitting to *ab initio* calculations is often needed.

As much as it is fundamental for the accuracy of the results, the quality of the force fields is also tightly connected to the timescale problem. Atomistic MD runs have now reached a few microseconds of biological time and, with the help of specialized machines, they have been extended up to the millisecond regime for small proteins in extreme cases [22, 23]. Longer simulations, however, highlighted a number of systematic inaccuracies of protein force fields that were not visible on shorter scales and led to the development of better parameters [24].

Notwithstanding the progress in specialized hardware specifically built to run atomistic simulations [17] and parallel codes able to run on graphics processing units (GPUs) [25], it is still difficult to observe a statistically significant number of events in most biological systems of interest (e.g. the binding or unbinding of a ligand, or the opening of a hidden druggable pocket) with plain MD.

Specifically, in ligand–target binding studies, the observation of numerous binding and unbinding trajectories is required to ensure that all relevant paths and target conformations have been explored and thus to allow for an accurate evaluation of the associated binding free energy. The development of an accurate method to evaluate this quantity is one of the top priorities of the CADD community. An MD-based method able to compute binding free energies accurately and quickly on a workstation would have a large impact on drug discovery pipelines, which to date still need to rely on less accurate and reliable methods, such as docking [26, 27].

To address the timescale problem and the other MD limitations related to it, a number of approaches have been developed by the computational chemistry community.

Coarse-graining, where a set of atoms are grouped together and parametrized as a single entity by specialized force fields, allows, for example, increasing the size of the systems and the timescales that can be simulated. Although examples of coarse-grained ligand binding studies can be found in the literature [28], these models are more commonly employed for the study of big multi-domain structures, transmembrane proteins within lipid bilayers, or self-assembly properties and are beyond the scope of this chapter [29–32].

Conversely, enhanced sampling methods are directly aimed at solving the timescale problem and can be used with atomistic, coarse-grained, and multiscale simulations.

The free energy landscape of even simple biological processes is characterized by a complex network of interconnected high-probability basins, separated by low-probability barriers and saddle points. The corrugated nature of this surface is what makes its full (ergodic) exploration rarely attainable with plain MD simulations. A large number of methods to accelerate the dynamics of the system and promote rare events have been thus developed to address the timescale problem.

The methods most used to study and quantify ligand–target binding can be roughly classified into five categories: tempering methods, multiple replica algorithms, endpoint methods, alchemical transformations, and collective variable-based methods [33–36].

3.3 Tempering Methods

Tempering methods artificially increase the temperature of the system, relying on the fact that the rate of free energy barrier crossing events is temperature dependent.

Simulated annealing (SA) takes this idea in its most basic implementation: an initial high-temperature run is followed by a slow cooling, allowing the system to escape initial metastable configuration and hopefully reach structures that are more physiologically relevant [37]. It is generally employed as an initial optimization step, running on Markov chain Monte Carlo rather than an MD engine and is strongly dependent on the cooling speed parameter.

As an improved version of this optimization algorithm, simulated tempering (ST) is also a Monte Carlo scheme that, however, is able to dynamically change the temperature while maintaining the system at equilibrium [38]. Continuous tempering molecular dynamics (CTMD) is a generalized version of ST that was recently developed by Lenner and Mathias. This approach works in a continuous temperature space and, in opposition to the stochastic nature of MC-based methods, integrates the deterministic equation of motions [39].

The most popular MD-based method in this category is however parallel tempering (PT), and it is discussed in more detail in the next section.

An interesting example illustrating the use of ST and CTMD in drug discovery was the case of peptide recognition on the orthosteric site of human immunoglobulin G. Monte Carlo-based ST was employed to identify three highly conserved residues Asn434, Ser254, and Tyr436, key to the peptide–target binding affinity [40].

A criticism concerning tempering approaches is that force fields are not parameterized to work outside physiological temperatures, which, when increased significantly, may lead to artifacts or unphysical behavior. They also rely on the assumption that ergodicity can be reached in the modified ensemble, which may not be the case, especially in the presence of entropic barriers. Indeed, increasing the temperature makes the entropic barrier more difficult to overcome, paradoxically making the exploration of the free energy surface more challenging [35, 36].

3.4 Multiple Replica Methods

A straightforward and convenient strategy to improve the conformational sampling is to simulate the system in multiple copies simultaneously. This trivially parallel nature of such simulations is well suited for large parallel computers, but the advantages of running multiple independent copies of the same model without biasing the dynamics (multiple walkers) are limited to the computational resources available, similar to running a single, long simulation. While multiple independent replicas can thus sometimes alleviate the sampling problems, they still need to be coupled with schemes to push the system across high-free energy barriers. This idea led to the development of a number of clever methods, employing on-the-fly communication and exchanges between replicas.

Among the first methods of this class to be developed was SWARM-MD, were a group (or swarm) of molecules or proteins are simulated in parallel starting from the same configuration. What distinguishes this method from other multiple walker techniques is the presence of a restraining potential that keeps the trajectory of the replicas compact in the conformational and dynamical space.

The copies are prevented from drifting too far from the average trajectory followed by the swarm, making the search for free energy pathways and minima cooperative [41].

In PT, a replica-exchange molecular dynamics (RE or REMD) method, starting from a system at standard room or physiological temperature, the subsequent copies are run at increasingly higher temperatures. Atomic coordinates are then periodically swapped between neighboring replicas, with a Metropolis acceptance probability based on their potential energies and temperatures: if the potential energy of the higher temperature is lower than the one of the lower temperature replica, the exchange is accepted; otherwise, the probability of acceptance is expressed by

$$p_{i,j} = \mathrm{e}^{\frac{1}{k_\mathrm{B}}(E_j - E_i)\left(\frac{1}{T_i} - \frac{1}{T_j}\right)}$$

where E_i is the potential energy of the ith replica, T_i is its temperature, and K_B is the Boltzmann constant. This allows the system to overcome potential barriers and access hidden conformations, while maintaining an equilibrium dynamics in the lowest temperature replica [42, 43].

Numerous variants of this simple but effective formula have been developed in recent years, tailored to solve specific biochemical problems, or, more generally, improve the algorithm. In solute tempering replica-exchange molecular dynamics (REST), for example, the Hamiltonian function of each replica is modified to make sure the acceptance probability is independent of the number of explicit water molecules in the system, improving the scalability of the method [44].

REMD is extremely versatile; the replicas are not necessarily limited by different temperatures. Modified Hamiltonians with exotic properties, as in the case of generalized RE [45] can also be very effective. On the basis of accelerated molecular dynamics (aMD), replica-exchange accelerated molecular dynamics (REXaMD), belongs to this group. In the original aMD method, the free energy landscape is modified by adding bias only to states below a certain energy threshold. This bias, effectively a Hamiltonian modification, raises these states, and makes the barriers separating them easier to overcome. In its REXaMD implementation, this bias is scaled to different levels for different replicas [46, 47] (Figure 3.1).

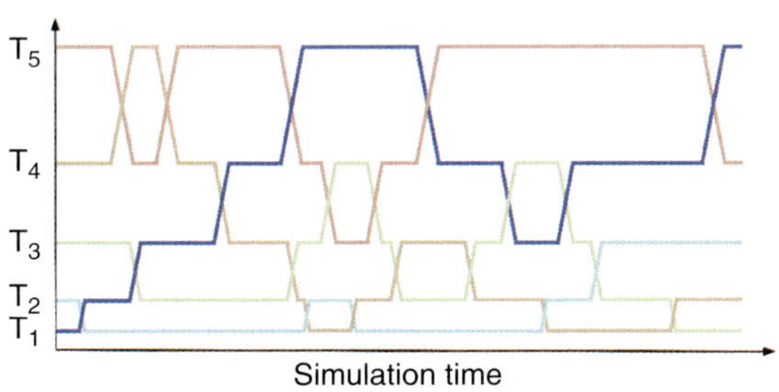

Figure 3.1 Schematic representation of a replica-exchange algorithm, where a number of identical copies of the same system are simulated concomitantly at different temperatures (parallel tempering). Exchanges between pairs of replica coordinates are attempted at regular intervals and accepted or rejected according to a probability dependent on the potential energies of the two replicas.

In 2016, Oleinikovas et al. developed SWISH (sampling water interfaces through scaled Hamiltonians), an efficient algorithm to improve the exploration of hidden (cryptic) pockets and test their druggability. It is based on a Hamiltonian RE method that gradually scales the interactions between water molecules and C-α and sulfur atoms of the solute. The scaling factor is gradually increased for different replicas, modulating the water–interface interactions through a λ parameter, and allowing highly hydrophobic, and possibly druggable, patches to open to the ligand. In the paper, the method has been initially to three exemplary systems: TEM1 β-lactamase, involved in antibiotic resistance; interleukin-2, which plays a role in the activation of T cells and in graft tissue rejection; and a validated anticancer target, Polo-like kinase-1. The results are indeed promising: SWISH was able to correctly identify known cryptic pockets (previously discovered by chance) on all three proteins. More recently, its has been successfully applied to four more pharmacologically relevant targets [48]. This method is particularly effective in driving the opening of hidden hydrophobic pockets that would otherwise go unnoticed by canonical structure-based methods. Moreover, combined with molecular probes with suitable physicochemical properties, it allows to characterize such sites and to test their druggability [49].

3.5 Endpoint Methods

Endpoint methods rely on the fact that, at variance from the kinetics of a reaction, the free energy is a state function, defined as the difference between the energy of two states (e.g. the bound and unbound states of a ligand to its target) and is independent of the paths joining them [50].

Molecular mechanics/Poisson–Boltzmann surface area (MM/PBSA) and its more approximate version molecular mechanics/generalized Born surface area (MM/GBSA) are two of the most widely used algorithms within this group. The difference in free energy between bound and unbound states is obtained through the calculation of an MM term, a solvation free energy, a conformational entropic contribution from normal modes and an electrostatic term obtained through either a generalized Born or Poisson–Boltzmann method. These scores are calculated over a few static snapshots of the bound state, while the unbound counterpart is obtained by trivially removing the ligand from its site [51, 52].

No conformational change in the target protein is taken into account in the original algorithms, often resulting in inaccurate free energy predictions [53, 54]. In spite of these obvious limitations, these approaches are often used to obtain a qualitative comparison between different target mutants or models, as recently shown for ion channels [55], or rank the affinities of different ligands to the same protein, as in the case of REV-ERBα, tested against 35 different agonists [56].

3.5.1 Alchemical Methods

Alchemical free energy calculations are among the most accurate and well-established endpoint methods in CADD.

The term "alchemical" was inspired by the fact that, *in silico*, a molecule can be "transmuted" to a noninteracting molecule (its interaction potentials are switched off) or to a molecule with different physicochemical properties through a series of unphysical intermediates steps. One of the first applications of such methods to compute the relative binding free energies of a number of congeneric ligands was reported by Tembre and McCammon in 1984 [57].

Nowadays, the most used alchemical methods with explicit solvent are thermodynamic integration (TI) [58] and free energy perturbation (FEP) [59].

In TI, a continuous parameter λ is introduced to define a series of intermediate nonphysical (alchemical) states between the endpoints ($\lambda = 0$ to $\lambda = 1$, which, for instance, may correspond to the unbound and bound states). The free energy difference between the endpoints is calculated by computing the average of the derivative of the free energy with respect to λ at each step.

$$\Delta G = \int_0^1 \left\langle \frac{\mathrm{d}H(\lambda, x)}{\mathrm{d}\lambda} \right\rangle_\lambda$$

TI provides rather accurate estimates of the free energy difference if the chosen alchemical change results in a smooth function $\langle {}^{\mathrm{d}H}/_{\mathrm{d}\lambda} \rangle$. However, if there are large and abrupt changes, e.g. when a molecule finally disappears and the Lennard-Jones potentials is turned off, the discretization error can also become rather large [60]. Thus, the accuracy crucially depends on the choice of the specific alchemical transformation [61, 62]. Once an effective alchemical pathway has been devised, the free energy difference is computed from the individual simulation windows run at different λ values by numerically solving the integral. A number of techniques have been applied to this end, with high-order integration methods allowing a larger distance between integration points [63]. TI can be also combined with methods that introduce external forces to drive the target toward desired conformations, such as targeted MD and steered MD [64, 65].

On the other hand, the name FEP refers specifically to the alchemical changes made to the identities of ligand substituents, such as transforming a bromine to a chlorine. In FEP, the free energy is usually computed using the Zwanzig relationship

$$\Delta G = -kT \, \ln \langle \exp^{-\beta[H_1(x) - H_0(x)]} \rangle_0$$

where the two Hamiltonians $H_0(x)$ and $H_1(x)$ are the original and the "perturbed" ones, respectively [59].

As in the case of MM/PBSA, the accuracy of FEP is also affected by its limited conformational sampling capabilities. If the target undergoes large conformational changes upon binding, it can be difficult to converge the binding free energy by FEP.

Indeed, the method greatly benefited from the combination with effective conformational sampling algorithms. A systematic application of a combination of FEP and enhanced sampling protocols to a large number of different ligand–target systems has shown a high level of accuracy [66].

3.6 Collective Variable-Based Methods

Collective variable (CV)-based methods require the definition of a limited set of $M \ll 3N$ variables $\boldsymbol{s}(\boldsymbol{r}, t)$ (where N is the number of atoms in the system) that are more or less complex functions of the atomic Cartesian coordinates $\mathbf{r}(t)$ and approximate the ideal reaction coordinate. For instance, in the case of ligand binding, a relevant CV is usually the distance of the ligand from the target's pocket. These methods typically use an additional potential to enhance the exploration of the configurational space and reconstruct the free energy landscape as a function of the selected CVs. A few of these methods focus on the calculation of mean forces, that is, the gradient of the free energy as measured at specific locations.

One of the first and possibly the best-known example of the CV-based sampling method is umbrella sampling (US) [67]. In this framework, a set of neighboring and partially overlapping windows are defined along the transition path of interest. In turn, a harmonic potential is applied to the system to restrain it within the chosen window, forcing it to explore both hard-to-sample and favorable configurations alike within the localized region. The information obtained from each window can then be used to reconstruct the unbiased probabilities and free energies through the weighted histogram analysis method (WHAM) [67, 68].

Approaching the problem from different perspectives, various groups developed a number of more or less successful CV-based methods, and reviewing them all is beyond the scope of this chapter.

Belonging to this group is also, for example, the adaptive biasing force (ABF) algorithm, which tries to achieve a uniform sampling on a flat free energy surface by adding external forces to the equations of motion. These are built on the fly to counterbalance the mean forces felt locally by the system along with an MD trajectory [69].

aMD is another good example. It adds a nonnegative boost potential to smoothen the high-energy barriers separating different low-energy conformational basins and thus accelerating the transition [46]. Gaussian accelerated molecular dynamics (GaMD or hyperdynamics) is a recent improvement on aMD, where the bias potential follows a near-Gaussian distribution, allowing for an improved reweighting with a cumulant expansion to the second order that solves the energetic noise affecting classical aMD [70].

Steering molecular dynamics (SMD) is a perhaps more straightforward approach, where harmonic forces are applied to (parts of) the system to pull it in a chosen direction and force it over the energy barriers that would otherwise impede the exploration of certain conformations. This method typically finds good use to investigate mechanical functions, such as muscle protein stretching or to pull a ligand from a superficial binding on a substrate [71, 72].

3.6.1 Metadynamics

A subset of CV-based sampling methods uses the intriguing idea of a biasing potential that adapts on the fly to the local corrugations of the free energy. Together with algorithms such as conformational flooding [73] and local elevation [74], metadynamics (MetaD) is the most widely known member of this group. Its efficiency in driving the exploration of rare events and the possibility of reconstructing the free energy surface at convergence has made this method a popular choice in recent years. The algorithm is implemented in most major MD codes and has evolved considerably from the original implementation of Laio and Parinello [75, 76].

MetaD requires, first of all, the definition of a set of CVs that approximate the reaction coordinate. The free energy in this new reduced CV space can be obtained by weighting all the corresponding conformations in $\boldsymbol{r}$ that map onto the different values of the CV $\boldsymbol{s}$ itself. This gives us the probability

$$P(\boldsymbol{s}) = \langle \delta(\boldsymbol{s}(\boldsymbol{r}) - \boldsymbol{s}) \rangle$$

where the angle brackets represent the Boltzmann ensemble average. For a system subject to a potential, $U(\boldsymbol{r}, t)$ leads to a free energy of the form

$$F(\boldsymbol{s}) = -\frac{1}{\beta} \log \left(\int e^{-\beta U(\boldsymbol{r},t)} \delta(\boldsymbol{s} - \boldsymbol{s}(\boldsymbol{r})) \, \mathrm{d}\boldsymbol{r} \right)$$

where $\beta = \frac{1}{k_B T}$, k_B is the Boltzmann constant and T is the temperature.

To improve the exploration of the CV space, a history-dependent potential is then built on the fly along the trajectory of the system. This potential is composed of the sum of Gaussian-shaped repulsive terms deposited during the run.

This limits the exploration of conformational regions that were previously visited, while at the same time gradually filling the underlying free energy basins. The system is thus "pushed" to escape the energy wells and cross barriers that would otherwise trap it if tackled with classical MD simulations. In this manner, the occurrence of rare events, namely, the transition to new unexplored states by crossing high-energy barriers is efficiently accelerated.

The Gaussian-built bias potential is a function of the values $\boldsymbol{s}(\boldsymbol{r}, t)$ of the CVs at the corresponding coordinates $\boldsymbol{r}(t)$ at the present time t and at all its past values $t' < t$, according to the expression

$$U_G(\boldsymbol{s}, t) = w \sum_{t'=\tau, 2\tau, \ldots, <t} e^{\left(-\frac{(s(r,t) - s(r,t'))^2}{2\sigma^2} \right)}$$

Here, w is the Gaussian height, σ is its width, and τ is the time step of deposition.

Once the convergence of the simulation is reached, the free energy surface is completely filled by the MetaD potential, leading to a flat surface and possible diffusivity in the space of the CVs.

Under these conditions, the bias potential itself provides an unbiased estimate of the shape of the underlying free energy, according to the limit

$$\lim_{t \to \infty} U_G(\boldsymbol{s}, t) \sim -F(\boldsymbol{s})$$

This unassuming relation is however extremely powerful, as it implies that the free energy, an equilibrium quantity, can be estimated by a nonequilibrium dynamics, as the potential changes with time at each addition of a new Gaussian term. The validity of this property, which was postulated heuristically, has been verified empirically for a general case, but it can be demonstrated rigorously for a system evolving under the action of a Langevin dynamics [76–78].

This implementation of MetaD, however, may lead to errors in the case of the free energy landscape even at infinite time, due to the uneven distribution of Gaussians deposited during the simulations. After the diffusivity of the CVs has been reached, the algorithm keeps adding Gaussians following the position of the system; a straightforward solution to this problem requires calculating the average of a number of profiles extracted at discrete intervals after convergence [79].

In order to solve this problem from its source, the well-tempered ensemble implementation has been developed. The deposition rate (the Gaussian height per unit time) is adaptively decreased locally in regions where the bias is accumulating [77].

The Gaussian height w is rescaled every step according to the equation

$$w = w_0 \mathrm{e}^{-\frac{U(s,t)}{k_B \Delta T}}$$

where w_0 is the value set at the beginning of the simulation and ΔT is a tunable parameter with the dimension of a temperature. This term allows tuning the speed of rescaling through the bias factor $^{T+\Delta T}/_{T}$ and can determine the extent of the exploration.

The effective sampling of the CVs is carried out at the higher temperature $T + \Delta T$, as it can be understood more intuitively by looking at the two extreme cases: with $\Delta T \rightarrow 0$, one recovers classical MD, while $\Delta T \rightarrow \infty$ deactivates the lowering of the height of the Gaussians and recovers plain MetaD.

$$U(\boldsymbol{s}, t \rightarrow \infty) = -\frac{\Delta T}{T + \Delta T} F(\boldsymbol{s}) + \text{const}$$

Decreasing the size of the Gaussians in regions that were already well explored improves the accuracy of the reconstructed free energy; and at variance from the original fixed-height Gaussian implementation, has been mathematically shown to converge to the correct free energy in the long term [80]. The trade-off is that as the deposition rate is decreased, the time required to converge the free energy is increased.

While with this approach it is easy to reconstruct the probability distribution and the free energy as a function of the set of chosen CVs, the addition of the bias potential also distorts the distribution of other degrees of freedom (DOF). Different from other methods such as US, the MetaD bias is time dependent, making the recovery of the full Boltzmann distribution of nonbiased DOF nontrivial.

Under specific conditions, however, it is still possible to extract the unbiased probability, through a reweighting of the bias potential [81].

Exploiting the relationship between the biased and unbiased potentials, as in

$$P(\boldsymbol{r}, t) = \mathrm{e}^{-\beta(U(s(\boldsymbol{r}),t)+c(t))} P_0(\boldsymbol{r})$$

where $P_0(\boldsymbol{r})$ is the unbiased Boltzmann distribution $e^{-\beta U(\boldsymbol{r})}/Z$ and $c(t)$ is the time-dependent bias offset

$$c(t) = -\frac{1}{\beta} \log\left(\frac{\int \mathrm{e}^{-\beta F(\boldsymbol{s})} \mathrm{d}s}{\int \mathrm{e}^{-\beta(F(\boldsymbol{s})+U(\boldsymbol{s},t))} \mathrm{d}s} \right)$$

This property can be exploited to reconstruct the free energy landscape as a function of unbiased CVs, assuming that they were fast enough to properly sample the range of interest within the simulation time [81].

Several successful examples of the use of MetaD in ligand binding have been reported in the literature, many of which have been extensively reviewed in Ref. [26]. Among the first notable cases, Gervasio and coworkers were able to correctly predict the ligand–target association free energies of several systems, in agreement with experimental results [82]. In this case, MetaD proved faster than US in reconstructing the full FES, but the use of an approximate force field, the relatively short length of the simulations, and the choice of a suboptimal CV affected the results.

Given its ability to increase the sampling of binding along physical association pathways, MetaD seems a natural choice for the study of ligand–target binding. Still, the complexity of some of these binding mechanisms, especially when conformational changes in the protein are required prior to the ligand approaching the site, can make the choice of CVs difficult.

Trial runs with suboptimal CVs can still be useful in exploring qualitative interactions and the environment in proximity to the pocket, but are not sufficient to the evaluation of a binding free energy with accuracy comparable to experimental data [55, 83, 84] (Figure 3.2).

Coarse MetaD was developed by Masetti et al. specifically to rank different ligands binding to a target [85]. The exploration of the free energy is limited to the transition state and the ligand is prevented from fully leaving the binding pocket avoiding the time-consuming exploration of fully solvated states and multiple unbinding and rebinding events. The success of the method is mostly due to the fact that the height of the transition state is, in most cases, correlated with the binding free energy. Indeed, the results were so promising that they inspired the development of a binding pose MetaD approach recently included in Schrodinger tools [86].

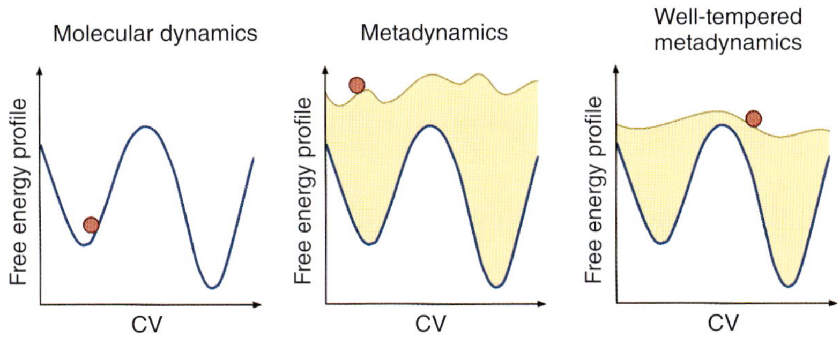

Figure 3.2 The effects of MetaD on a monodimensional toy model. In classical MD, the system (red sphere) is trapped in its initial conformational basin (blue profile). With metadynamics, the added bias (in yellow) fills the basins following the trajectory of the system, allowing it to escape its starting conformation and sample other areas of the free energy profile. In the well-tempered regime, the added bias does not overshoot the barriers and the free energy profile cast is more accurate.

A straightforward solution to improve the convergence time is to radically reduce the CV space that needs to be explored by means of a restraining potential. For instance, the number of solvated states accessible to the ligand when far from the target can be restricted by a potential shaped in such a way that it does not affect the binding region itself, but lowers the available volume of unbound states such that one can observe a statistically relevant number of in and out events, sufficient to improve on the accuracy of the binding free energy measured value.

This method has been applied successfully to study the behavior of a number of systems, from neurotransmitter binding to ion channels, to drugs targeting G protein-coupled reactors (GPCRs) [87–89], but it has limited applicability to simple mechanisms where the conformational changes are either minor or can be efficiently sampled by standard CVs without impacting too much on the MetaD calculation burden.

When the ligand is bound to follow a more complex path, with induced fit effects rather than conformational selection mechanisms at play in the protein, more advanced CV choices are required.

Path-based collective variables (pathCVs) represent a smart solution to this problem; the transition from an initial state A and a final state B is divided in n intermediated microstates l, where A and B are recovered when $l = 1$ and $l = 2$, respectively.

Two CVs, the progression along the path s and the distance from the path z, are then defined as nonlinear functions of the system's DOF $\boldsymbol{r}$, according to the following equations:

$$s(\boldsymbol{r}) = \frac{1}{P-1} \frac{\sum_{l=1}^{n} (l-1) \mathrm{e}^{-\lambda \|s(\boldsymbol{r}) - s(\boldsymbol{r}(t))\|^2}}{\sum_{l=1}^{n} \mathrm{e}^{-\lambda \|s(\boldsymbol{r}) - s(\boldsymbol{r}(t))\|^2}}$$

$$z(\boldsymbol{r}) = -\frac{1}{\lambda} \ln \left(\sum_{l=1}^{n} \mathrm{e}^{-\lambda \|s(\boldsymbol{r}) - s(\boldsymbol{r}(t))\|^2} \right)$$

Here, λ is a tunable parameter, inversely proportional to the square of the root-mean-square deviation (RMSD) between the path landmarks, while the term $\|s(\boldsymbol{r}) - s(\boldsymbol{r}(t))\|$ represents the distance from such configurations in a metric of choice (usually Euclidean) [90].

Complex conformational changes and movements, otherwise hard to describe through more basic geometrical CVs, can be sampled along the curvilinear path that connects the chosen snapshots and those that approximate the transition pathway.

PathCVs proved essential in explaining large-scale conformational rearrangements of cyclin-dependent kinase 5 (CDK5) [91], to the study of the binding mechanism of the β2-adrenergic receptor [92], and allowed an in-depth characterization of the dissociation mechanism of a potent inhibitor from the cyclooxygenase enzymes COX1 and COX2 [93, 94].

Even when using pathCVs, however, the introduction of a restraining potential may be advisable to reduce the exploration of high-z conformations, far from the path, in the solvated region (Figure 3.3).

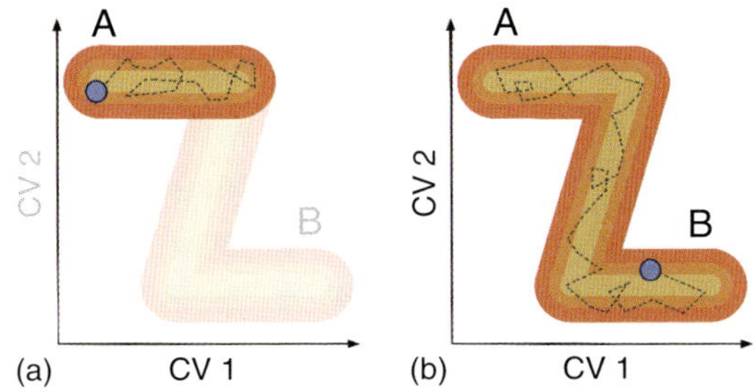

Figure 3.3 The importance of choosing the correct MetaD CVs on a bidimensional toy model. When using only partial or incorrect CVs (e.g. only CV 1 in (a)), the system (blue sphere) can be trapped in a limited sampling space and the reconstruction of the free energy surface may be incomplete or incorrect. When appropriate alternative CVs are added (e.g. CV 2 in (b)), the system can sample the free energy landscape in full.

3.7 Binding Kinetics

The link between kinetics and efficacy of compounds is becoming increasingly evident in recent years. While often ignored in the past, under the assumption that the binding affinity could quantify with accuracy the efficacy *in vivo*, the prediction of kinetic rates has now become a crucial step in the drug discovery pipeline. The concentration of drugs changes over time in the cellular environment, due to a variety of physiological and chemical mechanisms (such as distribution, adsorption, and excretion), all of which can be taken into account when working with the residence time (τ_{off}) instead, namely, the effective ligand–target binding time [82, 95]. Dealing with rate constants also allows improving selectivity and tuning residence/dissociation times and toxicity.

Computational methods represent a good opportunity to measure and investigate these quantities which are still elusive to most experimental alternatives.

To understand the kinetics of a compound, a number of binding events need to be observed between the ligand and its target. Averaging the association and dissociation times observed is a straightforward way to derive kinetic rates by means of MD-based simulations. However, as detailed in previous sections, reproducing rare binding events with classical MD may be well over the capabilities of most modern computers, requiring often the use of enhanced sampling methods.

Examples in the literature of computational studies of binding kinetics are limited and mostly focus on two strategies, employing either MSM or transition path sampling (TPS). MD-MSM methods have been successfully applied to small systems (such as benzamidine binding to trypsin), but they present remarkable limitations. Relying on short trajectories sampling the conformational basins of interest, the extracted rates tend to be affected by large errors, due to the poor sampling of full association and dissociation events. Still, an exceptional number of MD trajectories need to be generated, making this method quite expensive [96, 97].

At variance, TPS has been employed effectively to study the kinetic properties of a wide number of folding problems, from simple β-hairpin toy models [98] to the Trp-cage [99], the FBP28 WW domain [100], and the photoactive yellow protein (PYP) [101].

Given a number of favorable transition pathways from the bound to the unbound state and back, TPS builds a Monte Carlo random walk in the path space to generate the ensemble of all transition paths, from which one can then evaluate kinetic rates [102, 103]. This algorithm is powerful, but limitations in its implementation become apparent when dealing with pathways characterized by intermediate metastable states, and possibly multiple reaction channels, typical of some ligand–target binding problems.

Transition state partial-path transition interface sampling (TS-PPTIS) [104] is a reactive flux-like reformulation of PPTIS [105] and was developed specifically to treat systems with very diffusive barriers. In such cases, sampling the complete path in one go can be demanding in terms of computational resources; as an alternative, the pathway is separated in windows (each one defined by three interfaces) within which a shorter path can be independently sampled.

The rate is expressed as a reactive flux through the transition state TS (λ_0) $k = P_A(\lambda_0) \times R(\lambda_0)$, where the first term is derived from the free energy of the TS according to

$$P_A(\lambda_0) = \frac{e^{-\beta G(\lambda_0)}}{\int_{-\infty}^{\lambda_0} d\lambda \, e^{-\beta G(\lambda)}}$$

Instead of sampling entire trajectories from backward and forward in time starting from the TS, the transmission coefficient $R(\lambda_0)$, is obtained through successive approximations

$$R_0 = R^{TST}, R_1 = \frac{1}{2}R^{TST}(p_0^- p_0^{\pm} + p_0^+ p_0^{\mp}), \ldots$$

$$R_m = \frac{\frac{1}{2}R^{TST}(p_0^- p_0^{\pm} + p_0^+ p_0^{\mp})A_m \overline{A_m}}{p_0^{\pm} A_m + p_0^{\mp} \overline{A_m} + (1 - p_0^{\pm} - p_0^{\mp})A_m \overline{A_m}}$$

Here, R^{TST} is the transition state theory rate and is calculated as average crossing velocity at the λ_0 interface. This approximation requires PPTIS short-distance crossing probabilities ($p_0^+, p_0^-, p_0^{\pm}, p_0^{\mp}$) of interfaces measured at the barrier ($\lambda_{-m}, \ldots, \lambda_0, \ldots, \lambda_m$). A series of short trajectories are sampled within overlapping windows carefully positioned along the path barrier.

A_m can then be calculated through recursive relations, while p_0^- and p_0^+ are the standard conditional crossing probability without history dependence.

TS-PPTIS was recently applied to the study of the binding of imatinib to the proto-oncogene c-Src, a particularly difficult case, where both conformational selection and induced-fit mechanisms play a role, and predicted a k_{off} in good agreement with surface plasmon resonance experiments [106].

By taking advantage of the memory-loss assumption, this method provides a computationally efficient alternative to TPS and PPTIS, especially in cases where diffusive barriers require a relatively long time before the trajectories drop off the barrier into the stable states.

TPS and its derivatives still require the definition of reaction coordinates to separate the bound and unbound state and drive their transition. Reaction rates are independent of the specific form of the CVs employed, as long as the separation between reagent and product states is clearly respected. Less than ideal CVs, however, could hinder the convergence of the calculation, due to the presence of hysteresis or a low transmission coefficient. From this point of view, path collective variables represent the ideal choice when working with TPS-based methods, as they closely follow the transition pathway and are able to correctly take into account all the complex association and dissociation mechanisms involved.

Without relying on either MSM or TPS, Tiwary et al. recently proposed a method to recover binding and unbinding rates through infrequent deposition of MetaD bias [107]. The idea behind this approach builds on the assumption of processes characterized by transitions between two stable states through a dynamical bottleneck, rarely crossed. It thus requires, first of all, a simplified free energy landscape, where the process of interest is described by reagent and product states separated by a single well-defined barrier that can be described by simple CVs. Furthermore, the time spent by the system in the transition state needs to be much shorter than the bias deposition time set for the MetaD simulation.

Although effective when these assumptions are respected, its limited applicability makes this method less desirable when compared to TPS-based alternatives, especially when dealing with diffusive barriers or complex association/dissociation mechanisms involving multiple intermediate states.

An alternative approach was developed by Mollica et al. in 2015 [108]. In this case, the unbiased dissociation times are obtained by means of an Arrhenius-like formula applied to multiple partially restrained replicas that are run and stopped at the unbinding of the ligand [108].

The method has been applied to simulations enhanced through scaled potentials [109–111] on three test cases, a set of HSP90 inhibitors, adenosine-derived inhibitors targeting Grp78 and antagonists of the adenosine A_{2A} receptor complex, and more recently to a series of glucokinase inhibitors [108, 112]. The algorithm is sensitive to the choice of restraints which need to be applied to portions of the proteins: in all the test cases, in fact, no significant conformational changes were expected upon binding, allowing for a trivial choice of restraints. Moreover, the resulting dissociation rates are affected by large standard deviations due to the loss of information on the basins through the smoothing of the potential.

3.8 Conclusions

Recent progress in algorithms and force fields has made predictions based on enhanced sampling MD simulations increasingly accurate for CADD. In this chapter we have given an overview on a number of widely used algorithms and their successful applications, showing how enhanced sampling methods can be used to predict binding mechanisms, accurate relative and absolute binding free energies as well as binding kinetics, and target conformational free energy landscapes.

Their flexibility and increasing accuracy, together with the availability of inexpensive computing resources provided by cloud computing and GPUS, will beyond doubt make these MD-based methods a mainstay in drug discovery pipelines.

References

1 Alder, B.J. and Wainwright, T.E. (1957). Phase transition for a hard sphere system. *J. Chem. Phys.* 27: 1208.

2 Alder, B.J. and Wainwright, T.E. (1959). Studies in molecular dynamics. I. General method. *J. Chem. Phys.* 31: 459.

3 Rahman, A. (1964). Correlations in the motion of atoms in liquid argon. *Phys. Rev.* 136: A405–A411.

4 Stillinger, F.H. (1974). Improved simulation of liquid water by molecular dynamics. *J. Chem. Phys.* 60: 1545.

5 Warshel, A. and Levitt, M. (1976). Theoretical studies of enzymic reactions: dielectric, electrostatic and steric stabilization of the carbonium ion in the reaction of lysozyme. *J. Mol. Biol.* 103: 227–249.

6 McCammon, J.A., Gelin, B.R., and Karplus, M. (1977). Dynamics of folded proteins. *Nature* 267: 585–590.

7 Schames, J.R., Henchman, R.H., Siegel, J.S. et al. (2004). Discovery of a novel binding trench in HIV integrase. *J. Med. Chem.* 47: 1879–1881.

8 Summa, V., Petrocchi, A., Bonelli, F. et al. (2008). Discovery of raltegravir, a potent, selective orally bioavailable HIV-integrase inhibitor for the treatment of HIV-AIDS infection. *J. Med. Chem.* 51: 5843–5855.

9 Phillips, J.C., Braun, R., Wang, W. et al. (2005). Scalable molecular dynamics with NAMD. *J. Comput. Chem.* 26: 1781–1802.

10 Abraham, M.J., Murtola, T., Schulz, R. et al. (2015). GROMACS: high performance molecular simulations through multi-level parallelism from laptops to supercomputers. *SoftwareX* 1: 19–25.

11 Levitan, I., Singh, D.K., and Rosenhouse-Dantsker, A. (2014). Cholesterol binding to ion channels. *Front. Physiol.* 5: 65.

12 Brannigan, G., Hénin, J., Law, R. et al. (2008). Embedded cholesterol in the nicotinic acetylcholine receptor. *Proc. Natl. Acad. Sci. U.S.A.* 105: 14418–14423.

13 Purcell, E.K., Liu, L., Thomas, P.V., and Duncan, R.K. (2011). Cholesterol influences voltage-gated calcium channels and BK-type potassium channels in auditory hair cells. *PLoS One* (Dryer, S. E., Ed.) 6: e26289.

14 Szabò, I., Adams, C., and Gulbins, E. (2004). Ion channels and membrane rafts in apoptosis. *Pflugers Arch. Eur. J. Physiol.* 448 (3): 304–312.

15 Saunders, M.G. and Voth, G.A. (2013). Coarse-graining methods for computational biology. *Annu. Rev. Biophys.* 42: 73–93.

16 Simunovic, M., Evergren, E., Golushko, I. et al. (2016). How curvature-generating proteins build scaffolds on membrane nanotubes. *Proc. Natl. Acad. Sci. U.S.A.* 113: 11226–11231.

17 Shaw, D.E., Grossman, J.P., Bank, J.A. et al. (2014). Anton 2: raising the bar for performance and programmability in a special-purpose molecular dynamics supercomputer. *International Conference for High Performance Computing, Networking, Storage and Analysis, SC*, pp. 41–53. IEEE.

18 Lindorff-Larsen, K., Piana, S., Palmo, K. et al. (2010). Improved side-chain torsion potentials for the Amber ff99SB protein force field. *Proteins* 78: 1950–1958.
19 Shaw, D.E., Maragakis, P., Lindorff-Larsen, K. et al. (2010). Atomic-level characterization of the structural dynamics of proteins. *Science* 330: 341–346.
20 Robertson, M.J., Tirado-Rives, J., and Jorgensen, W.L. (2015). Improved peptide and protein torsional energetics with the OPLSAA force field. *J. Chem. Theory Comput.* 11: 3499–3509.
21 Harder, E., Damm, W., Maple, J. et al. (2016). OPLS3: a force field providing broad coverage of drug-like small molecules and proteins. *J. Chem. Theory Comput.* 12: 281–296.
22 Shaw, D.E., Bowers, K.J., Chow, E. et al. (2009). Millisecond-scale molecular dynamics simulations on Anton. *Proceedings of the Conference on High Performance Computing Networking, Storage and Analysis – SC'09*, p 1. New York, New York, USA: ACM Press.
23 Lindorff-Larsen, K., Maragakis, P., Piana, S., and Shaw, D.E. (2016). Picosecond to millisecond structural dynamics in human ubiquitin. *J. Phys. Chem. B* 120: 8313–8320.
24 Piana, S., Lindorff-Larsen, K., and Shaw, D.E. (2011). How robust are protein folding simulations with respect to force field parameterization? *Biophys. J.* 100: L47–L49.
25 Buch, I., Harvey, M.J., Giorgino, T. et al. (2010). High-throughput all-atom molecular dynamics simulations using distributed computing. *J. Chem. Inf. Model.* 50: 397–403.
26 Cavalli, A., Spitaleri, A., Saladino, G., and Gervasio, F.L. (2014). Investigating drug–target association and dissociation mechanisms using metadynamics-based algorithms. *Acc. Chem. Res.* 48: 277–285.
27 Saladino, G., Estarellas, C., and Gervasio, F.L. (2017). Recent progress in free energy methods. In: *Comprehensive Medicinal Chemistry III*, 34–50. Elsevier.
28 Negami, T., Shimizu, K., and Terada, T. (2014). Coarse-grained molecular dynamics simulations of protein–ligand binding. *J. Comput. Chem.* 35: 1835–1845.
29 Kamerlin, S.C.L., Vicatos, S., Dryga, A., and Warshel, A. (2011). Coarse-grained (multiscale) simulations in studies of biophysical and chemical systems. *Annu. Rev. Phys. Chem.* 62: 41–64.
30 Bond, P.J., Holyoake, J., Ivetac, A. et al. (2007). Coarse-grained molecular dynamics simulations of membrane proteins and peptides. *J. Struct. Biol.* 157: 593–605.
31 Orozco, M., Orellana, L., Hospital, A. et al. (2011). Coarse-grained representation of protein flexibility. Foundations, successes, and shortcomings. *Adv. Protein Chem. Struct. Biol.* 85: 183–215.
32 Wolkenhauer, O., Auffray, C., Brass, O. et al. (2014). Enabling multiscale modeling in systems medicine. *Genome Med.* 6: 21.

33 Leach, A.R. (2001). *Molecular Modelling: Principles and Applications.* Prentice Hall.

34 Adcock, S.A. and McCammon, J.A. (2006). Molecular dynamics: survey of methods for simulating the activity of proteins. *Chem. Rev.* 106: 1589–1615.

35 Abrams, C. and Bussi, G. (2014). Enhanced sampling in molecular dynamics using metadynamics, replica-exchange, and temperature-acceleration. *Entropy* 16: 163–199.

36 Bernardi, R.C., Melo, M.C.R., and Schulten, K. (2015). Enhanced sampling techniques in molecular dynamics simulations of biological systems. *Biochim. Biophys. Acta* 1850: 872–877.

37 Kirkpatrick, S., Gelatt, C.D., and Vecchi, M.P. (1983). Optimization by simulated annealing. *Science* 220: 671–680.

38 Marinari, E. and Parisi, G. (1992). Simulated tempering: a new monte carlo scheme. *Europhys. Lett.* 19: 451.

39 Lenner, N. and Mathias, G. (2016). Continuous tempering molecular dynamics: a deterministic approach to simulated tempering. *J. Chem. Theory Comput.* 12: 486–498.

40 Verkhivker, G.M., Bouzida, D., Gehlhaar, D.K. et al. (2002). Monte Carlo simulations of the peptide recognition at the consensus binding site of the constant fragment of human immunoglobulin G: the energy landscape analysis of a hot spot at the intermolecular interface. *Proteins Struct. Funct. Genet.* 48: 539–557.

41 Huber, T. and van Gunsteren, W.F. (1998). SWARM-MD: searching conformational space by cooperative molecular dynamics. *J. Phys. Chem. A* 102 (29): 5937–5943.

42 Hansmann, U.H.E. (1997). Parallel tempering algorithm for conformational studies of biological molecules. *Chem. Phys. Lett.* 281: 140–150.

43 Sugita, Y. and Okamoto, Y. (1999). Replica-exchange molecular dynamics method for protein folding. *Chem. Phys. Lett.* 314: 141–151.

44 Liu, P., Kim, B., Friesner, R.A., and Berne, B.J. (2005). Replica exchange with solute tempering: a method for sampling biological systems in explicit water. *Proc. Natl. Acad. Sci. U.S.A.* 102: 13749–13754.

45 Sugita, Y. and Okamoto, Y. (2000). Replica-exchange multicanonical algorithm and multicanonical replica-exchange method for simulating systems with rough energy landscape. *Chem. Phys. Lett.* 329: 261–270.

46 Hamelberg, D., Mongan, J., and McCammon, J.A. (2004). Accelerated molecular dynamics: a promising and efficient simulation method for biomolecules. *J. Chem. Phys.* 120: 11919–11929.

47 Fajer, M., Hamelberg, D., and McCammon, J.A. (2008). Replica-exchange accelerated molecular dynamics (REXAMD) applied to thermodynamic integration. *J. Chem. Theory Comput.* 4: 1565–1569.

48 Comitani, F. and Gervasio, F. L. (2018). Exploring cryptic pockets formation in targets of pharmaceutical interest with SWISH. *Journal of Chemical Theory and Computation.* 14 (6): 3321–3331.

49 Oleinikovas, V., Saladino, G., Cossins, B.P., and Gervasio, F.L. (2016). Understanding cryptic pocket formation in protein targets by enhanced sampling simulations. *J. Am. Chem. Soc.* 138: 14257–14263.

50 Durrant, J.D. and McCammon, J.A. (2011). Molecular dynamics simulations and drug discovery. *BMC Biol.* 9: 71.

51 Hou, T., Wang, J., Li, Y., and Wang, W. (2011). Assessing the performance of the MM/PBSA and MM/GBSA methods. 1. The accuracy of binding free energy calculations based on molecular dynamics simulations. *J. Chem. Inf. Model.* 51: 69–82.

52 Miller, B.R., McGee, T.D., Swails, J.M. et al. (2012). MMPBSA.py: an efficient program for end-state free energy calculations. *J. Chem. Theory Comput.* 8: 3314–3321.

53 Genheden, S., Cabedo Martinez, A.I., Criddle, M.P., and Essex, J.W. (2014). Extensive all-atom Monte Carlo sampling and QM/MM corrections in the SAMPL4 hydration free energy challenge. *J. Comput.-Aided Mol. Des.* 28: 187–200.

54 Rastelli, G., Del Rio, A., Degliesposti, G., and Sgobba, M. (2010). Fast and accurate predictions of binding free energies using MM-PBSA and MM-GBSA. *J. Comput. Chem.* 31: 797–810.

55 Comitani, F., Cohen, N., Ashby, J. et al. (2014). Insights into the binding of GABA to the insect RDL receptor from atomistic simulations: a comparison of models. *J. Comput.-Aided Mol. Des.* 28: 35–48.

56 Westermaier, Y., Ruiz-Carmona, S., Theret, I. et al. (2017). Binding mode prediction and MD/MMPBSA-based free energy ranking for agonists of REV-ERBα/NCoR. *J. Comput.-Aided Mol. Des.* 31: 755–775.

57 Tembre, B.L. and McCammon, J.A. (1984). Ligand–receptor interactions. *Comput. Chem.* 8: 281–283.

58 Kirkwood, J.G. (1935). Statistical mechanics of fluid mixtures. *J. Chem. Phys.* 3: 300.

59 Zwanzig, R.W. (1954). High-temperature equation of state by a perturbation method. I. Nonpolar gases. *J. Chem. Phys.* 22: 1420–1426.

60 Steinbrecher, T., Mobley, D.L., and Case, D.A. (2007). Nonlinear scaling schemes for Lennard–Jones interactions in free energy calculations. *J. Chem. Phys.* 127: 214108.

61 Krusemark, C.J. (2012). Drug design: structure- and ligand-based approaches. *Q. Rev. Biol.* 87: 176.

62 Chodera, J.D., Mobley, D.L., Shirts, M.R. et al. (2011). Alchemical free energy methods for drug discovery: progress and challenges. *Curr. Opin. Struct. Biol.* 21 (2): 150–160.

63 Resat, H. and Mezei, M. (1993). Studies on free energy calculations. I. Thermodynamic integration using a polynomial path. *J. Chem. Phys.* 99: 6052–6061.

64 Schlitter, J., Engels, M., and Krüger, P. (1994). Targeted molecular dynamics: a new approach for searching pathways of conformational transitions. *J. Mol. Graph.* 12: 84–89.

65 Carter, E.A., Ciccotti, G., Hynes, J.T., and Kapral, R. (1989). Constrained reaction coordinate dynamics for the simulation of rare events. *Chem. Phys. Lett.* 156: 472–477.

66 Wang, L., Wu, Y., Deng, Y. et al. (2015). Accurate and reliable prediction of relative ligand binding potency in prospective drug discovery by way of a modern free-energy calculation protocol and force field. *J. Am. Chem. Soc.* 137: 2695–2703.

67 Torrie, G.M. and Valleau, J.P. (1977). Nonphysical sampling distributions in Monte Carlo free-energy estimation: umbrella sampling. *J. Comput. Phys.* 23: 187–199.

68 Kumar, S., Rosenberg, J.M., Bouzida, D. et al. (1992). THE weighted histogram analysis method for free-energy calculations on biomolecules. I. The method. *J. Comput. Chem.* 13: 1011–1021.

69 Darve, E., Rodríguez-Gómez, D., and Pohorille, A. (2008). Adaptive biasing force method for scalar and vector free energy calculations. *J. Chem. Phys.* 128: 144120.

70 Miao, Y., Feher, V.A., and McCammon, J.A. (2015). Gaussian accelerated molecular dynamics: unconstrained enhanced sampling and free energy calculation. *J. Chem. Theory Comput.* 11: 3584–3595.

71 Grubmuller, H., Heymann, B., and Tavan, P. (1996). Ligand binding: molecular mechanics calculation of the streptavidin-biotin rupture force. *Science (80-.)* 271: 997–999.

72 Park, S. and Schulten, K. (2004). Calculating potentials of mean force from steered molecular dynamics simulations. *J. Chem. Phys.* 120: 5946–5961.

73 Grubmüller, H. (1995). Predicting slow structural transitions in macromolecular systems: conformational flooding. *Phys. Rev. E* 52: 2893–2906.

74 Huber, T., Torda, A.E., and van Gunsteren, W.F. (1994). Local elevation: a method for improving the searching properties of molecular dynamics simulation. *J. Comput.-Aided Mol. Des.* 8: 695–708.

75 Laio, A. and Parrinello, M. (2002). Escaping free-energy minima. *Proc. Natl. Acad. Sci. U.S.A.* 99: 12562–12566.

76 Laio, A. and Gervasio, F.L. (2008). Metadynamics: a method to simulate rare events and reconstruct the free energy in biophysics, chemistry and material science. *Rep. Prog. Phys.* 71: 126601.

77 Barducci, A., Bussi, G., and Parrinello, M. (2008). Well-tempered metadynamics: a smoothly converging and tunable free-energy method. *Phys. Rev. Lett.* 100: 20603.

78 Barducci, A., Bonomi, M., and Parrinello, M. (2011). Metadynamics. *Wiley Interdiscip. Rev. Comput. Mol. Sci.* 1: 826–843.

79 Micheletti, C., Laio, A., and Parrinello, M. (2004). Reconstructing the density of states by history-dependent metadynamics. *Phys. Rev. Lett.* 92: 170601.

80 Dama, J.F., Parrinello, M., and Voth, G.A. (2014). Well-tempered metadynamics converges asymptotically. *Phys. Rev. Lett.* 112: 240602.

81 Bonomi, M., Barducci, A., and Parrinello, M. (2009). Reconstructing the equilibrium Boltzmann distribution from well-tempered metadynamics. *J. Comput. Chem.* 30: 1615–1621.

82 Swinney, D.C. (2009). The role of binding kinetics in therapeutically useful drug action. *Curr. Opin. Drug Discovery Dev.* 12: 31–39.
83 Gervasio, F.L., Laio, A., and Parrinello, M. (2005). Flexible docking in solution using metadynamics. *J. Am. Chem. Soc.* 127: 2600–2607.
84 Comitani, F., Melis, C., and Molteni, C. (2015). Elucidating ligand binding and channel gating mechanisms in pentameric ligand-gated ion channels by atomistic simulations. *Biochem. Soc. Trans.* 43: 151–156.
85 Masetti, M., Cavalli, A., Recanatini, M., and Gervasio, F.L. (2009). Exploring complex protein–ligand recognition mechanisms with coarse metadynamics. *J. Phys. Chem. B* 113: 4807–4816.
86 Clark, A.J., Tiwary, P., Borrelli, K. et al. (2016). Prediction of protein–ligand binding poses via a combination of induced fit docking and metadynamics simulations. *J. Chem. Theory Comput.* 12: 2990–2998.
87 Limongelli, V., Bonomi, M., and Parrinello, M. (2013). Funnel metadynamics as accurate binding free-energy method. *Proc. Natl. Acad. Sci. U.S.A.* 110: 6358–6363.
88 Comitani, F., Limongelli, V., and Molteni, C. (2016). The free energy landscape of GABA binding to a pentameric ligand-gated ion channel and its disruption by mutations. *J. Chem. Theory Comput.* 12: 3398–3406.
89 Saleh, N., Ibrahim, P., Saladino, G. et al. (2017). An efficient metadynamics-based protocol to model the binding affinity and the transition state ensemble of G-protein-coupled receptor ligands. *J. Chem. Inf. Model.* 57: 1210–1217.
90 Branduardi, D., Gervasio, F.L., and Parrinello, M. (2007). From A to B in free energy space. *J. Chem. Phys.* 126: 54103.
91 Berteotti, A., Cavalli, A., Branduardi, D. et al. (2009). Protein conformational transitions: the closure mechanism of a kinase explored by atomistic simulations. *J. Am. Chem. Soc.* 131: 244–250.
92 Provasi, D., Artacho, M.C., Negri, A. et al. (2011). Ligand-induced modulation of the free-energy landscape of G protein-coupled receptors explored by adaptive biasing techniques. *PLoS Comput. Biol.* (Nussinov, R., Ed.) 7: e1002193.
93 Limongelli, V., Bonomi, M., Marinelli, L. et al. (2010). Molecular basis of cyclooxygenase enzymes (COXs) selective inhibition. *Proc. Natl. Acad. Sci. U.S.A.* 107: 5411–5416.
94 Bešker, N. and Gervasio, F.L. (2012). Using metadynamics and path collective variables to study ligand binding and induced conformational transitions. *Methods Mol. Biol.* 819: 501–513.
95 Copeland, R.A., Pompliano, D.L., and Meek, T.D. (2006). Drug–target residence time and its implications for lead optimization. *Nat. Rev. Drug Discovery* 5: 730–739.
96 Buch, I., Giorgino, T., and De Fabritiis, G. (2011). Complete reconstruction of an enzyme-inhibitor binding process by molecular dynamics simulations. *Proc. Natl. Acad. Sci. U.S.A.* 108: 10184–10189.

97 Plattner, N. and Noé, F. (2015). Protein conformational plasticity and complex ligand-binding kinetics explored by atomistic simulations and Markov models. *Nat. Commun.* 6: 7653.

98 Juraszek, J. and Bolhuis, P.G. (2009). Effects of a mutation on the folding mechanism of a β-hairpin. *J. Phys. Chem. B* 113: 16184–16196.

99 Juraszek, J. and Bolhuis, P.G. (2006). Sampling the multiple folding mechanisms of Trp-cage in explicit solvent. *Proc. Natl. Acad. Sci. U.S.A.* 103: 15859–15864.

100 Juraszek, J. and Bolhuis, P.G. (2010). (Un)Folding mechanisms of the FBP28 WW domain in explicit solvent revealed by multiple rare event simulation methods. *Biophys. J.* 98: 646–656.

101 Juraszek, J., Vreede, J., and Bolhuis, P.G. (2012). Transition path sampling of protein conformational changes. *Chem. Phys.* 396: 30–44.

102 Dellago, C. and Bolhuis, P.G. (2004). Activation energies from transition path sampling simulations. *Mol. Simul.* 30: 795–799.

103 Dellago, C. and Bolhuis, P.G. (2009). Transition path sampling and other advanced simulation techniques for rare events. In: *Advanced Computer Simulation Approaches for Soft Matter Sciences III*, vol. 221 (ed. C. Holm and K. Kremer), 167–233. Berlin: Springer-Verlag.

104 Juraszek, J., Saladino, G., Van Erp, T.S., and Gervasio, F.L. (2013). Efficient numerical reconstruction of protein folding kinetics with partial path sampling and pathlike variables. *Phys. Rev. Lett.* 110: 108106.

105 Moroni, D., Bolhuis, P.G., and van Erp, T.S. (2004). Rate constants for diffusive processes by partial path sampling. *J. Chem. Phys.* 120: 4055–4065.

106 Morando, M.A., Saladino, G., D'Amelio, N. et al. (2016). Conformational selection and induced fit mechanisms in the binding of an anticancer drug to the c-Src kinase. *Sci. Rep.* 6: 24439.

107 Tiwary, P., Limongelli, V., Salvalaglio, M., and Parrinello, M. (2015). Kinetics of protein–ligand unbinding: predicting pathways, rates, and rate-limiting steps. *Proc. Natl. Acad. Sci. U.S.A.* 112 (5): E386–E391.

108 Mollica, L., Decherchi, S., Zia, S.R. et al. (2015). Kinetics of protein–ligand unbinding via smoothed potential molecular dynamics simulations. *Sci. Rep.* 5: 11539.

109 Mark, A.E., van Gunsteren, W.F., and Berendsen, H.J.C. (1991). Calculation of relative free energy via indirect pathways. *J. Chem. Phys.* 94: 3808–3816.

110 Tsujishita, H., Moriguchi, I., and Hirono, S. (1993). Potential-scaled molecular dynamics and potential annealing: effective conformational search techniques for biomolecules. *J. Phys. Chem.* 97: 4416–4420.

111 Sinko, W., Miao, Y., de Oliveira, C.A.F., and McCammon, J.A. (2013). Population based reweighting of scaled molecular dynamics. *J. Phys. Chem. B* 117: 12759–12768.

112 Mollica, L., Theret, I., Antoine, M. et al. (2016). Molecular dynamics simulations and kinetic measurements to estimate and predict protein–ligand residence times. *J. Med. Chem.* 59: 7167–7176.

4

Markov State Models in Drug Design

Bettina G. Keller, Stevan Aleksić, and Luca Donati

Department of Biology, Chemistry, Pharmacy, Freie Universität Berlin, Takustr. 3, 14195 Berlin, Germany

4.1 Introduction

Starting from the lock-and-key model [1], models of ligand–target binding have been extended to acknowledge the role of conformational flexibility. In current models, the targets are assumed to fluctuate between several different conformations. In the conformational selection model [2], one of these conformations is the active state, i.e. the conformation which is assumed in complex with the ligand. The ligand then "selects" this conformation from the ensemble and stabilizes it by forming a complex. By contrast, in the induced-fit model [3], the active conformation is not sampled by the *apo*-target. Instead, the receptor and the ligand form an unspecific encounter complex. This weak complex then triggers a conformational rearrangement in the receptor, which leads to the fully formed complex. While examples of mechanisms have been found [4–7], most binding processes fall somewhere in between these two extremes.

The conformational selection and the induced-fit model neglect the conformational dynamics of the ligand, which is justified if the ligand is either rigid or its dynamics is fast compared to the dynamics of the target. However, this is not always a valid assumption. In particular, peptides and peptidomimetics exhibit a complex and often slow conformational dynamics. It is increasingly recognized that the flexibility of the target and the ligand and their mutual interaction are crucial factors in the ligand-binding process. Thus, to systematically vary the thermodynamic and kinetic properties of a drug molecule, not only binding affinities but also dynamics need to be taken into account.

Experimentally, it is difficult to characterize the full conformational ensemble and its dynamics. However, an increase in computer power combined with improved algorithms has rendered molecular dynamics (MD) simulations as useful tools in structure-based drug design [8]. With progress in distributed computing [9], special purpose computers [10], and graphics processing unit (GPU) devices [11], trajectories of several tens in microseconds are now accessible on a routine basis.

Biomolecular Simulations in Structure-Based Drug Discovery,
First Edition. Edited by Francesco L. Gervasio and Vojtech Spiwok.

A visual inspection of these trajectories is sometimes not feasible, due to the shear size of the data set, and almost always unrewarding, because it does not yield a quantitative description of the system.[1] While statistical analyses of the trajectory for the stationary properties of the system have been used routinely for decades, methods which yield a model of the dynamics have matured only recently. Markov State Models (MSMs) [12–17], in which the dynamics is approximated as a Markovian jump process between distinct microstates, are the most widely used dynamic models. MSMs have been used to improve ensemble docking, to optimize a specific conformation in a ligand, to identify cryptic allosteric sites, and to characterize ligand-binding processes as well as inactive-to-active transitions in signaling proteins. We do not aim at a comprehensive survey of the literature on this subject, nor do we focus on specific results. Our goal is to explain the different ways in which MSMs can be helpful in structure-based design.

4.2 Markov State Models

4.2.1 MD Simulations

MSMs [12–17] are constructed from the time series of MD, which can be generated by classical MD simulations. In an MD simulation, the molecular system is represented by N particles, where m_i denotes the mass and $\mathbf{r}_i(t) \in \mathbb{R}^3$ the three-dimensional coordinates of the ith particle at time t. The dynamics of the system is governed by a classical equation of motion:

$$m_i \frac{\partial^2 \mathbf{r}_i(t)}{\partial t^2} = -\nabla V(\mathbf{r}_1(t), \dots, \mathbf{r}_N(t)) + \text{thermostat} \quad \forall i \in [1, N] \tag{4.1}$$

where $V(\cdot)$ is a predefined potential energy function which determines how the particles interact. The thermostat term ensures that the simulation samples the canonical ensemble, and a wide range of algorithms have been proposed to achieve this task [18]. The MD simulation program then generates a trajectory of the particle positions

$$\mathbf{r}(t) = (\mathbf{r}_1(t), \mathbf{r}_2(t), \dots, \mathbf{r}_N(t)) \in \mathbb{R}^{3N} \tag{4.2}$$

by numerically integrating Eq. (4.1) (including the thermostat terms). Often, the momentum $\mathbf{p}_i(t) \in \mathbb{R}^3$ of each particle is additionally calculated, yielding a phase space trajectory

$$\mathbf{x}(t) = (\mathbf{r}_1(t), \mathbf{p}_1(t), \dots, \mathbf{r}_N(t), \mathbf{p}_N(t)) \in \mathbb{R}^{6N} \tag{4.3}$$

where $\Omega \subset \mathbb{R}^{6N}$ is the $6N$-dimensional phase space or, equivalently, state space of the system. For a detailed explanation of the MD simulation technique, see Refs. [19–21]. Reference [8] gives an overview of the current use of MD simulations in computational drug design.

---

1 Nonetheless, any good researcher would have a look at the raw data before proceeding with more complex analyses.

4.2.2 The Molecular Ensemble

An individual trajectory samples an extremely high-dimensional space and is usually far too complex to be comprehensible by visual inspection. It needs to be further analyzed to transform the trajectory *data* into *information* about the molecular system. To achieve this, one shifts the viewpoint from the dynamics of the individual system to the properties of an ensemble of systems.

In a thermostatted MD simulation, the trajectory samples the Boltzmann distribution

$$\pi(\mathbf{x}) = \frac{1}{Q} \exp\left(-\frac{1}{k_B T} H(\mathbf{x})\right) \tag{4.4}$$

where Q is the partition function, k_B is the Boltzmann constant, T is the temperature, and $H(\mathbf{x})$ is the classical Hamiltonian of the system. We assume that the dynamics is ergodic, i.e. each region of the state space has to be reachable from any other region of the state space. In practice, this means that the trajectory has to be long enough such that all relevant parts of the state space have been sampled sufficiently. In this case, the ensemble average of any observable $o(\mathbf{x})$ can be replaced by a time integral over the trajectory $\mathbf{x}(t)$

$$\langle o \rangle = \int_{\Omega} \pi(\mathbf{x}) o(\mathbf{x}) \mathrm{d}\mathbf{x} = \lim_{T_{\mathrm{trj}} \to \infty} \frac{1}{T_{\mathrm{trj}}} \int_{t=0}^{T_{\mathrm{trj}}} o(\mathbf{x}(t))\, \mathrm{d}t \tag{4.5}$$

where T_{trj} is the length of the trajectory. Equations (4.4) and (4.5) are the starting point for MD analysis methods, which yield information on the stationary properties of the system.

4.2.3 The Propagator

To gain insight into the dynamics of the system, one considers a time-dependent probability density $p_t(\mathbf{x})$ and the transition probability density $p(\mathbf{x}, \mathbf{y}; \tau)$, i.e. the probability density to find the system in state $\mathbf{y}$ at time $t + \tau$, given that it has been in state $\mathbf{x}$ at time t. Note that the transition probability density is parameterized by the lag time τ, which denotes the time interval at which the states of the system are evaluated. Different values of τ yield different transition probability densities. In addition, the transition probability density relies on two assumptions on the dynamics of the system. First, $p(\mathbf{x}, \mathbf{y}; \tau)$ is independent of t, which is the case if no time-dependent forces act on the system (time-homogeneous dynamics). Second, the transition probability density only depends on the current state $\mathbf{x}$ and not on any of the previous states of the system, i.e. the dynamics need to be Markovian. We make a third assumption, namely, that the dynamics fulfill detailed balance

$$\pi(\mathbf{x}) p(\mathbf{x}, \mathbf{y}; \tau) = \pi(\mathbf{y}) p(\mathbf{y}, \mathbf{x}; \tau) \tag{4.6}$$

That is, in an ensemble which is distributed according to the Boltzmann distribution $\pi(\mathbf{x})$, the number of transitions from state $\mathbf{x}$ to $\mathbf{y}$ within time τ is equal to the number of transitions in the opposite direction.

The transition probability density $p(\mathbf{x}, \mathbf{y}; \tau)$ defines the propagator $\mathcal{P}(\tau)$, a mathematical object that propagates the probability density function $p_t(\mathbf{x})$ forward in time

$$p_{t+\tau}(\mathbf{x}) = \mathcal{P}(\tau)p_t(\mathbf{x}) \tag{4.7}$$

Because we assumed Markovian dynamics, the probability density at $t + n\tau$ can be obtained by applying the propagator n times

$$p_{t+n\tau}(\mathbf{x}) = \mathcal{P}^n(\tau)p_t(\mathbf{x}) = \mathcal{P}(n\tau)p_t(\mathbf{x}) \tag{4.8}$$

When $n \to \infty$, the probability density tends to the Boltzmann probability density $\pi(\mathbf{x})$

$$\pi(\mathbf{x}) = \lim_{n\to\infty} \mathcal{P}^n(\tau)\pi(\mathbf{x}) \tag{4.9}$$

Note that the lag time τ defines the time resolution at which the time evolution of the probability density can be represented by $\mathcal{P}(\tau)$.

4.2.4 The Dominant Eigenspace

The advantage of introducing the propagator is that the slow dynamics of the system can be rationalized using a small set of its eigenfunctions $l_i(\mathbf{x})$ and eigenvalues $\lambda_i(\tau)$

$$\mathcal{P}(\tau)l_i(\mathbf{x}) = \lambda_i(\tau)l_i(\mathbf{x}) \tag{4.10}$$

the so-called dominant eigenspace. For ergodic and detailed-balanced dynamics, the eigenvalues are real-valued and lie on the interval $\lambda_i(\tau) \in]0, 1]$. The largest possible eigenvalue $\lambda_1(\tau)$ is unique and always exists. Its associated eigenfunction $l_1(\mathbf{x})$ is equal to the Boltzmann distribution

$$\pi(\mathbf{x}) = \mathcal{P}(\tau)\pi(\mathbf{x}) = \mathcal{P}(\tau)l_1(\mathbf{x}) \tag{4.11}$$

The eigenfunctions $l_1(\mathbf{x}), l_2(\mathbf{x}), \ldots$ form a complete basis of the Hilbert space of probability density functions. This means that any probability density function can be expressed as a linear combination of these eigenfunctions

$$p_t(\mathbf{x}) = \sum_{i=1}^{\infty} c_i\ l_i(\mathbf{x}) \tag{4.12}$$

where the eigenfunctions are ordered according to their associated eigenvalues. Inserting Eq. (4.12) into Eq. (4.8) yields

$$\begin{aligned} p_{t+n\tau}(\mathbf{x}) &= \mathcal{P}^n(\tau) \sum_{i=1}^{\infty} c_i\ l_i(\mathbf{x}) = \sum_{i=1}^{\infty} c_i\ \lambda_i^n(\tau)\ l_i(\mathbf{x}) \\ &\approx \pi(\mathbf{x}) + \sum_{i=1}^{M-1} c_i\ \lambda_i^n(\tau)\ l_i(\mathbf{x}) \end{aligned} \tag{4.13}$$

The time evolution of a probability density can be regarded as a superposition of dynamic modes $l_i(\mathbf{x})$ with amplitudes $c_i\ \lambda_i^n(\tau)$. Because $\lambda_i(\tau) \in]0, 1]$, the factor

$\lambda_i^n(\tau)$ decays exponentially

$$\lambda_i^n(\tau) = \exp(n \ln(\lambda_i(\tau))) = \exp\left(-\frac{n\tau}{t_i}\right) \tag{4.14}$$

where t_i is the implied timescale (its) of the ith mode

$$t_i = -\frac{\tau}{\ln(\lambda_i(\tau))} \tag{4.15}$$

Note that, for Markovian dynamics, the implied timescales are independent of τ. The smaller the $\lambda_i(\tau)$, the faster the $c_i\ \lambda_i^n(\tau)$ decays to zero. Thus, for sufficiently large $n\tau$, the fast decaying modes can be neglected, and the dynamics is determined by the slowest M modes. This works particularly well if there is a gap in the eigenvalue spectrum between $\lambda_M(\tau)$ and $\lambda_{M+1}(\tau)$ (Figure 4.1b). The first mode is the Boltzmann density, for which always $c_1 = 1$. And because $\lambda_1(\tau) = 1$, this mode does not decay ($t_1 = \infty$). Hence, for $n \to \infty$, Eq. (4.13) yields Eq. (4.9).

Figure 4.1 illustrates the dominant eigenspace of a diffusion on a one-dimensional potential energy surface (Figure 4.1a). An example trajectory is shown in Figure 4.2a. The eigenvalue spectrum (Figure 4.1b) has a gap between $\lambda_3(\tau)$ and $\lambda_4(\tau)$. Thus, besides the stationary Boltzmann distribution (Figure 4.1c), the system has two slow modes: $l_2(x)$ (Figure 4.1d) and $l_3(x)$ (Figure 4.1e). Note that the number of slow modes coincides with the number of minima in the potential energy surface. The modes represent kinetic exchange processes between regions in which the eigenfunction assumes a negative sign and those regions in which the eigenfunction assumes a positive sign. The mode $l_2(x)$ represents the exchange between the minimum on the right-hand side of the potential energy surface and the two other minima, i.e. the transition across the largest barrier in the system. The equilibration across this barrier takes place on a timescale of approximately $t_2 = 1260$ time steps. The mode $l_3(x)$ represents the equilibration between the central minimum and the two minima on the left and right side of the potential energy surface, which takes place on a

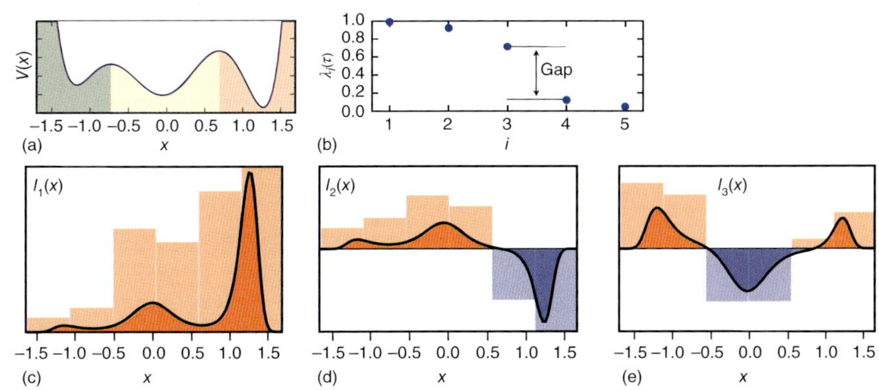

Figure 4.1 Dominant eigenspace of the propagator. (a) Energy function; (b) eigenvalue spectrum at $\tau = 100$ time steps; (c)–(e) dominant eigenfunctions; continuous lines, true eigenfunctions; histogram, MSM approximation.

timescale of approximately $t_3 = 300$ time steps. The modes thus contain a wealth of information on the energy minima of the system and the relative height of the barriers between them.

We remark that the theory of MSMs is usually not formulated in terms of the propagator, but in terms of the closely related transfer operator. In fact, for detailed-balanced dynamics, the transfer operator is the adjoint of the propagator. The two operators share the same eigenvalues, the eigenfunctions of the transfer operator $r_i(\mathbf{x})$ are related to the eigenfunctions of the propagator by $l_i(\mathbf{x}) = \pi(\mathbf{x})r_i(\mathbf{x})$. For the sake of brevity, we omit a discussion on the transfer operator here and point interested readers to Refs. [16, 22, 23].

4.2.5 The Markov State Model

The analytical expression of the propagator contains the potential energy function and is too complex to be analyzed directly. Instead, one approximates the propagator by a transition matrix $\mathbf{T}(\tau)$ whose elements can be estimated from an MD trajectory.

The state space Ω is discretized into a set of m non-overlapping microstates $A_1, A_2, \ldots, A_m$, such that $\bigcup_{i=1}^{m} A_i = \Omega$ and $A_i \cap A_j = \emptyset$ for $i \neq j$. The probability density $p_t(\mathbf{x})$ is then approximated as a probability vector $\mathbf{p}_t$ whose elements are obtained by integrating $p_t(\mathbf{x})$ over the individual microstates

$$[\mathbf{p}_t]_i = \int_{A_i} p_t(\mathbf{x})\, \mathrm{d}\mathbf{x} \tag{4.16}$$

Equation (4.7) can then be approximated as

$$\mathbf{p}^\top(t+\tau) = \mathbf{p}^\top(t)\mathbf{T}(\tau) \tag{4.17}$$

The elements of the transition matrix $\mathbf{T}(\tau)$ represent the transition probabilities between pairs of microstates, i.e the conditional probability of finding the system in state A_j at time $t+\tau$, given that it has been in state A_i at time t

$$T_{ij}(\tau) = \mathbb{P}[\mathbf{x}(t+\tau) \in A_j \mid \mathbf{x}(t) \in A_i] \tag{4.18}$$

The transition probabilities can be estimated from an MD trajectory $\mathbf{x}(t)$ as

$$\hat{T}_{ij}(\tau) = \frac{\hat{C}_{ij}(\tau)}{\sum_j \hat{C}_{ij}(\tau)} \tag{4.19}$$

$\hat{C}_{ij}(\tau)$ are the number of transitions between state A_i and state A_j within lag time τ which have been sampled by the trajectory. We use the "hat" to denote estimated properties. The transition count $\hat{C}_{ij}(\tau)$ is normalized by the total number of outgoing transitions from state A_i, thereby yielding a probability. Consequently, the transition matrix $\hat{\mathbf{T}}(\tau)$ is row-normalized, where each row vector $\hat{\mathbf{t}}_i(\tau)$ represents the probability distribution of reaching any microstate A_j in the state space within time τ, starting from state A_i.

Mathematically, $\hat{C}_{ij}(\tau)$ is an estimate of a correlation function

$$C_{ij}(\tau) = \mathrm{cor}(\mathbf{1}_{A_i}, \mathbf{1}_{A_j}; \tau) = \lim_{T_{\mathrm{trj}}\to\infty} \int_0^{T_{\mathrm{trj}}} \mathbf{1}_{A_i}(t)\, \mathbf{1}_{A_j}(t+\tau)\, \mathrm{d}t \tag{4.20}$$

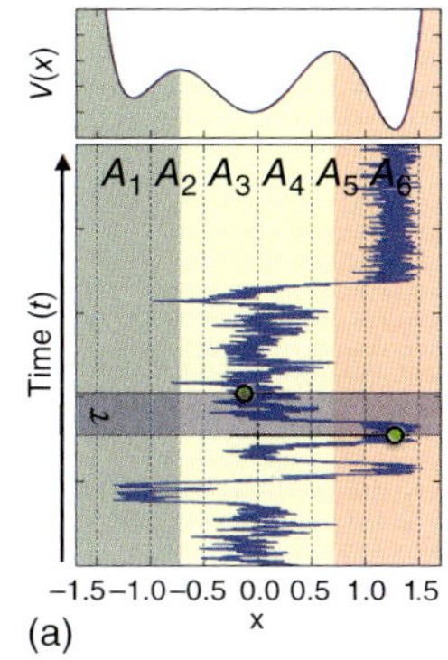

$j(t+\tau)$, $i(t)$

$C_{ij}^{(\tau)}$	1	2	3	4	5	6
1	0	0	210	64	55	29
2	23	2	86	56	0	26
3	233	69	423	313	86	216
4	102	52	168	119	24	155
5	0	20	59	44	4	56
6	0	0	83^{+1}	94	29	202

(b)

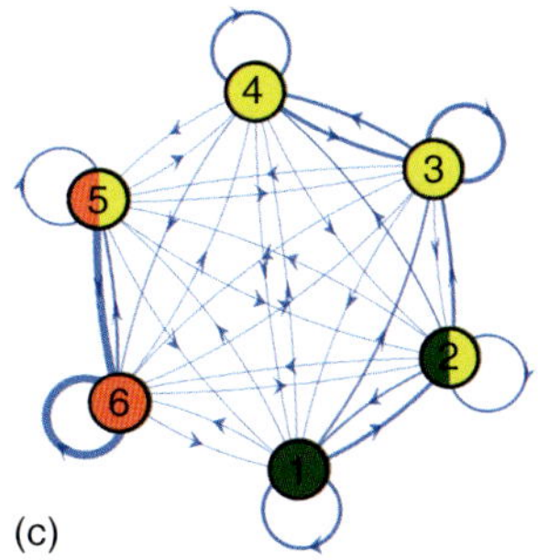

Figure 4.2 Construction of an MSM. (a) Example trajectory; (b) count matrix; (c) network representation of the transition matrix.

where we used the abbreviated notation $\mathbf{1}_{A_i}(t) = \mathbf{1}_{A_i}(\mathbf{x}(t))$. The function

$$\mathbf{1}_{A_i}(\mathbf{x}) = \begin{cases} 1 & \text{if } \mathbf{x} \in A_i \\ 0 & \text{if } \mathbf{x} \notin A_i \end{cases} \tag{4.21}$$

is the indicator function of microstate A_i. One can show that $T_{ij}(\tau) = \frac{C_{ij}(\tau)}{\sum_j C_{ij}(\tau)}$ is a valid discretization of the transfer operator and, by extension, the propagator, and that the discretized operator converges to the true operator if the microstates are infinitesimally small [22, 24].

Figure 4.2 illustrates the estimation of an MSM on a trajectory $x(t)$ which samples a one-dimensional potential energy function $V(x)$ (Figure 4.2a). The one-dimensional state space is discretized into six microstates $\{A_1, \ldots, A_6\}$ and a sliding window of length τ (blue box) is moved across the time axis. The two green dots denote the states at time t and $t + \tau$; in this case, $x(t) \in A_6$ and $x(t + \tau) \in A_3$. Consequently, the element $C_{63}(\tau)$ of the transition count matrix is increased by one (Figure 4.2b). Only the states at the beginning and at the end of the sliding window are relevant for the transition. The system might visit other states within the time window τ. The transition matrix can be visualized as a kinetic network, in which the edges correspond to the transition probabilities (Figure 4.2b).

Thus, an MSM replaces a dynamics which is continuous in space by a jump process between discrete microstates. However, the discretization introduces an error. In particular, although the continuous dynamics is Markovian (Eq. (4.8): $\mathcal{P}^n(\tau) = \mathcal{P}(n\tau)$), the discretized dynamics is not

$$\mathbf{T}^n(\tau) \approx \mathbf{T}(n\tau) \tag{4.22}$$

It is therefore crucial to validate each MSM.

Two tests for the deviation from Markovian behavior are commonly used: the Chapman–Kolmogorov test [16] and the implied timescale test [13]. The Chapman–Kolmogorov test starts from an initial distribution $\mathbf{p}_0$ which is restricted to a particular set of microstates, e.g. all microstates which belong to the central minimum in Figure 4.1 $S = \{A_3, A_4\}$:

$$[\mathbf{p}_0]_i = \begin{cases} \dfrac{\pi_i}{\sum_{j\in S}\pi_j} & A_i \in S \\ 0 & A_i \notin S \end{cases} \tag{4.23}$$

where $\pi_i = \int_{A_i} \pi(\mathbf{x})\,\mathrm{d}\mathbf{x}$. This distribution is propagated by a transition matrix $\mathbf{T}(\tau_0)$ n times; and in the resulting probability distribution, the elements corresponding to S are added to find the probability of being in this set of microstates after $n\tau_0$, given that the process has started in S

$$p_{\mathrm{MSM}}(S, S; n\tau_0) = \sum_{i\in S} [\mathbf{p}_0^\top \mathbf{T}^n(\tau_0)]_i = \sum_{i\in S} [\mathbf{p}_{n\tau_0}^\top]_i \tag{4.24}$$

This probability is compared to the corresponding probability estimated from the trajectory by counting the transitions $S \to S$ within time $n\tau_0$

$$p_{\mathrm{MD}}(S, S; n\tau_0) = \sum_{i\in S} \hat{T}_{ij}(n\tau_0) = \sum_{i\in S} \frac{\hat{C}_{ij}(\tau_0)}{\sum_j \hat{C}_{ij}(\tau_0)} \tag{4.25}$$

This test has to be carried out over a wide range of values for n. If the resulting curves $p_{\mathrm{MD}}(S, S, n\tau_0)$ and $p_{\mathrm{MSM}}(S, S, n\tau_0)$ are close enough, the model with lag time τ_0 is a valid model. The advantage of the Chapmann–Kolmogorov test is that it directly probes Eq. (4.22). The disadvantages are that the test has to be repeated for many values τ_0 and that its results tend to be sensitive to the choice of S.

The implied timescale test probes Eq. (4.15). One estimates MSMs at a range of lag times τ and plots the resulting implied timescales as a function of τ (Figure 4.3). For small values of τ, the implied timescales are monotonically increasing functions, but after a certain value τ_{Markov}, the implied timescales become approximately constant, thus indicating that for $\tau \geq \tau_{\mathrm{Markov}}$ the MSM is approximately Markovian [24]. Figure 4.3 shows the implied timescales associated with the second and third dominant eigenfunction, obtained from an MSM with 100 microstates (solid line) and an MSM with 6 microstates (dashed line). For the fine discretization, the implied timescales converge to a

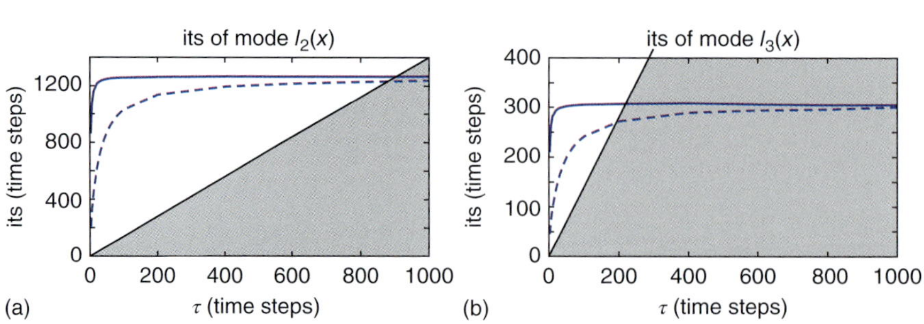

Figure 4.3 Implied timescale test for the one-dimensional diffusion process (a) its of mode $l_2(x)$, (b) its of mode $l_3(x)$, gray area: its $< \tau$, solid line: MSM with 100 microstates, dashed line: MSM with 6 microstates.

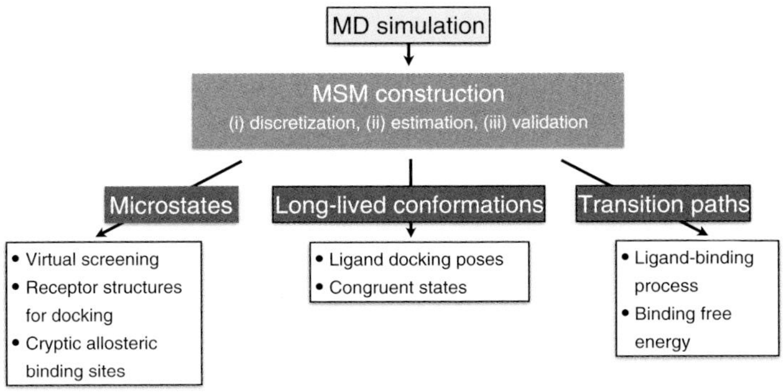

Figure 4.4 Workflow of MSM analyses for drug design.

constant function at much smaller values of τ than for the 6-microstate MSM. The approximation error of the 6-microstate MSM is also demonstrated in Figure 4.1c–e, which shows the MSM eigenvectors as a histogram. As a rule of thumb, the finer the discretization, the smaller the approximation error. However, one faces the following trade-off: the larger the number of microstates, the lower the average number of observed counts per transition. Thus, by decreasing the approximation error one tends to increase the statistical uncertainty [25, 26]. The challenge in the construction of MSMs is to balance these two errors.

To summarize, the construction of an MSM from an MD simulation consists of three fundamental steps (Figure 4.4): (i) discretization of the state space, (ii) estimation of the transition probabilities, and (iii) validation of the MSM. Several features of the MSMs can be utilized for rational drug design. From the microstates, one can extract representative structures of the conformational ensemble (Section 4.3). The microstates can be aggregated into long-lived conformations (Section 4.4), and the kinetic network can be analyzed (Section 4.5).

4.3 Microstates

Structure-based drug design methods, such as molecular docking to a known binding pocket and the detection of new binding pockets, require high-resolution structures of the target. However, targets are not entirely rigid, and minor fluctuations on the surface of the protein might heavily affect the shape or accessibility of any binding pocket. Using only a single-crystal structure arbitrarily limits the predictive power of any structure-based approach. MD simulations, in principle, cover the entire conformational ensemble of a target; however, using each frame of the trajectory as a starting point for a structure-based design is computationally prohibitive. Thus, a method which extracts meaningful representatives from the conformational ensemble is needed. The discretization of a validated MSM is particularly suited for this task, because the discretization yields a small number of microstates (compared to the number of frames) and mirrors the features of the free energy landscape. In particular, it ensures that slowly interconverting structures are assigned to different microstates. From each microstate, a representative structure is extracted and the set of representative structures is processed by the structure-based method of choice. In addition, the MSM readily provides the relative probability and the lifetime of each microstate.

In Ref. [27], this approach has been used for a docking study to MDM2. MDM2 is an interesting target for anticancer drugs because it binds to the tumor-suppressor p53 and thereby blocks its activity. Inhibiting MDM2 might thus restore p53 activity in cancer cells. However, the N-terminal lid region of MDM2 transiently covers its binding cleft, making MDM2 a particularly difficult receptor for docking studies. The authors conducted simulations of the *apo*-MDM2 and constructed an MSM with 2000 microstates. A set of 10 ligands, for which crystal structures in complex with MDM2 are available, was then docked to representatives of each MDM2 microstate. For each ligand, the 20 docking poses with best docking score were considered suitable docking poses. This approach recovered the correct docking pose for 8 of the 10 ligands from the conformational ensemble of the *apo*-MDM2, whereas the success rate of docking the set of ligands to any of the available receptor structures was considerably lower.

Knowledge of the conformational ensemble of the target is also important for the detection of binding sites. The group of Bowman analyzed MSM microstates to identify cryptic allosteric sites in TEM β-lactamase and two other receptors [28, 29]. Targeting allosteric sites is a promising avenue for several reasons. Allosteric sites drastically increase the range of possible targets, including proteins which were previously considered undruggable. The allosteric drugs might be effective at lower affinities, because they do not have to compete with a natural ligand. In addition, they might be highly specific, because allosteric sites are less evolutionarily conserved and therefore more unique than orthosteric sites. Finally, with allosteric drugs, one can potentially increase a desired activity of the target. Computational methods for the rational design of allosteric drugs have been reviewed in [30]. Unfortunately, many allosteric sites only open transiently in the conformational ensemble and are often not detectable in a crystal structure of the receptor; hence, the adjunct "cryptic."

TEM β-lactamase hydrolyzes β-lactam antibiotics and thereby confers antibiotic resistance. Based on extensive MD simulations, the conformational space of TEM β-lactamase was discretized into 5152 microstates. Representative structures from each microstate were then screened for pockets using the program LIGSITE [31]. Pockets close to the orthosteric binding site and those visible in the crystal structure were discarded, and the remaining pockets were clustered into merged pockets. The relative probability of a given pocket in the conformational ensemble was estimated as the sum of the relative probabilities π_i of all microstates A_i in which the pocket occurred. This yielded a considerable number of transiently open pockets. The authors then used a mutual-information analysis [32, 33] to identify those pockets which are coupled to the orthosteric site. The approach correctly identified a known allosteric binding site as the most accessible transient pocket, with a relative probability of 53% and the strongest coupling to the orthosteric site. In addition, it yielded a series of previously unknown allosteric sites with sufficient relative probability to be binding pockets. In a subsequent study [34], the group virtually screened a library of compounds against an ensemble of TEM β-lactamase structures which contained the cryptic pockets identified by the MSM analysis. Experimental screening of the hits from the virtual screening identified one inhibitor and two activators of TEM β-lactamase.

4.4 Long-Lived Conformations

Peptidomimetics and cyclic peptides mimic the folded structure of a protein and open up a route to inhibiting protein–protein interactions [35, 36]. However, their dynamics is complex and often not markedly faster than the dynamics of the target. Moreover, small modifications to the chemical structure can induce startling variations in the affinity or the bioavailability of the drug. To design these ligands, it is paramount to understand the dynamic equilibrium between the targeted long-lived conformation and the competing long-lived conformations.

Long-lived conformations consist of a set microstates which show high transition rates within the set and low transition rates to microstates outside of the set; see, for example, microstate 5 and 6 in Figure 4.2. The relative stability of a long-lived conformation is obtained by summing over Boltzmann populations π_i of each microstate in the set.

The Bayesian agglomerative clustering engine (BACE) algorithm [37] uses the observed transition counts to extract long-lived conformations from an MSM. For each pair of microstates, a Bayes factor is calculated which compares how likely the observed transitions have been generated from the same underlying distribution (i.e. both microstates belong to the same long-lived conformations) or from two different distributions (i.e the microstates belong to different long-lived conformations). By iteratively merging microstates according to the Bayes factor and recalculating the Bayes-factor matrix, the algorithm yields an aggregation of the microstates into long-lived conformations.

The Perron-cluster cluster analysis+ (PCCA+) [38] is based on the insight that the dominant MSM eigenvectors contain information on the slow dynamics and thus on the long-lived conformations. Each microstate corresponds to a point in the M-dimensional space of the dominant eigenvectors. One can show that, for dynamics which exhibits long-lived conformations, the microstates are arranged along a simplex in this space, where each vertex of the simplex corresponds to one long-lived conformation. PCCA+ identifies long-lived conformations by aggregating microstates in the vicinity of each vertex.

Although identifying long-lived conformations from an MSM has become a routine analysis, comparing long-lived conformations of different chemical constructs or in different external conditions remains a challenge. Of course, the long-lived conformations can be compared *a posteriori* by visual inspection or by calculating the RMSD. This, however, amounts to matching conformational sub-ensembles – a task which is often sensitive to the choice of distance measure. On the other hand, comparing single representatives of each conformation tends to be sensitive to the choice of the representative. A direct comparison of the long-lived conformations of two different systems is only possible if the

MSMs are built using a common discretization. For this, one concatenates the trajectories of both systems and decides on a discretization based on this combined data set. The count matrices are then estimated on this discretization from each data set separately. As a result, the eigenvectors of the MSMs can be compared directly. For example, a shift in the population of the ith microstate from one system to the other results in a shift in the ith element of the first (left) transition matrix eigenvector. Likewise, if the long-lived conformations change, the set of microstate which defines these long-lived conformations changes. Thus, one can match long-lived conformations in different systems by comparing their sets of microstates.

This approach has been used in Ref. [39] to compare a set of cyclic peptide hairpin mimetics. These peptides were designed to mimic a β-hairpin segment of LapD which forms an interface with LapG and thereby modulates the activity of LapG. The two proteins occur in the bacterium *Pseudomonas fluorescens* and are involved in bacterial biofilm formation. The objective of the study was to identify those molecules in which the hairpin structure was particularly stable, and to rationalize the stability relative to the competing conformations. A pre-screening of a larger set of constructs yielded four nine-residue constructs, each of which was covalently cross-linked by two linkers. These constructs had two design switches: a substituent in the terminal linker could be either a hydrogen (R=H) or a methyl group (R=CH_3), and the N-terminal valine was incorporated either in its D-isomer (D-Val) or its L-isomer (L-isomer), yielding four possible combinations. MSMs with a common discretization were constructed for the D-isomers and L-isomers, respectively, and revealed a native and a near-native conformation, whose populations could now be compared directly. Interestingly, in the D-isomers, the native structure was more for the R=CH_3 construct than for the R=H construct, whereas it was the opposite in the L isomers. This could be rationalized by a steric interaction between the residue R and the valine side chain.

In some cases, a crucial feature of a drug molecule relies on a shift in the stability of its long-lived conformations. For example, cyclosporin A, a cyclic peptide, and an immunosuppressive drug, crosses membranes by passive diffusion despite its high molecular weight and lipophilicity. However, the peptide is capable of forming a closed structure in which many backbone hydrogen bond donors and acceptors are engaged, and intramolecular hydrogen bonds are thus shielded from the solvent (Figure 4.5). It has thus been hypothesized that cyclosporin A crosses the membrane in the closed conformation and that a suitable design would hence be to stabilize this particular structure. We investigated this hypothesis [40] and simulated cyclosporin A in water and in chloroform as a proxy for the membrane interior. We then constructed MSMs based on a common discretization. The resulting long-lived conformations and hierarchy of the free energy landscape are shown in Figure 4.5a,b. Both simulations sampled the open and the closed conformation. In water, the closed conformation was neither energetically favorable nor particularly stable: it quickly interconverted with the long-lived conformations W2 and W5 (Figure 4.5a). Intriguingly, we could identify a half-open structure which was present and long-lived in both ensembles. In water, this structure is energetically more favored and more stable than in the closed structure, and it is highly populated and stable in chloroform. The existence of

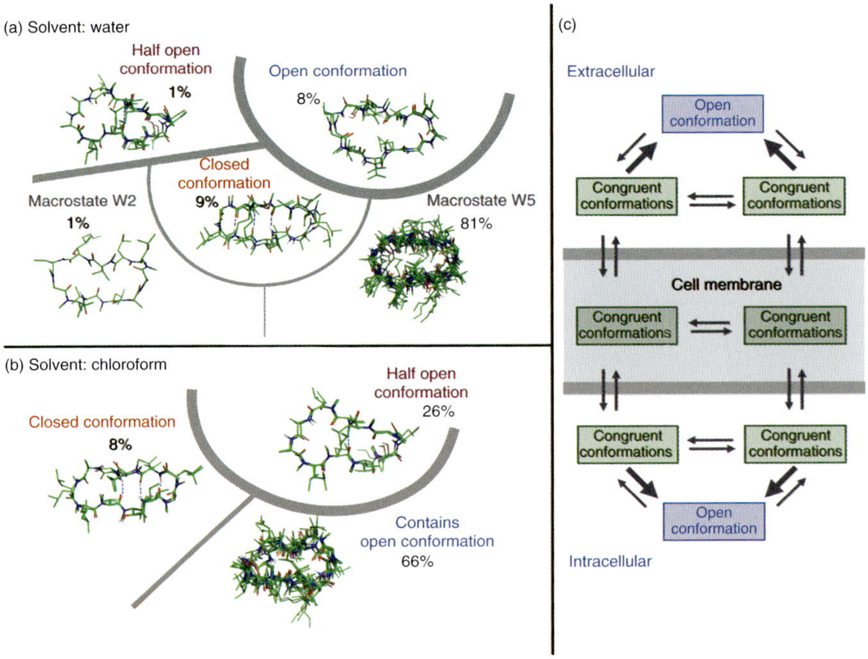

Figure 4.5 MSMs of Cyclosporin A (a) hierarchy of the free energy in water, (b) hierarchy of the free energy in chloroform, (c) updated hypothesis for the diffusion of cyclosporin A across a membrane. Source: Adapted with permission from Ref. [40]. Copyright 2016 American Chemical Society.

this third conformation was supported by NMR data. We thus proposed that there is actually an ensemble of congruent conformations which occur in polar and apolar solvent and can thus diffuse through a membrane (Figure 4.5a). An appropriate design strategy would then be to stabilize not a single structure but to enable the peptide to form as many congruent conformations as possible.

4.5 Transition Paths

Direct simulation and MSM analysis of ligand-binding processes have become available in recent years. Investigated systems include the wide range of target–receptor pairs: serine protease trypsin–benzamidine [41–43], DNA–cisplatin [44], LAO-binding protein–L-arginine [45], choline-binding protein–choline [46], MDM2–transactivation domain of p53 [47], and GPCR $S1P_1R$–lipid inhibitor ML056 [48]. Likewise, MSMs of transitions between inactive and active states of signaling proteins have been reported for the GPCR β_2 adrenergic receptor [49, 50], Src kinase [51] and the cyclic nucleotide-binding domain of protein kinase A [52]. What these systems have in common is a clearly defined initial state I, i.e. the unbound state or the inactive state, and an equally well-defined final state F, i.e. the bound state or the active state. Both the initial and the final state, usually consist of several microstates. MSMs can be used to characterize the transition paths and mean first passage times (MFPTs) between I and F.

The MFPT [53] of a microstate outside of the final state, $A_i \notin F$, is the average time it takes to get from this microstate to any microstate within the final state, $A_f \in F$. If A_j is a microstate on the way between A_i and the final state F, and if a transition $A_i \rightarrow A_j$ has been made, then MFPT(i) = τ + MFPT(j), i.e. the time τ it took to get to A_j plus the MFPT(j) from there. All possible transitions $A_i \rightarrow A_j$ are accounted for by this equation

$$\mathrm{MFPT}(i) = \sum_j T_{ij}(\tau) \cdot [\tau + \mathrm{MFPT}(j)] \tag{4.26}$$

where $T_{ij}(\tau)$ is the transition probability from A_i to A_j. We define that, for microstates which belong to the final state,

$$\mathrm{MFPT}(f) = 0 \quad \forall A_f \in F \tag{4.27}$$

Equation (4.26) can be set up for all the microstates. One obtains a system of linear equations which can be solved iteratively. The MFPT from initial to final state is then obtained by summing over the MFPTs of all microstates within I

$$\mathrm{MFPT}(I) = \frac{1}{\pi_I} \sum_{i \in I} \pi_i \, \mathrm{MFPT}(i) \tag{4.28}$$

MFPTs are particularly useful for binding processes, because they can be converted into association and dissociations rates

$$\begin{aligned} k_{\mathrm{on}} &= \frac{1}{\mathrm{MFPT}(\mathrm{unbound} \rightarrow \mathrm{bound})} \\ k_{\mathrm{off}} &= \frac{1}{\mathrm{MFPT}(\mathrm{bound} \rightarrow \mathrm{unbound})} \end{aligned} \tag{4.29}$$

and thus be compared to experiment.

Contrary to rates, transition paths are difficult to elucidate experimentally. Using transition path theory [54, 55], these paths can be extracted. The flux $f_{ij}(\tau)$ between two microstates A_i and A_j is the total probability to observe a transition $A_i \rightarrow A_j$ within τ. When considering all possible trajectories, $f_{ij} = \pi_i T_{ij}(\tau)$. From these trajectories, one extracts the reactive trajectories, i.e. those trajectories which go directly from the initial state I to the final state F without hitting I in between. This yields an effective flux,

$$f_{ij}^{\mathrm{effective}} = q_i^- \cdot \pi_i T_{ij}(\tau) \cdot q_j^+ \tag{4.30}$$

where q_i^- is the probability that the trajectory came from the initial state I, when it reached A_i, rather than from the final state F, and q_j^+ is the probability that the trajectory will proceed to the final state F after leaving A_j. q_i^-, and q_j^+ are the called backward and forward committor, respectively. Both properties can be calculated from the transition matrix by solving a system of linear equations. In the present set of trajectories, many contain recrossing events, i.e. sequences like $A_i \rightarrow A_j \rightarrow A_i \rightarrow A_j \ldots$ This yields a high effective flux between A_i and A_j, which, however, does not mean that many reactive trajectories pass through A_i and A_j. One therefore reduces the set of trajectories further and removes all trajectories which contain recrossing events or detours. The probability of observing a transition $A_i \rightarrow A_j$ in this set of trajectories is the net flux

$$f_{ij}^{+}(\tau) = \max\{0, f_{ij}^{\text{effective}}(\tau) - f_{ji}^{\text{effective}}(\tau)\} \tag{4.31}$$

For detailed-balanced processes, and if the microstate are arranged such that $q_i^+ \leq q_j^+$, Eq. (4.31) can be rearranged as

$$f_{ij}^{+}(\tau) = \pi_i T_{ij}(\tau)(q_j^+ - q_i^+) \tag{4.32}$$

Representing the net flux matrix as a network graph usually yields a very intuitive representation of the transition paths from I to F; see Refs. [43, 45]

In addition to MFPTs and transition paths, many authors have extracted long-lived conformations from their MSMs and have found that ligand-binding processes, as well as activation processes are not purely diffusive but proceed via a series of metastable states [41–47, 49, 51, 52]. Also, binding free energies have been estimated from MSMs [41–47]. The results are comparable to or slightly better than the molecular mechanics Poisson–Boltzmann surface area method (MMPB-SA) [42, 44], but not as good as the estimates from a free energy perturbation [42].

4.6 Outlook

In our account, we have not touched on the question of how to find a suitable discretization. In practice, the actual discretization is preceded by a dimensionality reduction of the trajectory, either by manually choosing relevant coordinates or by means of a dimensionality reduction method. Many of the studies we mentioned used the time-lagged independent component analysis (tICA) [56, 57] to find a suitable low-dimensional representation of the system. Similar to the principle component analysis, tICA is a coordinate transformation. It yields an optimal slow subspace which decorrelated (at lag time τ) from the remaining coordinates. This subspace is then usually discretized into non-overlapping microstates and a conventional MSM is constructed.

Besides this, there are several advanced discretizations, such as the core-set method [58, 59] and the variational approach to molecular dynamics [60, 61]. These methods abandon the use of discrete microstates and instead use functions of the conformational space as ansatz functions for the discretization of the transfer operators. This significantly reduces the approximation error of the model and thus yields a more accurate description of the conformational dynamics.

Dynamical reweighting methods are another important development [62–65]. So far, MSMs have been estimated from direct simulations of the conformational transitions. With dynamical reweighting methods, one can estimate a dynamical model from a biased simulation, such as a metadynamics simulation [66] or a replica-exchange MD [63]. This drastically increases the range of molecular time and length scale, for which dynamical models can be constructed.

User-friendly software packages [67, 68] have made MSMs accessible to a wide range of academic researchers. This has led to innovative ways to use this technique in structure-based drug design. Currently, the workflows of an MSM analysis have not been automized to the point where they can be used routinely in pharmaceutical companies. However, as the accuracy of dynamics models further improves and the sampling becomes less of an issue, taking the conformational flexibility into account is likely to become the norm in structure-based drug design.

Acknowledgments

This research has been funded by Deutsche Forschungsgemeinschaft (DFG) through CRC 765 "Multivalency as a chemical organization and action principle," Project C10, and through CRC1114 "Scaling Cascades in Complex Systems," Project B05.

References

1 Fischer, E. (1894). Einfluss der configuration auf die wirkung der enzyme. II. *Ber. Dtsch. Ges.* 27 (3): 3479–3483.
2 Monod, J., Wyman, J., and Changeux, J.-P. (1965). On the nature of allosteric transitions: a plausible model. *J. Mol. Biol.* 12: 88–118.
3 Koshland, D.E., Namethy, G., and Filmer, D. (1966). Comparison of experimental binding data and theoretical models in proteins containing subunits. *ACS Biochem.* 5 (1): 365–385.
4 Berger, C., Weber-Bornhauser, S., Eggenberger, J. et al. (1999). Antigen recognition by conformational selection. *FEBS Lett.* 450 (1–2): 149–153.
5 Miller, Y., Ma, B., and Nussinov, R. (2010). Polymorphism in Alzheimer Aβ amyloid organization reflects conformational selection in a rugged energy landscape. *Chem. Rev.* 110 (8): 4820–4838.
6 Williamson, J.R. (2000). Induced fit in RNA-protein recognition. *Nat. Struct. Mol. Biol.* 7 (10): 834–837.
7 Johnson, K.A. (2008). Role of induced fit in enzyme specificity: a molecular forward/reverse switch. *J. Biol. Chem.* 283 (39): 26297–26301.
8 De Vivo, M., Masetti, M., Bottegoni, G., and Cavalli, A. (2016). Role of molecular dynamics and related methods in drug discovery. *J. Med. Chem.* 59 (9): 4035–4061.
9 Shirts, M. and Pande, V.S. (2000). Screen savers of the world unite! *Science* 290 (5498): 1903–1904.
10 Shaw, D.E., Maragakis, P., Lindorff-Larsen, K. et al. (2010). Atomic-level characterization of the structural dynamics of proteins. *Science* 330 (6002): 341–346.
11 Stanley, N. and De Fabritiis, G. (2015). High throughput molecular dynamics for drug discovery. *In Silico Pharmacol.* 3: 1–4. https://www.ncbi.nlm.nih.gov/pmc/articles/PMC4339319/

12 Schütte, Ch., Fischer, A., Huisinga, W., and Deuflhard, P. (1999). A direct approach to conformational dynamics based on hybrid Monte Carlo. *J. Comput. Phys.* 151: 146–168.
13 Swope, W.C., Pitera, J.W., and Suits, F. (2004). Describing protein folding kinetics by molecular dynamics simulations. 1. Theory†. *J. Phys. Chem. B* 108: 6571–6581.
14 Chodera, J.D., Singhal, N., Pande, V.S. et al. (2007). Automatic discovery of metastable states for the construction of Markov models of macromolecular conformational dynamics. *J. Chem. Phys.* 126 (15): 155101.
15 Buchete, N.-V. and Hummer, G. (2008). Coarse master equations for peptide folding dynamics. *J. Phys. Chem. B* 112: 6057–6069.
16 Prinz, J.-H., Wu, H., Sarich, M. et al. (2011). Markov models of molecular kinetics: generation and validation. *J. Chem. Phys.* 134: 174105.
17 Keller, B., Hünenberger, P., and van Gunsteren, W.F. (2011). An analysis of the validity of Markov state models for emulating the dynamics of classical molecular systems and ensembles. *J. Chem. Theory Comput.* 7: 1032–1044.
18 Hünenberger, P.H. (2005). Thermostat algorithms for molecular dynamics simulations. In: *Advances in Polymer Science*, 105–149. Berlin, Heidelberg: Springer-Verlag.
19 Frenkel, D. and Smit, B. (2001). *Understanding Molecular Simulation*, 2e. Elsevier.
20 Leach, A.R. (2001). *Molecular Modelling: Principles and Applications*, 2e. Pearson Education.
21 Allen, M.P. and Tildesley, D.J. (1991). *Computer Simulation of Liquids*. Oxford Science Publications, reprinted edition, 1987.
22 Schütte, Ch., Huisinga, W., and Deuflhard, P. (1999). Transfer Operator Approach to Conformational Dynamics in Biomolecular Systems. ZIB Report, 36, SC99–36.
23 Prinz, J.-H., Keller, B., and Noé, F. (2011). Probing molecular kinetics with Markov models: metastable states, transition pathways and spectroscopic observables. *Phys. Chem. Chem. Phys.* 13: 16912–16927.
24 Frank Noé, M.S. and Schütte, Ch. (2010). On the approximation quality of Markov state models. *Multiscale Model. Simul.* 8: 1154–1177.
25 Noé, F. (2008). Probability distributions of molecular observables computed from Markov models. *J. Chem. Phys.* 128 (24): 244103.
26 Chodera, J.D. and Noé, F. (2010). Probability distributions of molecular observables computed from Markov models. II. Uncertainties in observables and their time-evolution. *J. Chem. Phys.* 133 (10): 105102.
27 Mukherjee, S., Pantelopulos, G.A., and Voelz, V.A. (2016). Markov models of the apo-MDM2 lid region reveal diffuse yet two-state binding dynamics and receptor poses for computational docking. *Sci. Rep.* 6 (1): 1038.
28 Bowman, G.R. and Geissler, P.L. (2012). Equilibrium fluctuations of a single folded protein reveal a multitude of potential cryptic allosteric sites. *Proc. Natl. Acad. Sci. U.S.A.* 109: 11681–11686.
29 Bowman, G.R., Bolin, E.R., Hart, K.M. et al. (2015). Discovery of multiple hidden allosteric sites by combining Markov state models and experiments. *Proc. Natl. Acad. Sci. U.S.A.* 112 (9): 2734–2739.

30 Wagner, J.R., Lee, C.T., Durrant, J.D. et al. (2016). Emerging computational methods for the rational discovery of allosteric drugs. *Chem. Rev.* 116 (11): 6370–6390.
31 Hendlich, M., Rippmann, F., and Barnickel, G. (1997). LIGSITE: automatic and efficient detection of potential small molecule-binding sites in proteins. *J. Mol. Graph. Modell.* 15 (6): 359–363.
32 Keller, B., Gattin, Z., and van Gunsteren, W.F. (2010). What stabilizes the $3_1$4-helix in β3-peptides? A conformational analysis using molecular simulation. *Proteins Struct. Funct. Bioinf.* 78 (7): 1677–1690.
33 Hanske, J., Aleksić, S., Ballaschk, M. et al. (2016). Intradomain allosteric network modulates calcium affinity of the C-type lectin receptor Langerin. *J. Am. Chem. Soc.* 138: 12176–12186.
34 Hart, K.M., Moeder, K.E., Ho, C.M.W. et al. (2017). Designing small molecules to target cryptic pockets yields both positive and negative allosteric modulators. *PLOS ONE* 12 (6): e0178678.
35 Arkin, M.R., Tang, Y., and Wells, J.A. (2014). Small-molecule inhibitors of protein-protein interactions: progressing toward the reality. *Chem. Biol.* 21 (9): 1102–1114.
36 Rakers, C., Bermudez, M., Keller, B.G. et al. (2015). Computational close up on protein–protein interactions: how to unravel the invisible using molecular dynamics simulations? *WIREs Comput. Mol. Sci.* 5 (5): 345–359.
37 Bowman, G.R. (2012). Improved coarse-graining of Markov state models via explicit consideration of statistical uncertainty. *J. Chem. Phys.* 137 (13): 134111.
38 Deuflhard, P. and Weber, M. (2005). Robust Perron cluster analysis in conformation dynamics. *Linear Algebra Appl.* 398: 161–184.
39 Razavi, A.M., Wuest, W.M., and Voelz, V.A. (2014). Computational screening and selection of cyclic peptide hairpin mimetics by molecular simulation and kinetic network models. *J. Chem. Inf. Model.* 54 (5): 1425–1432.
40 Witek, J., Keller, B.G., Blatter, M. et al. (2016). Kinetic models of cyclosporin A in polar and apolar environments reveal multiple congruent conformational states. *J. Chem. Inf. Model.* 56 (8): 1547–1562.
41 Buch, I., Giorgino, T., and De Fabritiis, G. (2011). Complete reconstruction of an enzyme-inhibitor binding process by molecular dynamics simulations. *Proc. Natl. Acad. Sci. U.S.A.* 108 (25): 10184–10189.
42 Takahashi, R., Gil, V.A., and Guallar, V. (2013). Monte Carlo free ligand diffusion with Markov state model analysis and absolute binding free energy calculations. *J. Chem. Theory Comput.* 10 (1): 282–288.
43 Plattner, N. and Noé, F. (2015). Protein conformational plasticity and complex ligand-binding kinetics explored by atomistic simulations and Markov models. *Nat. Commun.* 6: 7653–7663.
44 Lucas, M.F., Cabeza de Vaca, I., Takahashi, R. et al. (2014). Atomic level rendering of DNA-drug encounter. *Biophys. J.* 106 (2): 421–429.
45 Silva, D.-A., Bowman, G.R., Sosa-Peinado, A., and Huang, X. (2011). A role for both conformational selection and induced fit in ligand binding by the LAO protein. *PLoS Comput. Biol.* 7 (5): e1002054.

46 Gu, S., Silva, D.-A., Meng, L. et al. (2014). Quantitatively characterizing the ligand binding mechanisms of choline binding protein using Markov state model analysis. *PLoS Comput. Biol.* 10 (8): e1003767.
47 Zhou, G., Pantelopulos, G.A., Mukherjee, S., and Voelz, V.A. (2017). Bridging microscopic and macroscopic mechanisms of p53-MDM2 binding using molecular simulations and kinetic network models. *Biophys. J.* 113 (4): 785–793.
48 Stanley, N., Pardo, L., and De Fabritiis, G. (2016). The pathway of ligand entry from the membrane bilayer to a lipid G protein-coupled receptor. *Sci. Rep.* 6 (1). doi: 10.1038/srep22639.
49 Bai, Q., Pérez-Sánchez, H., Zhang, Y. et al. (2014). Ligand induced change of β2 adrenergic receptor from active to inactive conformation and its implication for the closed/open state of the water channel: insight from molecular dynamics simulation, free energy calculation and Markov state model analysis. *Phys. Chem. Chem. Phys.* 16 (30): 15874–15885.
50 Kohlhoff, K.J., Shukla, D., Lawrenz, M. et al. (2014). Cloud-based simulations on Google Exacycle reveal ligand modulation of GPCR activation pathways. *Nat. Chem.* 6: 15–21.
51 Shukla, D., Meng, Y., Roux, B., and Pande, V.S. (2014). Activation pathway of Src kinase reveals intermediate states as novel targets for drug design. *Nat. Commun.* 5: 3397.
52 Malmstrom, R.D., Kornev, A.P., Taylor, S.S., and Amaro, R.E. (2015). Allostery through the computational microscope: cAMP activation of a canonical signalling domain. *Nat. Commun.* 6: 7588.
53 Singhal, N., Snow, C.D., and Pande, V.S. (2004). Using path sampling to build better Markovian state models: predicting the folding rate and mechanism of a tryptophan zipper beta hairpin. *J. Chem. Phys.* 121 (1): 415.
54 Vanden-Eijnden, E. (2006). Transition path theory. In: *Computer Simulations in Condensed Matter Systems: From Materials to Chemical Biology*, Vol. 1, 453–493. Berlin Heidelberg: Springer-Verlag.
55 Noé, F., Schütte, C., Vanden-Eijnden, E. et al. (2009). Constructing the equilibrium ensemble of folding pathways from short off-equilibrium simulations. *Proc. Natl. Acad. Sci. U.S.A.* 106: 19011–19016.
56 Pérez-Hernández, G., Paul, F., Giorgino, T. et al. (2013). Identification of slow molecular order parameters for Markov model construction. *J. Chem. Phys.* 139: 015102.
57 Schwantes, C.R. and Pande, V.S. (2013). Improvements in Markov state model construction reveal many non-native interactions in the folding of NTL9. *J. Chem. Theory Comput.* 9 (4): 2000–2009.
58 Schütte, Ch., Noé, F., Lu, J. et al. (2011). Markov state models based on milestoning. *J. Chem. Phys.* 134: 204105.
59 Lemke, O. and Keller, B.G. (2016). Density-based cluster algorithms for the identification of core sets. *J. Chem. Phys.* 145: 164104.
60 Nüske, F., Keller, B.G., Pérez-Hernández, G. et al. (2014). Variational approach to molecular kinetics. *J. Chem. Theory Comput.* 10: 1739–1752.

61 Vitalini, F., Noé, F., and Keller, B.G. (2015). A basis set for peptides for the variational approach to conformational kinetics. *J. Chem. Theory Comput.* 11: 3992–4004.
62 Tiwary, P. and Parrinello, M. (2013). From metadynamics to dynamics. *Phys. Rev. Lett.* 111: 230602.
63 Wu, H., Paul, F., Wehmeyer, C., and Noé, F. (2016). Multiensemble Markov models of molecular thermodynamics and kinetics. *Proc. Natl. Acad. Sci. U.S.A.* 113: E3221–E3230.
64 Donati, L., Hartmann, C., and Keller, B.G. (2017). Girsanov reweighting for path ensembles and Markov state models. *J. Chem. Phys.* 146 (24): 244112.
65 Donati, L. and Keller, B.G. (2018). Girsanov reweighting for metadynamics simulations. *J. Chem. Phys.* 149 (7): 072335.
66 Casasnovas, R., Limongelli, V., Tiwary, P. et al. (2017). Unbinding kinetics of a p38 MAP kinase type II inhibitor from metadynamics simulations. *J. Am. Chem. Soc.* 139: 4780–4788.
67 Harrigan, M.P., Sultan, M.M., Hernández, C.X. et al. (2017). MSMBuilder: statistical models for biomolecular dynamics. *Biophysical Journal* 112 (1): 10–15. January
68 Scherer, M.K., Trendelkamp-Schroer, B., Paul, F. et al. (2015). PyEMMA 2: a software package for estimation, validation, and analysis of Markov models. *J. Chem. Theory Comput.* 11 (11): 5525–5542.

5

Monte Carlo Techniques for Drug Design: The Success Case of PELE

Joan F. Gilabert[1], Daniel Lecina[1], Jorge Estrada[1], and Victor Guallar[1,2]

[1] *Barcelona Supercomputing Center (BSC), Life Science Department, Jordi Girona 29, 08034 Barcelona, Spain*
[2] *ICREA, Passeig Lluís Companys 23, 08010 Barcelona, Spain*

5.1 Introduction

The Monte Carlo (MC) method dates back to the 1940s. While recovering from a surgery, Stanislaw Ulam tried to compute the probability of successfully finishing the Canfield solitaire game. After initial attempts looking for the combinatorial solution, he conceived an easier method that consisted of repeatedly simulating different games and counting the number of successful draws [1]. In other words, the idea was using statistical sampling to obtain a numerical approximation for a problem with a difficult analytical solution. Soon after, in the context of the Manhattan project and the first electronic computers, MC techniques were quickly applied to different problems in physics and afterwards became increasingly popular in a multitude of scientific fields, from biology to engineering [2–4].

Nowadays, MC techniques cover a broad spectrum of computational methodologies that make use of random numbers to perform numerical calculations. In drug design, these techniques are typically used to sample the Boltzmann distribution, building a Markov Chain Monte Carlo (MCMC) [5] (mainly using the Metropolis algorithm [6]), or to find bound structure conformations. Protein–ligand complexes are modeled using an all-atom force field paradigm [7], which yields a highly dimensional phase space that does not allow direct estimates. Luckily, the relative probabilities between any pair of states are known, the ratio of Boltzmann weights, which permits the construction of an MCMC [8]. From this point of view, MC represents an alternative to molecular dynamics (MD) [9]. In the latter, Newton's equations of motion are integrated in small time increments, in the order of the femtosecond, to produce trajectories. One of the main drawbacks of this kind of simulation is the heavy computational cost associated to model the long timescales that correspond to protein–ligand binding and unbinding. On the contrary, MCMC does not produce time trajectories (i.e. dynamics); the algorithm draws samples in the energy landscape according to the corresponding statistical ensemble. This opens the door to design strategies that may yield, at least in theory, a more efficient sampling, since subsequent snapshots can be more uncorrelated compared to the MD counterpart, e.g. with

Biomolecular Simulations in Structure-Based Drug Discovery,
First Edition. Edited by Francesco L. Gervasio and Vojtech Spiwok.

side-chain rotations, backbone movements or ligand displacements. The main obstacle of MC simulations, however, is the difficulty in generating uncorrelated poses with a significant statistical weight, given the ruggedness of the energy landscape that is defined by the force field. Protein and ligand flexibility is one of the causes behind this. Considering coordinated atom displacements help mitigate the problem, for example with the use of internal coordinates, which take into account coordinated displacements in torsions [10–12]. However, random proposals typically lead to clashes and are usually, but not exclusively, reduced to small variations of the starting structure.

5.1.1 First Applications

The precursors of MC sampling are condensed matter simulations, such as the study of water carried out by Barker and Watts in 1969 [13]. Several instances of studies involving statistical and thermodynamic properties of water solutions of methane can be found; some examples are the works of Dashevsky and Sarkisov in 1974 [14], Owicki and Scheraga in 1977 [15] and by Swaminathan et al. in 1978 [16]. In a similar manner, in the late 1960s the first MC simulations of chain molecules were reported [17], including the work of Hagler et al. studying the conformation of oligopeptides in 1978 [18] or the study of bovine pancreatic trypsin by Wako and Scheraga in 1981 [19].

5.1.2 Free Energy Calculations

As computers started growing in power, the possibilities of molecular simulations expanded. Larger and more complex systems could be tackled, and new methods attempted to increase the efficiency of simulations. This permitted obtaining free energy estimations [20, 21], mainly using alchemical methods such as free energy perturbation. An example is the pioneering MCLiquids of Jorgensen's laboratory, used to study liquid hydrogen fluoride, ammonia, or ethanol, which later on evolved into BOSS (Biomolecular and Organic Simulation System) and MCPRO (MC for Proteins) [12]; both programs are still in active development, being used to perform molecular mechanics simulations and in the development of the OPLS force field [22]. Other free energy calculation methods, such as thermodynamic integration, have been used with MC. It can be seen in different works of Essex's group [23], and is included in their software package, ProtoMS [24], which specializes in explicit solvent calculations, for example, making use of Grand Canonical MC [25] or JAWS [26]. The Sire software (http://siremol.org), a collection of C++/Python building blocks allowing to build custom molecular simulation applications, also emerged from ProtoMS.

5.1.3 Optimization

MC methods have been largely used in protein optimization problems, often in combination with enhanced sampling methods such as simulated annealing [27] or replica-exchange ideas, where several replicas are kept at slightly different conditions and exchanges between such copies are attempted periodically. Two examples of the latter are replica exchange with solute tempering (REST) [28] and replica-exchange thermodynamic integration (RETI) [29]. Optimization problems have also been tackled using more sophisticated random schemes, such as the biased probability MC of Abagyan [30], based on generating the random step with an optimal distribution of proposals; the force-biased MC [31], where the acceptance probability becomes one (or nearly one) by using first energy derivatives; or with collective variable motions [32].

A different family of methodologies originated from the MC-minimization technique by Li and Scheraga [33]. This method combined a typical MC random displacement, in this case randomly moving one dihedral angle, followed by an energy minimization. The method identifies at each step a local minimum, and thus the exploration is constrained to different local minima. An important contribution was then developed in 1989 by Still and coworkers: the MacroModel software [34], a simple MC method combining random moves, minimizations, and an implicit solvation generalized Born surface area (GBSA) model. This technique is currently included in the Schrödinger suite of programs, being one of the most used molecular MC programs of all times. More recently, these techniques have led to the development of several MC methods with minimization schemes to perform conformational sampling of proteins, for example, the basin-hopping method of Wales [35], which can be used in combination with Wang–Landau sampling [36] to obtain the complete energy landscape, the activation–relaxation technique of Mousseau and coworkers [37] or our Protein Energy Landscape Exploration (PELE), which combines random movements with protein structure prediction algorithms to enhance the exploration [38].

5.1.4 MC and MD Combinations

Efforts have also been made to combine MC with MD. In Hamiltonian MC (also known as hybrid MC) [39], velocities are added in order to guide proposals and improve the acceptance. Recently, Chodera and coworkers presented nonequilibrium candidate Monte Carlo (NCMC) [40], where proposal moves are constructed via nonequilibrium dynamics to generate new configurations. These are built in a series of perturbation mixed with propagation moves, enhancing 2 orders of magnitude the performance in finding binding modes of toluene in a mutated T4 lysozyme [41].

In Table 5.1 we summarize the most representative software packages that readily allow running MC simulations in relevant scenarios for drug design. Some application examples include several studies using the MCPRO or BOSS tools from Jorgensen Labs. Besides their own applications studies, for example, the computationally aided discovery of anti-HIV agents [42], we find additional studies such as the inhibitor design on tyrosine kinase [43]. In addition, MCPRO+, a version under Schrödinger commercial use, has been also used in the drug design studies for the acetylcholine receptor [44] and the Bcr-Abl kinase [45]. We also find several studies using ProtoMS, more focused on obtaining the calculation of free energies [23] and the role of water in drug recognition and binding [25, 26]. Commercial software, such as ICM and Macromodel, include several reference and benchmark studies [46–49].

Table 5.1 Main MC software used in drug development modeling.

Software	Developers	Short description
ProtoMS http://www.essexgroup.soton.ac.uk/ProtoMS/	Essex Lab.	Relative and absolute free energy calculations and water placement with GCMC and JAWS methodology. GNU General Public License
BOSS http://zarbi.chem.yale.edu/software.html	Jorgensen Lab.	General purpose molecular modeling system that performs molecular mechanics calculations, Metropolis MC statistical simulations, and semiempirical quantum mechanics calculations. Free for nonprofit/commercial for profit
MCPRO http://zarbi.chem.yale.edu/software.html	Jorgensen Lab.	MC statistical mechanics simulations of peptides, proteins, and nucleic acids; it was derived from BOSS. Free for nonprofit/commercial for profit
PELE https://pele.bsc.es, http://www.nostrumbiodiscovery.com	Guallar Lab.	MC specialized in mapping protein–ligand interactions. Free for nonprofit/commercial for profit
Sire http://siremol.org	Christopher Woods, et al.	C++/Python rewrite of ProtoMS. It now includes many more modeling modules. GNU General Public License
ICM http://www.molsoft.com/icm_pro.html	Molsoft, LLC	Biased probability MC. Prediction of the three-dimensional structure of peptides and proteins. Commercial
Tinker https://dasher.wustl.edu/tinker	Ponder Lab.	General molecular modeling program including an MC minimization technique. Free of charge, registered under US Copyright Law
Macromodel https://www.schrodinger.com/macromodel	Schrödinger, LLC	Mixed MC algorithms and other molecular modeling methods. Commercial Schrodinger also distributes a (commercial) MCPRO+ version

The remaining sections of this chapter explore in detail the PELE program, providing first the main characteristics of the technique, followed by an analysis of the different application studies in mapping protein–ligand interactions.

5.2 The PELE Method

PELE, originally described by Borrelli et al. in 2005 [38] and whose latest version is implemented in C++, follows a heuristic MC approach, generating new conformational proposals using vibration modes of the macromolecule (protein or nucleic acid) with translations and rotations of the ligand, and relaxing the system with structure prediction methods, so that the probability of acceptance in the Metropolis test remains high.

5.2.1 MC Sampling Procedure

The overall MC procedure, shown in Figure 5.1, involves a ligand plus a receptor perturbation, followed by a relaxation step comprising a side-chain sampling and a minimization step. Typical simulations involve from tens to hundreds of processors for thousand MC steps. We describe now briefly each of these steps.

5.2.2 Ligand Perturbation

The ligand, formed by a rigid core (at least one atom) and a set of rotatable side chains, is perturbed by translating and rotating it. PELE creates a list of perturbation poses (typically between 1 and 20), and then chooses the one with the lowest system energy. Each pose is generated through a random procedure, where a number of translations and rotations are tried, until a combination is found that does not show steric clashes after a clash-relieving process; clashes of up to 30% of the van der Waals interatomic distance between atoms are allowed, since the relaxation phase will probably relieve them. When searching for the binding site, large translations are usually selected (around 6 Å), while once in the binding site, small adjustments of ~1 Å (or less) are used; different translation ranges are often applied depending on the ligand solvent accessible surface area (SASA) at each iteration. In addition, to enhance the sampling of rare events, the translational vector may be kept constant in a random direction for a given number of MC simulation steps. Also, for those simulations that require exploring a restricted volume of the search space (as it is done in binding pose refinement), perturbations that place the ligand center of mass outside a predefined box are rejected.

5.2.3 Receptor Perturbation

After perturbing the ligand, the macromolecule backbone is perturbed following normal modes calculated using the Anisotropic Network Model (ANM, see Ref. [50]) or a principal component analysis (PCA) from different structures (derived from X-ray structures, MD, etc.). This phase is typically executed every four MC steps [51].

In ANM, if a protein, the macromolecule is modeled as an elastic network of α-carbon (CA) nodes using the power contact model of [52]. For nucleic acids, atoms P, C2, and C4′ are used as nodes of the network, using the exponential contact model of Setny and Zacharias [53]. A combination of modes is selected to represent the given vibration from which to select the perturbation to try. By default, the six lowest frequency modes are used, since they are the ones more closely related to conformational movements [50]. It is possible to use a single mode, or to mix several ones randomly, with a preferred mode being given more weight, or weighting each mode according to its frequency. Besides, a given combination of modes can be kept during several simulation steps, changing sense

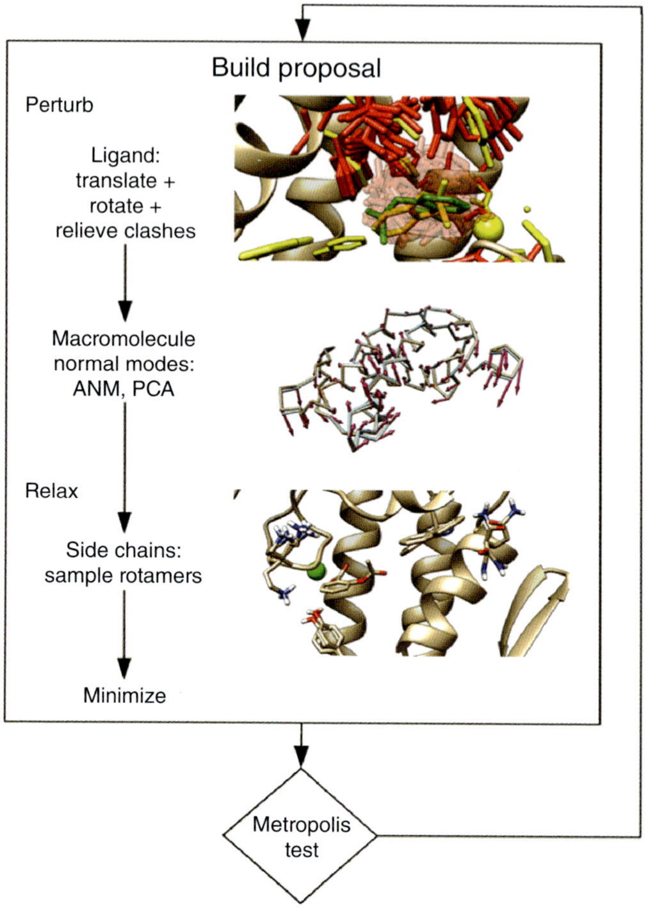

Figure 5.1 Process flow in a PELE simulation: Each simulation step builds a new conformation proposal by perturbing the ligand and the macromolecule, then relaxing it to an energetically favorable conformation by sampling side-chain rotamers and by minimizing the whole system; finally, the proposal is either accepted or rejected following the Metropolis test. The top image shows the initial conformation (yellow), all the candidate ligand perturbations obtained in a given step by translating and rotating the ligand (transparent red) and relieving existing clashes by moving side chains (solid red), and the selected perturbation (green). The middle image shows the first ANM mode as magenta vectors placed at the atoms of the α-carbons trace for the starting protein conformation (brown), and the trace obtained after applying the ANM mode (cyan). The bottom image shows the representative rotamers of the generated clusters for several side chains.

periodically, to allow a more thorough sampling of a given vibrational movement. To apply the movement for all atoms, and not only CAs, harmonic constraints are placed on the latter to their target coordinates, and an all-atom energy minimization is performed, using the multiscale truncated Newton algorithm. To avoid reverting the ligand perturbation, a constraint on the center of mass of the ligand's core atoms is typically applied; in this way, the ligand perturbation can be coupled to backbone reorganization.

5.2.4 Side-Chain Adjustment

The side-chain readjustment is done for all the side chains in the macromolecule that surround the ligand, and for the ligand itself. In addition, one might include those side chains that increased their energy the most during the perturbation phase (regardless of their location in the receptor). The sampling follows the algorithm of Jacobson et al. [54], where the problem becomes manageable by considering only rotamers as possible conformations for the residue side chains and, above all, by placing the best rotamer for each side chain (after clustering) in a residue by residue way, keeping all the other residues fixed at their current rotamers. This is iterated until a new round of placement for all residues does not significantly change the conformation (or a maximum number of iterations are exceeded). If (even with the allowed overlap) there is no rotamer without clashes for some side chain, then this phase fails and the MC step is rejected. Once the side chains have been readjusted, a partial minimization affecting only the side-chain atoms is performed.

5.2.5 Minimization

The last relaxation step continues through a multiscale truncated Newton minimization [55], the same algorithm as used in the minimizations done in previous phases. Along the minimization, the atoms representing the nodes in the ANM phase are (typically) weakly constrained, so that this minimization does not undo the perturbation phase. Convergence criteria is significantly relaxed, ~0.1 kcal/mol/Å, allowing to approximately recover the detailed balance condition (lost when introducing the relaxation step) [56].

5.2.6 Coordinate Exploration

PELE introduces two further techniques to enhance conformational sampling. In the original one, *the spawning approach* introduced in Borrelli et al. [38], a reaction coordinate (root-mean-square distance (RMSD), distances, etc.) is defined along with a range criteria and a sense (increase or decrease). For a given processor and MC step, if the current reaction coordinate value falls outside of the range from the best achieved value (along all processors and steps), then this processor will abandon its current conformation and receive the one corresponding to the best value, which is constantly updated. This kind of strategy is particularly useful when studying binding or escape trajectories.

More recently, an enhanced strategy has been introduced, *adaptive PELE* [57], aiming at eliminating the bias introduced in the spawning approach. In this approach, standard PELE is run for a small number of MC steps (~4–20), in what constitutes an epoch. All generated conformations in the epoch (from all processors) are then clustered using the ligand RMSD, where we also consider clusters from previous epochs. Each cluster is assigned a reward value that favors (by default) clusters less explored, that is inversely proportional to the cluster population. Additional rewards might be defined from physical/structural properties such as reaction coordinates, ligand exposure to solvent, etc. Then, a new epoch is started using a set of selected conformations, based on the reward function, as new seeds. Thus, the overall procedure is designed to enhance the sampling of poorly sampled regions. In addition, for an even faster biased exploration, the spawning and the adaptive techniques can be combined.

5.2.7 Energy Function

By default, PELE uses the OPLS 2005 energy function and parameters [58, 59], but also implements the AMBER99sbBSC0 force field [60], more suitable for nucleic acids. For the implicit solvent model, PELE has three options: the OBC model [61] with a nonpolar term following the ACE model [62]; the surface generalized Born (SGB) model [63], modeling the nonpolar term as in [64]; and the variable dielectric generalized Born model [65], using the same nonpolar model as SGB. PELE allows placing specific discrete water molecules that are deemed important, which will be moved during the different energy minimizations in a simulation step (the next release will include a water perturbation/relaxation step). As in most molecular mechanics codes, to gain speed negligible contributions to the energy, such as those caused by non-bonded interactions between far apart atoms, are discarded using a set of distance cutoffs.

5.3 Examples of PELE's Applications

We present here examples of application studies using PELE.

5.3.1 Mapping Protein Ligand and Biomedical Studies

PELE was originally designed to map ligand migration pathways: its first applications aimed at finding exit pathways starting from ligand-bound crystallographic structures. In a few hours of a small computational cluster (four cores), for example, it could map how a palmitate fatty acid molecule escaped the fatty acid–binding protein active site, without any additional knowledge or bias [38]. Many additional initial studies centered on "simple" ligand migration studies involving globin enzymes, following a trend (at that time) using different biased MD studies, such as the random expulsion MD or steered MD. Studied systems included truncated hemoglobins TrHbN [66] and TrHBO [67], where the involvement of multiple internal cavities was addressed. In addition, the fast performance of PELE allowed the first migration study on a (tetramer) human hemoglobin, identifying differences in ligand migration between the α and the β subunits and between their tense and relaxed states [68].

By expanding the exploration to a few more cores (tens to hundreds), we realized that we could perform a complete nonbiased exploration, mapping the entire surface active site search (from the solvent) and binding; similar studies, at that time, could only be performed using the special purpose (and \$10 M cost) Anton computer (see, for example, the pioneer study by Shaw Labs. [69]). Figure 5.2 shows an example of such an exhaustive search using PELE's web server on the

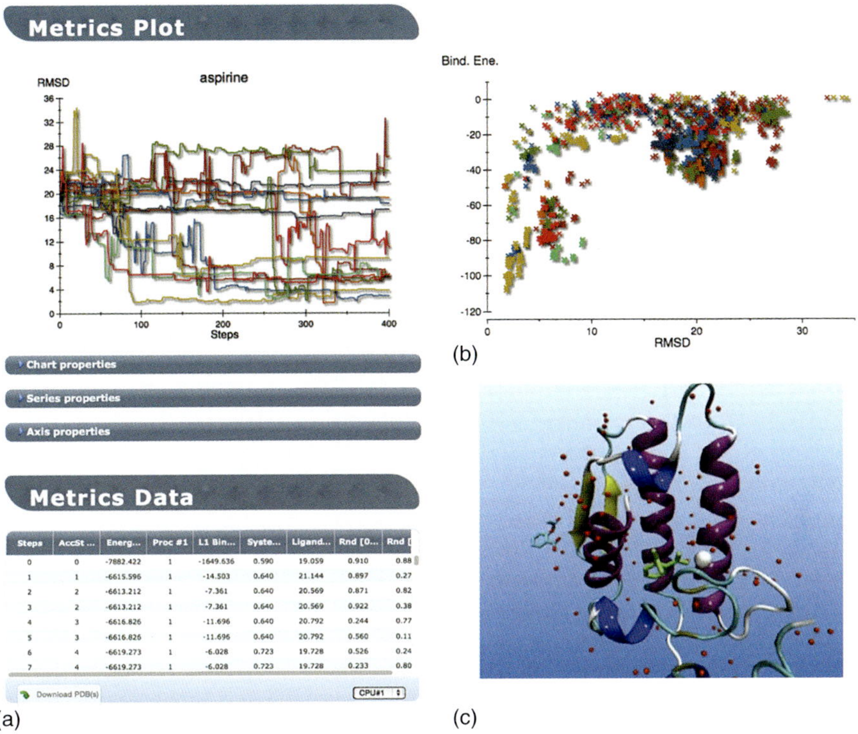

Figure 5.2 Example of active site search and binding on the aspirin–phospholipase A2 complex performed by the PELE web server (using 16 processors and 4 h). Panel (a) shows a screen capture of the server with the evolution of the ligand RMSD (all heavy atoms) to the crystal-bound complex (PDB ID: 1OXR) along the MC simulation. Crystal-like models, with RMSD < 2 Å, are found in ~90 MC steps, requiring approximately 1 h of wall clock simulation. Panel (b), also extracted from the server, shows the molecular mechanics protein–ligand interaction energy with respect to the same RMSD, indicating that the modeled bound complexes (<2.0 Å) can clearly be identified as the best affinity models; we should emphasize that no crystallographic information is used along the simulation but only to analyze the results. Panel (c) shows the initial molecular model, with the ligand on the bulk solvent (~21 Å RMSD); notice that we have also superposed in green the ligand active site bound structure (underlying the presence of a calcium ion). For more information, please visit https://pele.bsc.es/pele.wt/examples.

aspirin–phospholipase A2 complex. Additional studies involved quite complex systems such as prolyl oligopeptidase [70] and the binding of porphyrin into Gun4 [71]. Analogous simulations have been also produced for a DNA receptor, where the non-covalent binding of cisplatin was modeled with PELE's MC techniques [72].

Such modeling capabilities shifted our attention to biomedical/pharmaceutical studies, attracting, in addition, the interest from some pharmaceutical companies. Studies included research on cancer targets such as mTOR [73] and BCL-2 [74], on the glycosylation disorder through the study of the human phosphomannomutase2 receptor [75], and on diverse nuclear hormone receptors (NHRs), in collaboration with AstraZeneca [76, 77]. To enhance ligand screening, we have recently coupled PELE with advanced analysis techniques such as Markov State Models (MSM), approximating absolute binding free energies in a fast manner [77, 78].

In most cases, however, modeling the receptor–drug interactions does not require such extensive study; rearrangements of the ligand at the already known binding site suffice. Thus, we placed particular emphasis on protocols mapping the active site induced fit. Our initial benchmark study showed significant improvement from that of state-of-the-art commercial software, providing accurate cross-docking and apo docking [79]. Applied biomedical studies in this line, for example, included the prediction of drug resistance in the HIV-1 protease receptor, where we were able to blindly identify high-resistance patients using only their viruses' sequence data [80]. We should underline that this effort involved modeling 15–30 mutations in each receptor of 42 different patients.

PELE's position at the forefront of the induced-fit modeling techniques is clear if we look at the results of the latest Community Structure–Activity Resource (CSAR) competition (http://www.csardock.org/). The CSAR competition is a community-wide blind evaluation of docking and scoring methods performed by the University of Michigan (and participated by GlaxoSmithKline). In the summary article from the organizers [81], we find that PELE is the software most highlighted: "…In particular, submission J (Glide-XP2/PELE7) deserves special recognition for docking well to all targets. This group cross-docked all the ligands against all the given crystal structures. To see robust performance across an ensemble of protein structures for multiple targets is an impressive accomplishment…"

As stated, a recent enhanced conformational sampling has been introduced by means of an adaptive sampling [57]. This procedure improves about 1 order of magnitude the exploration performed by PELE. Such enhancement, for example, allowed to map complex ligand binding mechanisms, such as in G protein–coupled receptors (GPCRs) or in NHRs in only 3 hours using a commodity computing cluster (~32 computing cores), achieving interactive rates (~5 minutes) in difficult induced-fit cases.

These advances motivated PELE's transfer of the technology to Nostrum Biodiscovery (www.nostrumbiodiscovery.com), a recent small and medium-sized enterprise (SME) dedicated to drug development by means of supercomputation, which currently performs the commercial applications of the technique; academic (nonprofit) usage, however, is still possible through the public website https://pele.bsc.es, where hardware and software resources are available [82].

5.3.2 Enzyme Characterization

Additional applied studies have centered on modeling enzymatic mechanisms and engineering; the same techniques applied in mapping protein–drug interactions can be used in the study of substrate recognition by enzymes. While these studies often require combining PELE with mixed quantum mechanics and molecular mechanics (QM/MM) techniques, biochemical activity can be (largely) addressed by predicting the enzyme–substrate complex with simpler molecular mechanics methods. A substrate-bound model for the aryl alcohol oxidase with *p*-anisyl alcohol, for example, was developed in 2011 by mapping the substrate entrance in the apo crystallographic structure, which, due to the small size cavity, did not allow direct docking studies [83]. This model, which has recently been confirmed by crystallographic studies [84], allowed to select mutations for mapping the enzymatic mechanism and, more importantly, to improve the oxidation of secondary alcohols [85]. Similar simulations on toluene 4-monooxygenase allowed designing an improved (2.1×) variant for 2-phenyethanol oxidation [86].

A complete substrate migration study exploring the entire enzyme surface was used to compare two manganese peroxidases: one active toward 2,2′-azino-bis(3-ethylbenzothiazoline-6-sulphonic acid) (ABTS) and another one with no activity. Simulations identified the oxidation active site and allowed to introduce two solvent exposed mutations activating the ABTS-inactive manganese peroxidase achieving high turnover rates [87].

In another set of enzymatic studies on laccases, PELE's induced-fit procedure was coupled with QM/MM spin density single-point calculations, achieving excellent prediction of oxidation rates on different substrates [88]. This protocol was later expanded using an *in silico*-directed evolution protocol, allowing to improve activity and to modify the specificity on laccases [89].

In summary, in this chapter we aimed at providing first a quick overview of the development and applications of MC techniques in drug design. Then, we centered on delivering a deeper analysis of the techniques and capabilities in PELE, our own MC code that provides a quick and accurate manner to map protein–ligand interactions.

Acknowledgments

Along the development of PELE in the past years, we gratefully acknowledge financial support from the European Union (in particular from the ERC program) and from the Catalan and Spanish Governments. In addition, we thank all present and past members from the EAPM Lab at BSC for their dedication and work.

References

1 Special Issue, Stanislaw Ulam 1909-1984 (1987). In Los Alamos Sci.

2 Vitalis, A. and Pappu, R.V. (2009). Methods for Monte Carlo simulations of biomacromolecules. *Annu. Rep. Comput. Chem.* 5: 49–76.

3 Robert, C.P. (2016). Monte Carlo methods. In: *Wiley StatsRef: Statistics Reference Online* (ed. N. Balakrishnan, T. Colton, B. Everitt, et al.). Wiley.

4 Liu, J.S. (2004). *Monte Carlo Strategies for Scientific Computing*. New York: Springer.

5 Hastings, W.K. (1970). Monte Carlo sampling methods using Markov chains and their applications. *Biometrika* 57: 97–109.

6 Metropolis, N., Rosenbluth, A.W., Rosenbluth, M.N. et al. (1953). Equation of state calculations by fast computing machines. *J. Chem. Phys.* 21: 1087–1092.
7 Ponder, J.W. and Case, D.A. (2003). Force fields for protein simulations. *Adv. Protein Chem.* 66: 27–85.
8 Frenkel, D. and Smit, B. (2002). *Understanding Molecular Simulation. From Algorithms to Applications.* San Diego, CA: Academic Press.
9 Paquet, E. and Viktor, H.L. (2015). Molecular dynamics, Monte Carlo simulations, and Langevin dynamics: a computational review. *Biomed. Res. Int.* 2015: 183918.
10 Favrin, G., Irbäck, A., and Sjunnesson, F. (2001). Monte Carlo update for chain molecules: biased Gaussian steps in torsional space. *J. Chem. Phys.* 114: 8154–8158.
11 Yun, M.R., Lavery, R., Mousseau, N. et al. (2006). ARTIST: an activated method in internal coordinate space for sampling protein energy landscapes. *Proteins* 63: 967–975.
12 Jorgensen, W.L. and Tirado-Rives, J. (2005). Molecular modeling of organic and biomolecular systems using BOSS and MCPRO. *J. Comput. Chem.* 26: 1689–1700.
13 Barker, J.A. and Watts, R.O. (1969). Structure of water; a Monte Carlo calculation. *Chem. Phys. Lett.* 3: 144–145.
14 Dashevsky, V.G. and Sarkisov, G.N. (1974). The solvation and hydrophobic interaction of non-polar molecules in water in the approximation of interatomic potentials: the Monte Carlo method. *Mol. Phys.* 27: 1271–1290.
15 Owicki, J.C. and Scheraga, H.A. (1977). Monte Carlo calculations in the isothermal-isobaric ensemble. 2. Dilute aqueous solution of methane. *J. Am. Chem. Soc.* 99: 7413–7418.
16 Swaminathan, S., Harrison, S.W., and Beveridge, D.L. (1978). Monte Carlo studies on the structure of a dilute aqueous solution of methane. *J. Am. Chem. Soc.* 100: 5705–5712.
17 Lal, M. (1969). 'Monte Carlo' computer simulation of chain molecules. I. *Mol. Phys.* 17: 57–64.
18 Hagler, A.T., Stern, P.S., Sharon, R. et al. (1979). Computer simulation of the conformational properties of oligopeptides. Comparison of theoretical methods and analysis of experimental results. *J. Am. Chem. Soc.* 101: 6842–6852.
19 Wako, H. and Scheraga, H.A. (1981). Use of distance constraints to fold a protein. *Macromolecules* 14: 961–969.
20 Lee, W.-G., Gallardo-Macias, R., Frey, K.M. et al. (2013). Picomolar inhibitors of HIV reverse transcriptase featuring bicyclic replacement of a cyanovinylphenyl group. *J. Am. Chem. Soc.* 135: 16705–16713.
21 Michel, J. and Essex, J.W. (2008). Hit identification and binding mode predictions by rigorous free energy simulations. *J. Med. Chem.* 51: 6654–6664.
22 Jorgensen, W.L., Maxwell, D.S., and Tirado-Rives, J. (1996). Development and testing of the OPLS all-atom force field on conformational energetics and properties of organic liquids. *J. Am. Chem. Soc.* 118: 11225–11236.

23 Michel, J. and Essex, J.W. (2010). Prediction of protein–ligand binding affinity by free energy simulations: assumptions, pitfalls and expectations. *J. Comput.-Aided Mol. Des.* 24: 639–658.
24 Woods, C.J., Bodnarchuk, M., Genheden, S. et al. (2017). ProtoMS 3.3. http://protoms.org.
25 Ross, G.A., Bodnarchuk, M.S., and Essex, J.W. (2015). Water sites, networks, and free energies with grand canonical Monte Carlo. *J. Am. Chem. Soc.* 137: 14930–14943.
26 Bodnarchuk, M.S., Viner, R., Michel, J., and Essex, J.W. (2014). Strategies to calculate water binding free energies in protein–ligand complexes. *J. Chem. Inf. Model.* .
27 Kirkpatrick, S., Gelatt, C.D. Jr.,, and Vecchi, M.P. (1983). Optimization by simulated annealing. *Science* 220: 671–680.
28 Cole, D.J., Tirado-Rives, J., and Jorgensen, W.L. (2014). Enhanced Monte Carlo sampling through replica exchange with solute tempering. *J. Chem. Theory Comput.* 10: 565–571.
29 Woods, C.J., Essex, J.W., and King, M.A. (2003). The development of replica-exchange-based free-energy methods. *J. Phys. Chem. B* 107: 13703–13710.
30 Abagyan, R. and Totrov, M. (1994). Biased probability Monte Carlo conformational searches and electrostatic calculations for peptides and proteins. *J. Mol. Biol.* 235: 983–1002.
31 Neyts, E.C., Thijsse, B.J., Mees, M.J. et al. (2012). Establishing uniform acceptance in force biased Monte Carlo simulations. *J. Chem. Theory Comput.* 8: 1865–1869.
32 Noguti, T. and Gō, N. (1985). Efficient monte carlo method for simulation of fluctuating conformations of native proteins. *Biopolymers* 24: 527–546.
33 Li, Z.Q. and Scheraga, H.A. (1987). Monte Carlo-minimization approach to the multiple-minima problem in protein folding. *Proc. Natl. Acad. Sci. U.S.A.* 84: 2633–2636.
34 Chang, G., Guida, W.C., and Still, W.C. (1989). An internal-coordinate Monte Carlo method for searching conformational space. *J. Am. Chem. Soc.* 111: 4379–4386.
35 Evans, D.A. and Wales, D.J. (2003). Free energy landscapes of model peptides and proteins. *J. Chem. Phys.* 118: 3891–3897.
36 Wang, F. and Landau, D.P. (2001). Efficient, multiple-range random walk algorithm to calculate the density of states. *Phys. Rev. Lett.* 86: 2050–2053.
37 El-Mellouhi, F., Mousseau, N., and Lewis, L.J. (2008). Kinetic activation-relaxation technique: An off-lattice self-learning kinetic Monte Carlo algorithm. *Phys. Rev. B* 78: 153202.
38 Borrelli, K.W., Vitalis, A., Alcantara, R., and Guallar, V. (2005). PELE: protein energy landscape exploration. A novel Monte Carlo based technique. *J. Chem. Theory Comput.* 1: 1304–1311.
39 Duane, S., Kennedy, A.D., Pendleton, B.J., and Roweth, D. (1987). Hybrid Monte Carlo. *Phys. Lett. B* 195: 216–222.

40 Nilmeier, J.P., Crooks, G.E., Minh, D.D.L., and Chodera, J.D. (2011). Nonequilibrium candidate Monte Carlo is an efficient tool for equilibrium simulation. *Proc. Natl. Acad. Sci. U.S.A.* 108: E1009–E1018.
41 Samuel, G., Nathan, L.M., Patrick, G. et al. (2018). Binding modes of ligands using enhanced sampling (BLUES): rapid decorrelation of ligand binding modes using nonequilibrium candidate Monte Carlo. *J. Phys. Chem. B* 122: 5579–5598.
42 Jorgensen, W.L. (2016). Computer-aided discovery of anti-HIV agents. *Bioorg. Med. Chem.* 24: 4768–4778.
43 Lovering, F., Aevazelis, C., Chang, J. et al. (2016). Imidazotriazines: Spleen Tyrosine Kinase (Syk) inhibitors identified by Free-Energy Perturbation (FEP). *ChemMedChem* 11: 217–233.
44 Wang, N.X., Watson, G.B., Loso, M.R., and Sparks, T.C. (2016). Molecular modeling of sulfoxaflor and neonicotinoid binding in insect nicotinic acetylcholine receptors: impact of the Myzus beta1 R81T mutation. *Pest Manage. Sci.* 8: 1467–1474.
45 Peng, Z., Maxwell, D.S., Sun, D. et al. (2014). Imatinib analogs as potential agents for PET imaging of Bcr-Abl and c-KIT expression at a kinase level. *Bioorganic Med. Chem.* 22: 623–632.
46 Neves, M.A., Totrov, M., and Abagyan, R. (2012). Docking and scoring with ICM: the benchmarking results and strategies for improvement. *J. Comput.-Aided Mol. Des.* 26: 675–686.
47 Cross, J.B., Thompson, D.C., Rai, B.K. et al. (2009). Comparison of several molecular docking programs: pose prediction and virtual screening accuracy. *J. Chem. Inf. Model.* 49: 1455–1474.
48 Tubert-Brohman, I., Sherman, W., Repasky, M., and Beuming, T. (2013). Improved docking of polypeptides with glide. *J. Chem. Inf. Model.* 53: 1689–1699.
49 Salameh, B.A., Cumpstey, I., Sundin, A. et al. (2010). 1H-1,2,3-Triazol-1-yl thiodigalactoside derivatives as high affinity galectin-3 inhibitors. *Biorg. Med. Chem.* 18: 5367–5378.
50 Atilgan, A.R., Durell, S.R., Jernigan, R.L. et al. (2001). Anisotropy of fluctuation dynamics of proteins with an elastic network model. *Biophys. J.* 80: 505–515.
51 Cossins, B.P., Hosseini, A., and Guallar, V. (2012). Exploration of protein conformational change with PELE and meta-dynamics. *J. Chem. Theory Comput.*.
52 Eyal, E., Yang, L.W., and Bahar, I. (2006). Anisotropic network model: systematic evaluation and a new web interface. *Bioinformatics* 22: 2619–2627.
53 Setny, P. and Zacharias, M. (2013). Elastic network models of nucleic acids flexibility. *J. Chem. Theory Comput.* 9: 5460–5470.
54 Jacobson, M.P., Kaminski, G.A., Friesner, R.A., and Rapp, C.S. (2002). Force field validation using protein side chain prediction. *J. Phys. Chem. B* 106: 11673–11680.
55 Zhu, K., Shirts, M.R., Friesner, R.A., and Jacobson, M.P. (2007). Multiscale optimization of a truncated Newton minimization algorithm and application to proteins and protein–ligand complexes. *J. Chem. Theory Comput.* 3: 640–648.

56 Lecina, D. (2017). A study on protein–ligand interactions using a Monte Carlo procedure. PhD thesis. Universitat de Barcelona, Barcelona.

57 Lecina, D., Gilabert, J.F., and Guallar, V. (2017). Adaptive simulations, towards interactive protein-ligand modeling. *Sci. Rep.* 7: 8466.

58 Banks, J.L., Beard, H.S., Cao, Y. et al. (2005). Integrated modeling program, applied chemical theory (IMPACT). *J. Comput. Chem.* 26: 1752–1780.

59 Kaminski, G.A., Friesner, R.A., Tirado-Rives, J., and Jorgensen, W.L. (2001). Evaluation and reparametrization of the Opls-Aa force field for proteins via comparison with accurate quantum chemical calculations on peptides. *J. Phys. Chem. B* 105: 6474–6487.

60 Pérez, A., Marchán, I., Svozil, D. et al. (2007). Refinement of the AMBER force field for nucleic acids: improving the description of α/γ conformers. *Biophys. J.* 92: 3817–3829.

61 Onufriev, A., Bashford, D., and Case, D.A. (2004). Exploring protein native states and large–scale conformational changes with a modified generalized born model. *Proteins Struct. Funct. Bioinf.* 55: 383–394.

62 Schaefer, M., Bartels, C., and Karplus, M. (1998). Solution conformations and thermodynamics of structured peptides: molecular dynamics simulation with an implicit solvation model. *J. Mol. Biol.* 284: 835–848.

63 Ghosh, A., Rapp, C.S., and Friesner, R.A. (1998). Generalized born model based on a surface integral formulation. *J. Phys. Chem. B* 102: 10983–10990.

64 Gallicchio, E., Zhang, L.Y., and Levy, R.M. (2002). The SGB/NP hydration free energy model based on the surface generalized born solvent reaction field and novel nonpolar hydration free energy estimators. *J. Comput. Chem.* 23: 517–529.

65 Zhu, K., Shirts, M.R., and Friesner, R.A. (2007). Improved methods for side chain and loop predictions via the protein local optimization program: variable dielectric model for implicitly improving the treatment of polarization effects. *J. Chem. Theory Comput.* 3: 2108–2119.

66 Marti, M.A., Bidon-Chanal, A., Crespo, A. et al. (2008). Mechanism of product release in NO detoxification from Mycobacterium tuberculosis truncated hemoglobin N. *J. Am. Chem. Soc.* 130: 1688–1693.

67 Guallar, V., Lu, C., Borrelli, K. et al. (2009). Ligand migration in the truncated hemoglobin-II from mycobacterium tuberculosis: the role of G8 tryptophan. *J. Biol. Chem.* 284: 3106–3116.

68 Lucas, M.F. and Guallar, V. (2012). An atomistic view on human hemoglobin carbon monoxide migration processes. *Biophys. J.* 102: 887–896.

69 Shan, Y., Kim, E.T., Eastwood, M.P. et al. (2011). How does a drug molecule find its target binding site? *J. Am. Chem. Soc.* 133: 9181–9183.

70 Kotev, M., Lecina, D., Tarragó, T. et al. (2015). Unveiling prolyl oligopeptidase ligand migration by comprehensive computational techniques. *Biophys. J.* 108: 116–125.

71 Kopečná, J., Cabeza de Vaca, I., Adams, N.B.P. et al. (2015). Porphyrin binding to Gun4 protein, facilitated by a flexible loop, controls metabolite flow through the chlorophyll biosynthetic pathway. *J. Biol. Chem.* 290: 28477–28488.

72 Cabeza de Vaca, I., Lucas, M.F., and Guallar, V. (2015). New Monte Carlo based technique to study DNA–ligand interactions. *J. Chem. Theory Comput.* 11: 5598–5605.

73 Espona-Fiedler, M., Soto-Cerrato, V., Hosseini, A. et al. (2012). Identification of dual mTORC1 and mTORC2 inhibitors in melanoma cells: prodigiosin vs. obatoclax. *Biochem. Pharmacol.* 83: 489–496.

74 Hosseini, A., Espona-Fiedler, M., Soto-Cerrato, V. et al. (2013). Molecular interactions of prodiginines with the BH3 domain of anti-apoptotic Bcl-2 family members. *PLoS One* 8.

75 Andreotti, G., Cabeza de Vaca, I., Poziello, A. et al. (2014). Conformational response to ligand binding in phosphomannomutase2: insights into inborn glycosylation disorder. *J. Biol. Chem.* 289: 34900–34910.

76 Edman, K., Hosseini, A., Bjursell, M.K. et al. (2015). Ligand binding mechanism in steroid receptors; from conserved plasticity to differential evolutionary constraints. *Structure* 23: 2280–2290.

77 Grebner, C., Lecina, D., Gil, V.A. et al. (2017). Exploration of binding mechanisms in nuclear hormone receptors by Monte Carlo simulations and X-ray derived motion modes. *Biophys. J.* 112: 1147–1156.

78 Takahashi, R., Gil, V.A., and Guallar, V. (2014). Monte Carlo free ligand diffusion with Markov state model analysis and absolute binding free energy calculations. *J. Chem. Theory Comput.* 10: 282–288.

79 Borrelli, K.W., Cossins, B., and Guallar, V. (2010). Exploring hierarchical refinement techniques for induced fit docking with protein and ligand flexibility. *J. Comput. Chem.* 31: 1224–1235.

80 Hosseini, A., Alibés, A., Noguera-Julian, M. et al. (2016). Computational prediction of HIV-1 resistance to protease inhibitors. *J. Chem. Inf. Model.* 56: 915–923.

81 Carlson, H.A., Smith, R.D., Damm-Ganamet, K.L. et al. (2016). CSAR 2014: a benchmark exercise using unpublished data from pharma. *J. Chem. Inf. Model.* 56: 1063.

82 Madadkar-Sobhani, A. and Guallar, V. (2013). PELE web server: atomistic study of biomolecular systems at your fingertips. *Nucleic Acids Res.* 41: W322–W328.

83 Hernandez-Ortega, A., Borrelli, K., Ferreira, P. et al. (2011). Substrate diffusion and oxidation in GMC oxidoreductases: an experimental and computational study on fungal aryl-alcohol oxidase. *Biochem. J.* 436: 341–350.

84 Carro, J., Martinez, M., Medina, M. et al. (2017). Protein dynamics promote hydride tunnelling in substrate oxidation by aryl-alcohol oxidase. *Phys. Chem. Chem. Phys.* 42: 28666–28675.

85 Hernandez-Ortega, A., Lucas, F., Ferreira, P. et al. (2012). Role of active site histidines in the two half-reactions of the aryl-alcohol oxidase catalytic cycle. *Biochemistry* 51: 6595–6608.

86 Hosseini, A., Brouk, M., Lucas, M.F. et al. (2014). Atomic picture of ligand migration in toluene 4-monooxygenase. *J. Phys. Chem. B* 119: 671–678.

87 Acebes, S., Fernandez-Fueyo, E., Monza, E. et al. (2016). Rational enzyme engineering through biophysical and biochemical modeling. *ACS Catal.* 6: 1624–1629.

88 Lucas, M.F., Monza, E., Jørgensen, L.J. et al. (2017). Simulating substrate recognition and oxidation in laccases: from description to design. *J. Chem. Theory Comput.* 13: 1462–1467.

89 Giacobelli, V.G., Monza, E., Fatima Lucas, M. et al. (2017). Repurposing designed mutants: a valuable strategy for computer-aided laccase engineering – the case of POXA1b. *Catal. Sci. Technol.* 7: 515–523.

6

Understanding the Structure and Dynamics of Peptides and Proteins Through the Lens of Network Science

Mathieu Fossépré[1,2], *Laurence Leherte*[1], *Aatto Laaksonen*[2,3,4], *and Daniel P. Vercauteren*[1]

[1] *University of Namur (UNamur), Unité de Chimie Physique, Théorique et Structurale, NAmur MEdicine and Drug Innovation Center (NAMEDIC), NAmur Research Institute for LIfe Science (NARILIS), Laboratoire de Physico-Chimie Informatique, 61 Rue de Bruxelles, 5000 Namur, Belgium*
[2] *Stockholm University, Arrhenius Laboratory, Division of Physical Chemistry, Svante Arrhenius väg 10-18, 10691 Stockholm, Sweden*
[3] *Uppsala University, Department of Chemistry–Ångström Laboratory, 75121 Uppsala, Sweden*
[4] *"Petru Poni" Institute of Macromolecular Chemistry, Centre of Advanced Research in Bionanoconjugates and Biopolymers, Aleea Grigore, Ghica-Voda, 41A, 700487 Iasi, Romania*

6.1 Insight into the Rise of Network Science

A network (or graph) is a set of nodes (or vertices) interconnected with edges (or links) to establish a pattern of relationships between the nodes. As such, networks are expected to describe the structure, dynamics, and behavior of various complex systems [1]. Helped by the progress in computer development, network science has become an increasingly important field [2, 3]. Indeed, network science is now a highly interdisciplinary field that emerged in a broad range of disciplines such as social sciences [4], computer science [5, 6], physics [7], chemistry [8, 9], geology [10], biology [11, 12], or medicine [13, 14]. For instance, networks are used to represent road networks [15] or Internet topologies [16], to understand the dynamics of social networks [17–19], financial markets [20, 21], the structure and modularity of the human brain [22–24], or metabolic pathways in biochemistry [25–27]. For further information on the impact and wide range of applications of networks, readers are invited to consult the book of Barabási [2] or reviews from Newman [28] or Barabási [29], wherein numerous examples of real-world networks are given.

In the field of drug design, networks are increasingly applied to model different levels of the living organisms. Notably, the cell functional organization can be modeled with networks to study the interactions between sets of proteins and to decipher complex biological pathways [11, 30, 31]. Drug discovery strategies are influenced by the network paradigm of cell organization to detect related targets in the multi-target drug design approach [32, 33]. At the molecular level, atoms and bonds can easily be represented by vertices and edges [34]. Therefore, the so-called molecular graphs are often considered to compare the connectivity of

Biomolecular Simulations in Structure-Based Drug Discovery,
First Edition. Edited by Francesco L. Gervasio and Vojtech Spiwok.

molecules [35]. Computers can, for example, easily enumerate chemical isomers once the molecules are modeled with a graph. The approach is also used in computer processing to rapidly search databases of molecular structures, to superimpose molecular structures, or to align three-dimensional molecular fields, all being various common tasks for drug designers. In addition, the era of -omics science implies an explosion of biological data, bringing the need for tools to automate information analyses [36]. In this sense, text mining and network tools can be combined to construct molecular interaction networks that represent complex pathways [37]. As a consequence, a drug designer is more likely to use networks to emphasize complex patterns of biological interactions [38].

As with the development of network sciences, the progress of computer technology has also allowed a colossal development of molecular simulations such as molecular dynamics (MD) applied to biological materials [39–41]. Hence, MD simulations of biomolecules are more complex and significant as a daily routine tool to complement and to emphasize experimental results [42]. However, increasing the complexity and size of the systems as well as the length and timescale of MD simulations can make them more difficult to apprehend. It is therefore natural that networks are also considered to clarify the complex dynamics of biological systems as observed with MD simulations. The recognition of the biomolecular simulations research field reached a new level when the 2013 Nobel Prize in Chemistry was attributed to Karplus, Levitt, and Warshel for the "development of multiscale models for complex chemical systems" [43]. The Prize outlined the importance of multiscale modeling that consists in considering different resolutions on the same system, either in parallel or in serial. Hence, one of the great challenges for the future of MD, in general, is not simply to use ever more powerful computational resources to perform simulations of ever-larger systems on longer timescales [44] but rather to develop ways to mix different spatial resolutions. Biological phenomena can indeed occur at very different timescales, i.e. from femtoseconds to hours, often in a concerted way [45, 46]. Therefore, coarse-grained (CG) models of proteins, in which a particle represents a group of atoms, are becoming more and more popular [47, 48]. Many applications of CG models efficiently combined with all-atom (AA) simulations in multiscale modeling strategies are now reported in literature [47, 49]. With nodes of a network being part of a complex system, it is obvious that such nodes can represent different levels of abstraction of a protein structure, in the same sense that particles of a CG model can gather different numbers of atoms. The so-called elastic network models (ENMs) allow to represent protein structures and coarse-grain their dynamics [50].

This review focuses on the usefulness of networks to understand structural, dynamical, and functional aspects of small peptides and proteins, and how network science can enhance drug design approaches. It is divided into six sections. Section 6.2 presents general principles of networks and the most used topological properties encountered to inspect protein structures [51]. In the same section, some recent applications to highlight properties of a (static) protein structure are reported. Section 6.3 concerns the use of networks to unveil complex dynamical properties of proteins as observed with simulations of biomolecules, and more particularly MD simulations [52]. Section 6.4 focuses on ENMs and reports several successful applications [50]. In Section 6.5, we address how the intrinsic hierarchical properties of complex networks, related to their high degree of clustering [53], make them interesting to highlight the modular nature of proteins [54]. Our particular contribution in the field of network science applied from small molecules to protein structures is summarized in Section 6.6. Future challenges and promising perspectives are discussed in Section 6.7. Notably, future developments of dynamical mesoscale (or highly CG) models of proteins based on the modularization techniques used in network sciences are also presented.

6.2 Networks of Protein Structures: Topological Features and Applications

6.2.1 Topological Features and Analysis of Networks: A Brief Overview

The purpose of this first subsection is to introduce the concept and the topological features of a network with examples related to protein structures. A graph, or a network, is rather simple to visualize. A graph is a set of nodes or vertices joined together by a set of lines called edges or links (Figure 6.1). The mathematical and more rigorous definition of graph is: *a graph G is an ordered pair G = (V, E), where V is a finite, non-empty set of objects called vertices, and E is a (possibly empty) set of unordered pairs of distinct vertices, i.e. 2-subsets of V called edges* [56]. The properties of graphs are fundamentally different according to the characteristics of the edges. Graphs in which edges have a direction are called directed graphs or digraphs. Although graphs in the context of biochemical pathways are often directed to indicate the series of events of a given biochemical process [57], graphs related to protein structures are, in general, undirected [51]. Graphs of protein structures are indeed used to describe chemical interactions between particles, without the need to consider directionality. In the following, only undirected graphs are considered.

Edges can also be weighted (or not) to lead to different degrees of importance of specific chemical interactions. For instance, the strength of the non-covalent interactions between residues can be a criterion to determine the weight of edges [58, 59]. Information issued from weighted residue interaction networks (RINs) can then be implied in a protein–protein docking score [60]. Interestingly, Karain and Qaraeen compared weighted and unweighted protein structure networks, MD simulations being used to define edge weights with regard to the fraction of contacts between residues [61]. Graph weighting is therefore an important concept for describing specific chemical interactions as it can replace the physical distance between particles. Particularly, the more separated the particles, the weaker their interaction. The edge weight modeling of such an interaction is thereby lower [62].

Representing protein structures as nodes and edges is now a popular tool, as recently reviewed [63, 64]. There are numerous methods of constructing RINs for proteins [51, 65, 66]. Even though RINs can be built from an AA structure [67], they are usually built from a CG structure with nodes gathering a group of atoms. RINs are often built at a resolution of one CG site per residue, with variations on

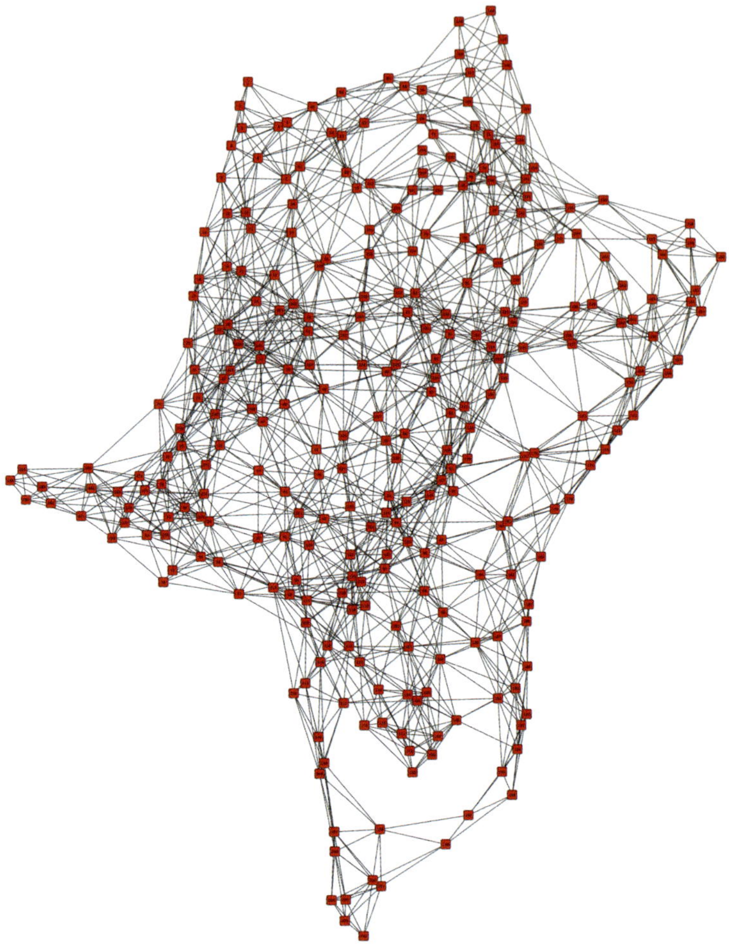

Figure 6.1 Network built from the Cα atoms of the μ-opioid receptor, with 288 nodes (red dots) and 1775 edges (black lines). Each node represents one residue and each edge is a link between a pair of nodes separated by a cutoff distance shorter than 9.1 Å. Such a network is used to describe the μOR dynamics at a coarse-grained level of spatial resolution [55].

the location of the CG sites. For instance, RINs built using the software NAPS [68] are built from the coordinates of Cα and Cβ atoms, the geometric center, or the center of mass (COM) of each residue. Variations in the CG coordinates obviously have a strong influence on the resulting graph topology.

Networks can be characterized by several topological measures to understand how nodes interact with each other [69]. A first level of analysis consists in treating the network as a whole entity. Hence, most basic properties are its *size*, noted N in the following, corresponding to the number of nodes, and its *general degree* E related to the number of edges. The first interesting property is the *density*, which is defined as the ratio between the actual number of edges and the maximum number of possible ones, i.e. a situation wherein each node is linked to all the others. Intuitively, the density indicates if the network can be modularized into *communities*. Community in a network means that its nodes can be grouped into sets of nodes that are densely connected internally [70]. The higher the density, the lower the probability to have communities, as the nodes are less isolated in a subgroup of nodes. As discussed in Section 6.5, the identification of communities in RINs is of great interest to reveal modularity of proteins.

Similar to the network density is the *average degree*, noted $\langle k \rangle$, where k is the degree of a node, i.e. the number of edges connected to it. Several network properties are based on the optimization of the shortest path between nodes. It is clear that the understanding and determination of such properties progressed a lot thanks to the advent of computer resources, allowing treatments of substantial networks. Determining the shortest path between two nodes is therefore the first step to calculate the average path length and network diameter. The *average shortest path length* is the sum of the shortest paths involved in each pair of nodes divided by the total number of pairs. This measure is an indicator of the average number of steps required to go from one node to another. The more the nodes are connected, the smaller the average shortest path, thus indicating a high degree of communication between nodes. As shown by Atilgan and Baysal [71], the average shortest path of a given residue in a protein is correlated with its thermal fluctuation. Indeed, residues located at the core of the protein are closer to the gravity center of the structure, and their average shortest path is therefore smaller than the ones associated with the residues located on the surface. Residues in the core are highly packed and are less sensitive to thermal fluctuations. Highly connected residues are known to be crucial for the protein stability [58, 72]. In contrast, residues that are characterized both by a low average shortest path and are highly conserved are important for allosteric communication [73, 74]. Related to the average path length is the network *diameter*. It corresponds to the longest shortest path found in a network once all shortest paths have been computed. The *clustering coefficient* is another measure to indicate how the nodes communicate indirectly with each other via the other nodes. For a node, it corresponds to the probability that its connected neighbors are also connected to each other. Intuitively, it can be easily depicted with the common everyday life sentence "friends of my friends are my friends." It is calculated as the ratio between the number of existing edges that connect neighbors of a node to the maximum possible number of such links. The clustering coefficient is often calculated for all nodes of the network to give the average clustering coefficient of the network.

The network properties described are all concerned with the network as a whole entity. However, a second level of network analysis consists in treating the local properties of each node, i.e. through *centrality measures*. Such measures are related to the relative importance of a node within the structure of the network. In a RIN, those measures allow the identification of functional residues that can, for instance, shed some light on the transduction mechanism along the membrane axis of membrane receptor structures [75]. Hence, clusters of important residues involved in ligand binding or in transduction, through long-range interactions between the clusters, can possibly be detected. Using centrality measures also helps in the detection of important residues such as catalytic ones in enzymes as recently reviewed [76]. The word "importance" can have different meanings according to the type of centrality measure that is considered. Many centrality measures in the field of network science [77, 78] can indeed be combined [76]. Moreover, depending on the application field, centrality measures do not present the same degree of efficiency [79, 80]. Many functional properties of proteins can be studied with centrality measures. For instance, mutations of residues, as detected by centrality measures, reduce the thermal stability of some enzymes [81]. This is a prime example of how RINs could orient experimental studies. Similarly, when residues involved in the binding sites of polymerase are mutated, they adopt lower values of centrality, showing that evolution of RINs upon mutations can highlight drug resistance [82]. Several examples illustrating the way most used centrality measures can bring functional information by inspecting protein structures are reported in the next section.

6.2.2 Centrality Measures and Protein Structures

The most intuitive notion of centrality is the *degree*: the node with the largest number of neighbors is the most important (Figure 6.2; Table 6.1). In a RIN, a residue with a high degree is, in general, located not only in the binding site of a protein but also in regions along the receptor structure involved in signal transduction [84, 85]. Interestingly, it was shown that there is a strong correlation between the centrality degree and the hydrophobicity of a residue. Indeed, the most connected residues are among the most hydrophobic ones, i.e. valine, isoleucine, leucine, and phenylalanine [86]. In contrast, poorly connected residues are among the most polar ones, i.e. glutamine, glutamate, aspartate, lysine, and arginine. The centrality degree is a purely local measure. As a consequence, it is not sufficient to capture the important nodes. For instance, in network B of Figure 6.2, it can be intuitively seen that the node with a centrality degree of 2, i.e. the one between two other nodes with a centrality degree of 3, is the most central one in the sense that it is the closest to all other ones. Hence, if a node has the ability to be a mediator between groups of nodes, or if a node is located in the network in such a way that information originating from anywhere else is likely to reach it, the centrality degree is often insufficient for detecting the importance of such nodes. Other more sophisticated measures of centrality are related to the rest of the network.

Centrality measures are used to highlight the role of individual amino acids within a complex RIN. Among them, *betweenness* and *closeness* centralities are the most known. Betweenness of a node is defined as the number of shortest paths in the network passing through it (Figure 6.2; Table 6.1). That centrality measure is a way to detect which nodes connect important groups of nodes, without necessarily having a high centrality degree; a node can be characterized by both a low centrality degree and a high betweenness. Such a node is therefore considered as important if it plays the role of a communicator between the others. In network A of Figure 6.2, the highest degree and betweenness values are attributed to the same nodes. However, network B of Figure 6.2 shows that nodes with the highest centrality degrees are not always the ones with the highest betweenness. The betweenness centrality was, for example, used by

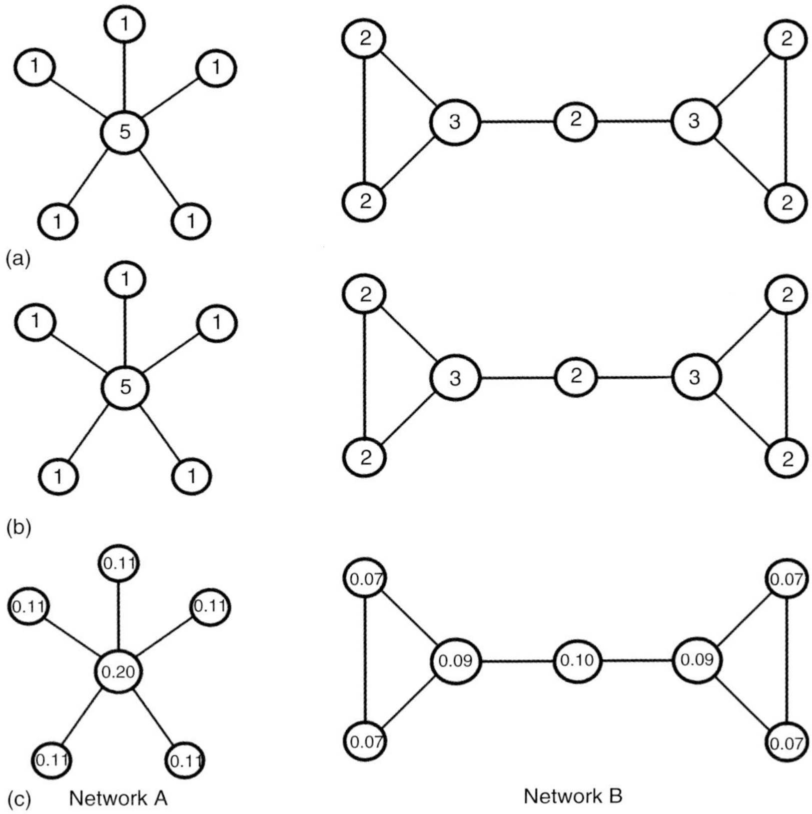

Figure 6.2 Illustration of the centrality degree (a), the betweenness centrality (b), and the closeness centrality (c) for two simple networks, i.e. network A with five nodes and five edges, and network B, with seven nodes and eight edges. The centrality measure of each node is indicated inside the circles representing the nodes. The size of the nodes is proportional to the importance of its centrality measure. Source: Adapted from [83].

Vendruscolo et al. [87] to show that residues with large betweenness values can correspond to conserved residues that are critical for the folding of overall native structure [87, 88]. This measure also permits to spot key residues involved in protein–protein interactions [89].

Closeness is another centrality measure to detect nodes which are neither "between" others, as measured by the betweenness, nor highly connected, as measured by the degree. Rather, they are considered to be "in the middle of" other nodes, i.e. not too far from them in terms of connectivity (Figure 6.2; Table 6.1). Closeness is the reverse concept of *remoteness*, which is computed as the sum of all node–node distances. In protein structures, it was shown that residues with high closeness centrality are critical to explain protein function [90]. Indeed, closeness values of active site residues are, in general, significantly higher than those of the other residues [84], as it is the case for the centrality degree. An interesting example is the comparison of inactive and active GPCR β2AR structures [91]. It was detected that residues with the highest closeness centrality

Table 6.1 Formulae of centrality indices: degree C_{deg}, betweeness C_{spb}, closeness C_{clo}, eccentricity C_{ecc}, radiality C_{rad}, and stress C_{str} [56].

Centrality index	Formula
Degree	$C_{deg}(v) := deg(v)$
Betweenness	$C_{spb}(v) := \sum_{s \neq v \in V} \sum_{t \neq v \in V} \delta_{st}(v)$
Closeness	$C_{clo}(v) := \frac{1}{\sum_{w \in V} \text{dist}(v, w)}$
Eccentricity	$C_{ecc}(v) := \frac{1}{\max\{\text{dist}(v, w) : w \in V\}}$
Radiality	$C_{rad}(v) := \frac{\sum_{w \in V} (\Delta_G + 1 - \text{dist}(v, w))}{n - 1}$
Stress	$C_{str}(v) := \sum_{s \neq v \in V} \sum_{t \neq v \in V} \sigma_{st}(v)$

Let $G(V, E)$ be an unweighted and undirected network where V and E are the node and edge sets of the network G. deg(v) is the number of edges linked to a node v. $\delta_{st}(v)$ is the ratio between $\sigma_{st}(v)$, i.e. the number of shortest paths between s and t that pass through the vertex v, and σ_{st}, a normalization number that denotes the number of shortest paths between s and t passing through v and that have the same length. dist(v, w) is the length of the shortest path between v and w. max [dist(v, w)] is the maximum distance, which is the length of the longest shortest path between v and all the other vertices in the network. Δ_G is the diameter of the graph. n is the number of edges.

are actually located in the middle of the seven transmembrane domains. More particularly, high closeness residues are mostly located in the middle of the H2, H3, H6, or H7 α-helices in both inactive and active β2AR structures. A difference of closeness centrality values of such inactive and active β2AR structures was observed at the cytosolic end of H6. Interestingly, this region is related to the most important conformational changes of a GPCR due to its activation mechanism, and corresponds to the location where the G protein docks with GPCR. In general, the centrality detected for the inactive structure is higher than the value obtained for the corresponding residues in the active structure. Importantly, the presence of a ligand can also affect the centrality of residues. Such results show that the conformational rearrangements of the receptor, due to the presence of agonists or blockers, are related to their corresponding RIN structures. Closeness centrality can therefore serve as an indicator of allosteric conformational changes in complex membrane proteins such as GPCRs. Other GPCRs, like dopamine receptors, were also inspected in the light of RINs to detect the role of central residues. A structural clustering of a few receptor segments was also carried out to highlight their allosteric nature [92].

In some cases, various centrality measures can converge to the detection of similar residues. The use of RINs showed, for instance, that residues with higher degree, closeness, and betweenness are related to heme-binding sites. Such an analysis, combined with bioinformatics and experimental results, proved that RINs generated from different conformational states present divergent topological features, providing a way to detect heme-binding sites [93].

Other less common RIN centrality measures are also encountered: *eccentricity*, *radiality*, and *stress* (Table 6.1). The key step to evaluate the eccentricity of a node is to detect the longest path among the set of all shortest paths passing through it. Eccentricity therefore corresponds to the reciprocal of the distance of the detected longest shortest path (Table 6.1). Hence, a high eccentricity value, i.e. close to 1, means that the considered node is close to all the others. Radiality is quite different to evaluate. First, as for eccentricity, each shortest path between a node and all others is computed. The length of each path is subtracted from the value of the network diameter and the resulting values are summed up. The radiality is then calculated by dividing the last number obtained by $(n-1)$, n being the total number of nodes (Table 6.1). Radiality is a centrality measure to detect the tendency of a node to be surrounded (or isolated) in the network. Indeed, the network diameter is the maximal possible distance involved by a pair of nodes. Thus, the difference occurring in the radiality definition is high when the path lengths dist(v, w) are short. In return, low values of radiality are obtained when the paths are long. Hence, a high radiality means that a node is, on average, closer to the others, while a low radiality is associated with a node that is peripheral to the others. Finally, the last centrality measure used in our work is stress. It corresponds to the number of shortest paths that pass through a given node (Table 6.1).

All the aforementioned studies tend to show that much structural and functional information can be extracted from RINs of single-protein structures. Besides that, network theory is also an efficient tool to highlight interactions involved in structures of protein dimers [94]. Hot spot residues, involved in dimer interfaces, are detectable with elaborate descriptors based on network topology features [95]. Databases of protein–protein interactions in which interactions between monomers are represented by networks are now available, for instance, in PIMAdb [96]. Such information can therefore guide biologists to design point mutation experiments.

Comparison of RINs of entire sets of protein structures was also proposed, in the same philosophy as protein sequences are aligned for comparison using bioinformatics tools. Comparisons of networks provide a way to throw off sequence data, focusing directly on patterns of connectivity among networks [97]. Zhang et al. studied how RINs evolve within protein families and the way they can complement sequence and structural information [98]. In addition, the way the RIN of a given protein is designed can strongly influence the centrality of residues [55]. In this study, we emphasized the dependence of the residues with the highest closeness centrality on their location along the μ-opioid receptor (μOR) structure according to the scheme connectivity of the μOR RIN. In Figure 6.3 is reported the location of the 10 most central residues on the μOR structure, according to different cutoff values [55]. Central residues, in terms of closeness, are located in the same region of μOR if the scheme connectivity is defined with a cutoff-based rule, whatever the cutoff value is. These particular residues can, however, be different, and ranked differently, as stated by us [55].

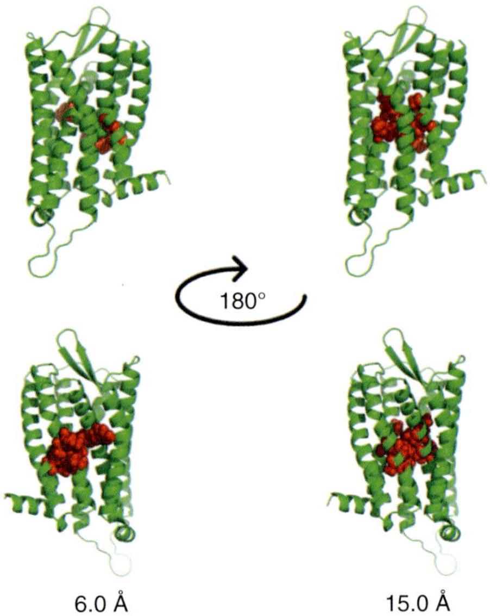

Figure 6.3 View of the 10 most central residues, as measured by the closeness centrality, for two selected cutoff RINs of μOR designed in [55]. μOR residues are represented at an all-atom resolution by red spheres. Source: Adapted from [55].

The multiple uses of RINs emphasize the richness in the architectures, the folding, and the topology of protein structures. DNA molecules were also modeled using RINs to design a variety of nanoscale structures presenting complex topologies [99, 100]. In parallel, emerging concepts, such as "protein calligraphy" [101], combined with the ability to control the design of increasingly complex supramolecular architectures, involve structures that are more difficult to apprehend. It is hence likely that inspecting features of complex bio- and/or nanomaterials through the light of networks will become more and more popular in the near future. Beyond the scope of this review, let us mention that networks have been shown to be useful to study RNA structures [102]. In the field of nanotechnology, artificial nucleic acid structures have been naturally described with networks, and graph theory has been used to design complex nanoscale structures from nucleic acids [103]. Also, network tools were employed to reveal the interactions between complex self-assembly of biomolecules, as for protein–DNA interactions [104].

6.2.3 Software

The use of networks to understand protein structures has been greatly favored by the availability of powerful software (Table 6.2). Some of them are completely general and can be used for any type of networks, related to protein structures or not; others are specific to RINs. The general software Gephi [109] is one of the most used open-source network analysis and visualization package. For graph visualization, Graphviz presents numerous features thanks to the implementation of DOT, a simple graph description language [112]. Other network analysis software specifically designed to treat biological problems are also available. At the biochemical pathway level, several visualization and network analysis tools were recently reviewed [123]. Among them, Cytoscape is

Table 6.2 Software for analysis and visualization of networks.

Software	Description	Platform	References
Bio3D	A free general R package used for the analysis of protein structures, sequences, and MD trajectories. Bio3D contains interesting tools for network correlation analysis of protein structures or MD simulations	GNU/Linux, MacOS X, Windows	[105]
CentiBiN	A free biocentric software for the calculation and visualization of centrality measures in complex biological networks	Windows	[106]
CompNet	A free tool for the comparison of biological networks with an easy-to-use graphical interface	GNU/Linux, Windows	[107]
Cytoscape	An open source platform for network analysis and visualization, with functions specific to molecular interaction networks. Cytoscape can be enhanced with many JAVA plugins, providing additional features such as centrality measures or network modularization tools	GNU/Linux, MacOS X, Windows	[108]
Gephi	A free and open-source visualization and network analysis software. Gephi is adapted to various types of networks with diverse sizes	GNU/Linux, MacOS X, Windows	[109]
GraphCrunch	A free software that focuses on the modeling, comparison, alignment, and clustering of protein–protein interaction networks	GNU/Linux, Windows	[110]
Graph-tool	A free Python module for the analysis of graphs. Data structures are implemented in C++, offering therefore a high level of performance within the framework of Python, a simple scripting language	GNU/Linux, MacOS X, Windows	[111]
Graphviz	An open-source and free visualization package. Graphviz implements an internal language that allows multiple features for graph representation	GNU/Linux, Mac, Windows	[112]
InFlow	A stand-alone commercial software that is generally used for social network analyses and visualization, but it can be used for any type of networks. It is easy to use for nontechnical people	Windows	[113]
NeAT	A user-friendly web server for the analysis and clustering of networks	Any OS with browsing capabilities	[114]

Table 6.2 (Continued)

Software	Description	Platform	References
NetMiner	A stand-alone commercial software for visualization and analyses of network data. NetMiner embeds an internal Python-based script engine offering the users the choice to operate NetMiner with their own programs or scripts	Windows	[115]
Pajek	A visualization and analysis software adapted to large networks. Pajek is free for noncommercial use	Windows	[116]
RING	A fast web server for the generation of Residue Interaction Networks of PDB protein structures. Output RINs can be read in Cytoscape or in Pymol	Any OS with browsing capabilities	[117, 118]
Statnet	A free library implemented as a command line R package. Statnet focuses on statistical modeling of network data	GNU/Linux, MacOS X, Windows	[119]
Tulip	A free framework with advanced functions for graph visualization. Tulip is a C++ library that should therefore be implemented in a user application	GNU/Linux, MacOS X, Windows	[120]
UCINET	A software for the analysis of network data, mainly social ones, but it can be used for general networks. Free for 90 days. UCINET contains NetDraw, a powerful network visualization tool	Windows	[121]
xiNET	An online visualization tool for exploring cross-link network maps	Any OS with browsing capabilities	[122]

The list includes stand-alone programs, plugins, libraries, and web servers, giving the user the choice of an appropriate framework according to its technical capabilities.

probably the most popular open-source software for the exploration of biological networks [124]. Cytoscape is generally used to study networks at the metabolic pathway level, but it becomes more and more popular to study RINs. Cytoscape plugins such as NetworkAnalyzer and RINalyzer allow exploring RINs in a convenient way. Indeed, the two tools together constitute a complete toolbox to characterize the topology and the centrality properties of RINs. RINalyzer permits to construct RINs from a protein structure stored as a Protein Data Bank (PDB) file, to not only interactively visualize RINs with several representation options but also to compute centrality measures to highlight important residues [125]. RINalyzer is also interfaced with UCSF Chimera, a molecular visualization software, to allow direct applications of RIN representations to 3D protein structures. NetworkAnalyzer allows the calculation of various parameters related to network topology such as shortest path length distribution, node degree distribution, etc. [125].

Comparing RINs is an important task. It is a way to either detect conformational changes when RINs are related to the same protein, some examples having been mentioned earlier, or to compare different protein structures. The recently introduced tool CompNet [107] is a versatile software, with an intuitive graphic interface, dedicated to align RINs and therefore to evaluate the similarity between multiple networks. Network analysis tools are now integrated within commonly used visualization software. As an example, xPyder is a plugin implemented in PyMOL that allows the visualization of RIN properties directly on the 3D structure of a protein [126]. Several webservers for user-friendly applications of network analysis tools are available. Among them, RING [117, 118] and xiNET [122] are accessible to scientists not accustomed to molecular visualization packages. At last, we mention that several libraries are now implemented in script languages such as graph-tool, an efficient Python module for graph analysis. Several other packages are also implemented in R, such as statnet or Bio3D [105]. There is thus a large panel of network analysis and visualization software, from general to protein-specific tools, from user-friendly webservers to programming-demanding algorithms. Such a diversity of network analysis packages will certainly make the use of network science more and more popular in the understanding of protein structures.

In this section, we have focused on the interest to apply RINs to static biomolecular structures. In the next section, we report studies in which network science principles are also useful to understand dynamical information issued from experimental or computational studies.

6.3 Networks of Protein Dynamics: Merging Molecular Simulation Methods and Network Theory

6.3.1 Molecular Simulations: A Brief Overview

As is the case for network sciences, the availability of powerful computers led to a tremendous development of simulations of biological materials. Nowadays, simulations of biomolecules are indeed largely considered to highlight dynamics of proteins at various microscopic lengths and timescales [41, 127–129]. Among the biomolecular modeling approaches, MD proved to bring numerous benefits to the understanding of protein dynamics. MD simulations therefore constitute a useful tool to complement experiments in rational drug design strategies, as reviewed by several researchers [130, 131]. MD simulations even appear to be essential in cases where experimental techniques such as X-ray crystallography, nuclear magnetic resonance (NMR) spectroscopy, or cryoelectron microscopy encounter difficulties in handling large biomolecules [132]. Historically, MD simulation techniques were first applied to simple hard sphere systems [133], the computer power at that time being insufficient to treat more realistic systems. To our knowledge, one of the first one-microsecond AA MD simulations of the folding of a small peptide was reported in 1998 [134]. More recently, another milestone was achieved with the 1-ms simulation of a small protein [135], thanks to the development of the supercomputer ANTON, a specific hardware adapted to MD, by Shaw's group [136]. The simulated timescale in protein MD has thus progressed a lot, following the well-known Moore's law stating that computer performance is doubling approximately every two years [137]. Opportunities with AA MD simulations are thus now open to study complex mechanisms in biology and, very importantly, to more easily compare simulations with experimental observations [138, 139], as well as to predict protein dynamics properties. However, increasing the complexity, the length, and the timescale of MD simulations can make them more difficult to apprehend. Networks are thus more and more used to understand complex dynamics of biological systems as they allow highlighting complex and subtle conformational changes in the context of MD simulations [52].

6.3.2 How Can Network Science Help in the Analysis of Molecular Simulations?

Allostery is one of the most important dynamical properties that can be handled with networks [140]. Network properties like the shortest path length can be exploited to understand dynamics correlations between distant regions of a protein [141, 142]. For instance, the allosteric properties of the apo and bound conformational states of Hsp70 chaperone proteins were emphasized by comparing the centrality measures of RINs issued from MD simulations [143, 144]. Similarly, the allosteric mechanisms of kinase proteins were detected by building RINs from MD simulation trajectories [145]. Tse et al. compared various network parameters such as the residue centrality to show that functional sites far from the catalytic domain have a great influence on the regularity function of the catalytic domain. Also, network analysis helped relate the allosteric effects of specific subdomains with the catalytic binding regions of kinases [146]. Recently, allosteric effects of G proteins were revealed thanks to the association of long MD simulations and RINs by partitioning their 3D structure [147]. Studying MD simulations with RINs is without doubt a promising approach in drug design to reveal alternative binding pockets for drugs.

RINs were also considered to characterize conformational changes in MD simulations. Particularly, the comparison of wild and mutated proteins through network theory is a promising approach to understand the importance of mutations. For example, RINs were applied to study the variation of the overall conformational landscape of HIV1 due to mutational effects [148]. Rhodopsin, a member of the well-known GPCR protein family involved in vision, was also investigated by combining MD simulations of mutants with the analysis of RINs [149]. Mariani et al. studied how the communication between different parts of mutated rhodopsin differs by comparing diverse features of graphs (node numbers, hubs, etc.) of wild-type and mutated proteins. Similarly, topological properties of HIV1 mutant RINs allowed to detect different communication abilities between residues, showing how drug resistance can be emphasized with network theory and MD techniques [150].

It is of great interest for a drug designer to characterize the fluctuations of residues due to conformational changes as it helps design drugs targeting a specific conformational subspace of a protein. In that sense, MD simulations of RIN representations of Hsp90 chaperones allowed to detect which parts of the proteins are rigid [151]. The comparison of RINs issued from MD conformations could indeed highlight preserved modules in flexible proteins. Network approaches are also efficient to detect rare but functionally relevant conformational states of disordered proteins, as reviewed by Csermely et al. [152]. Indeed, RIN features of functional and nonfunctional conformations of disordered proteins strongly differ. This is actually the observation from which RINs are foreseen as guides in the conformational sampling of proteins [153].

Combining MD simulations and network approaches also helped in the understanding of dimer or oligomer protein dynamics [154]. Baltoumas et al. showed that the stability and dynamics of various GPCR dimer/oligomers could be tracked and that important residues involved in interface interactions could be detected. It is actually of great importance for the design of drugs that target dimer entities.

Finally, network principles can also be used to interpret flexibility properties of complex MD data such as in the case of μOR [155]. In this sense, network modularization techniques, described in Section 6.5, helped understand the modularity of μOR in terms of the motions of a few modules [155].

6.3.3 Software

There are several promising software that automatize RIN inspection and analysis directly from a large set of MD conformations (Table 6.2). Among them, Bio3D, implemented as an R package, is probably the most elaborated one [105]. Bio3D considers data obtained from MD simulations to identify protein parts with correlated motions. Such a tool allows the construction of a weighted network, wherein the connections between nodes, representing the residues, depend on the contacts between the residues observed during MD. Interestingly, Bio3D can consider several proteins to combine the comparisons of structural dynamics with evolutionary information [156]. An online version of Bio3D, i.e. Bio3D-web, was recently released for researchers not familiar with the R software [157]. Bio3D was notably considered to reveal how the communication of a GPCR, i.e. the M2 muscarinic receptor, is influenced by its conformational states, thus highlighting its allosteric regulation [158]. Again, this study is a nice example of how correlations between subtle protein conformational changes of distant parts can be detected by describing MD motions with RINs. One can also cite MDN [159], a recently introduced webserver with a user-friendly interface, which was successfully applied to detect correlated motions between residues in relation with the allosteric properties of the coat protein of bacteriophage MS2 [160].

The aforementioned studies are related to protein structures issued from MD simulations. So far, applications of network tools to study experimentally obtained protein structures have not been often reported. Indeed, obtaining alternative conformations of proteins is still difficult. However, there are successful stories in which NMR studies were complemented by network science. Kurzbach et al. recently reported a graph analysis of NMR data of protein interactions [161, 162], an approach that enables the detection of multiple binding sites.

Studies reported in this section involved networks that are constructed from motions generated with MD data. In the next section, we explain how the structure of RINs can be the major ingredient guiding the dynamics of the protein.

6.4 Coarse-Graining and Elastic Network Models: Understanding Protein Dynamics with Networks

6.4.1 Coarse-Graining: A Brief Overview

As discussed in Section 6.3, AA MD techniques are widely considered to study the dynamical properties of proteins. Although the presented methods can interestingly complement experimental results, they require heavy computational resources for large systems. Hence, there are still limitations to the study of large spatial and temporal scales. However, multiple biological mechanisms occur at the micro- or even millisecond time scale, a phenomenological scale still out of reach for most reasonable computational resources. To apprehend the dynamics of very large systems, CG models, wherein atoms are merged, are now largely used [48]. Very large biological systems, including notably solvent and ions, as well as very long timescales can then be considered, as the number of simulated particles is smaller. Consequently, the energy function is smoother, allowing time steps of about 10–30 fs, compared to AA MD with time step values of 1–2 fs, as the number of degrees of freedom is smaller. Smoothing the energy function also implies an accelerated sampling of the conformational states of a biological system versus AA MD over an equivalent timescale [163].

There are numerous CG models and techniques to represent the various components of biomolecular systems, i.e. ions, lipids, proteins, DNA/RNA, and water. As general overviews, let us mention [164–171]. The variety of CG models is, notably, due to the diversity in the coarsening degree, i.e. the number of atoms that are merged into each bead. In contrast to AA models, CG models can be designed at different levels of resolution. In proteins, one-bead models may, for example, consist in only one particle per residue. Figure 6.4 illustrates a CG representation of the μOR structure depicted at such a resolution. Potential functions for one-bead models are, in general, parameterized with structural and/or geometrical features extracted from a reference or a set of structures generated by computational or experimental approaches. Therefore, these models are somehow biased by the accuracy of the reference structures. A good structural accuracy can nevertheless be achieved at a modest computational cost, as illustrated through the study of the opening motion in HIV-1 protease simulated with a one-bead model [172], and through dynamics studies of large systems such as ribosomes in which one residue represents one nucleotide [173]. Most one-bead models of proteins may be limited as they do not consider the specificity of local interactions due to the side chains. Developments of two-bead models (Figure 6.5), involving a bead on the residue side chains, were considered, for example, by Bahar and Jernigan, for the simulation of globular proteins [178]. Such models are, in general, parameterized from structural data extracted from atomistic simulations [179]. Two-bead models can also be extended with one

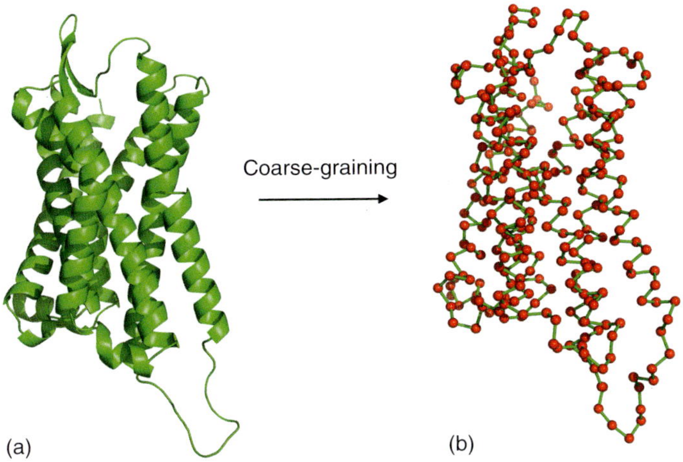

Figure 6.4 All-atom representation involving 4728 atoms (a) and coarse-grained representation involving 288 beads (b) of the μOR structure. Each CG site (red spheres), located on Cα atoms, stands for one residue.

additional grain for long side chains such as lysine [180]. At a finer resolution, CG models with several beads, i.e. four to six beads, are needed to study more precisely the conformations of the backbone and/or the side chains of a protein. At such a coarse-graining level, a description of particular interactions such as hydrogen bonds involved in the formation of secondary structure elements, is possible. As an example, the simulation of the formation of α-helices with four- and six-bead models was proposed by Smith and Hall [181, 182] and Irbäck et al. [183], respectively. CG models based on several beads to describe the side chains are also useful to consider the interactions between residues for processes such as the formation of protein complexes. PRIMO/PRIMONA is one of these models [174]. It consists in a mapping of one to five CG interaction sites for the side chain, whereas only one CG site is used to represent the residue backbone (Figure 6.5). One of the most popular models belonging to that family of CG models is MARTINI [167], with a mapping of four heavy atoms to one CG interaction side, leading to one to five beads describing the side chain according to the type of residue (Figure 6.5). In conventional MD force fields (FFs), such as Amber99, the atoms of a protein bear an electric charge that is involved in the calculation of electrostatic properties. In this sense, we have developed the so-called reduced point charge models (RPCMs) for proteins [176, 177, 184, 185]. In these studies, the electrostatic properties are evaluated from a limited number of point charges, which are located at the extrema, or critical points (CPs) (cf. Section 6.6.1), of a smoothed charge density (CD) distribution function (Figure 6.5). Such models thus consist of a reduced number of charged interaction sites, which, interestingly, are not always located on atoms, and can be used to generate MD trajectories of protein systems. Depending on the method used to generate the charge values and the MD solvation conditions, the protein conformational sampling can be enhanced [176, 185].

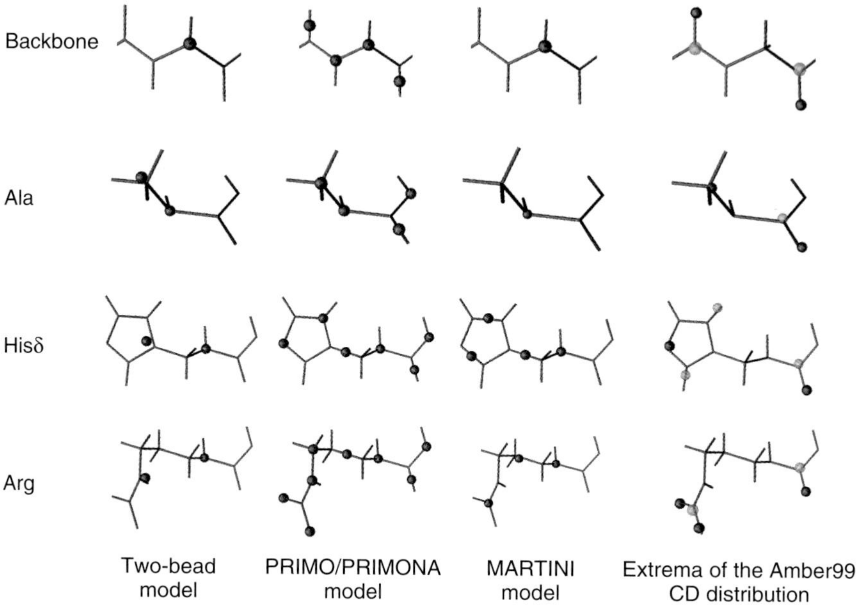

Figure 6.5 Two-bead, PRIMO/PRIMONA [174], MARTINI [167] models, and extrema of the smoothed Amber99 charge density (CD) distributions at a smoothing factor of 1.70 bohr^2 [175–177] of selected amino acid residues. In the last case, black and gray spheres are obtained from the analysis of the negative and positive CD, respectively.

Coarser descriptions gathering several residues per bead were also introduced in literature. To our knowledge, one of the earliest studies was proposed by Doruker et al. [186] with very low resolutions models, i.e. from one bead per 10 residues to one bead per 40 residues. Such models are reported as highly CG models. Even lower resolutions, the so-called ultra-CG models, were proposed by, for instance, the research group of Voth [187]. A CG model of the HIV-1 CA protein dimer, composed of 440 residues, was reduced to four ultra-CG sites, each of them being related to the COM of a domain. The coarse-graining of the actin filament (375 residues) also led to four CG sites, each one corresponding to a subdomain [188, 189]. Readers interested in ultra-CG processes can refer to [188, 190, 191]. Figure 6.6 reports diverse levels of coarse-graining of μOR [55].

Various mapping schemes can be applied for a selected CG model level. For a resolution of one CG per residue, one can, for instance, locate the CG interaction sites on the Cα, on the Cβ, and also on the COM or on the geometric center of the residues, as is the case for RINs. It is important to note that the different mapping schemes can influence the parameterization procedure of the CG FF, as reviewed by Tozzini [170].

In the following section, we present the so-called CG ENMs, i.e. networks of interactions constructed from pairs of residues. Their relevance in the study of dynamical processes of proteins is also discussed.

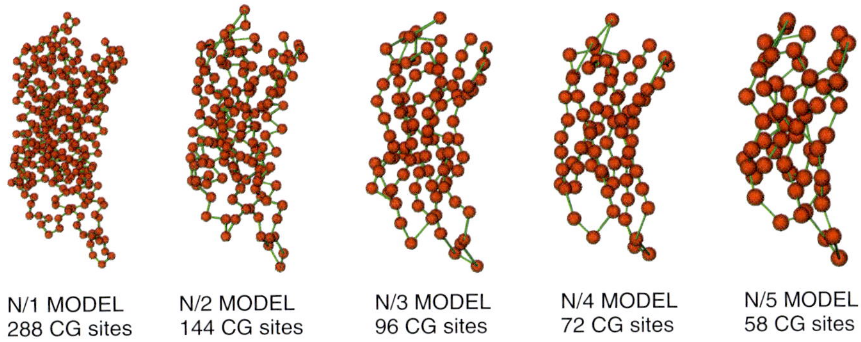

Figure 6.6 Various levels of coarse-graining of the μOR structure, from one particle per residue (N/1 model) to one particle per five residues (N/5 model). Coarse-grained particles are located on Cα atoms.

6.4.2 Elastic Network Models: General Principles

ENMs describe a protein as a network of connected beads interacting via a harmonic potential. The most usual mathematical expression of a CG ENM potential consists in the following potential energy function (E):

$$E = \frac{1}{2}\sum_{i<j}^{N} k_{ij}(d_{ij} - d_{ij,0})^2 \tag{6.1}$$

where N is the number of CG particles (or beads); d_{ij} and $d_{ij,0}$ are the distance between the coordinates at a given time and the equilibrium distance between particles i and j, respectively, and k_{ij} is the force constant associated with the bond occurring between i and j. ENM parameterization thus involves the determination of the connection scheme, i.e. the topology of the network interactions, and the definition of the forces between the connected beads. The force constant adjusts the stiffness of the bonds occurring between the connected beads. Numerous versions of CG ENMs were proposed in literature by introducing various spring constant forms, for example, distance-dependent exponential decay and inverse power distance–dependent functions [62].

The first protein ENMs designed at a CG resolution were introduced by Hinsen, who considered each residue as one bead located on its Cα atom [192]. CG ENMs at lower resolutions, i.e. from one bead per two residues to one bead per 40 residues, were studied by Doruker et al. [186]. ENMs at a CG resolution and at the atomistic scale can be combined to yield hybrid resolution models, and as such are useful to highlight the dynamics of a limited part of a receptor without retaining all details of its surroundings [193]. CG ENMs have, for example, been considered to reproduce backbone structural flexibility, slow dynamics, and low-frequency motions of proteins in relation to their biological function. They were widely applied to membrane proteins as reviewed by Bahar et al. [194]. Among the potential applications of CG ENMs to important therapeutic targets, we mention the interesting work reported by Isin et al. regarding GPCRs [195]. In the paper, a mechanobiological approach, wherein a force constraint is applied to the binding site of rhodopsin, allows to drive the propagation of conformational changes that induce the active state of rhodopsin. Interestingly, the recent publication of several new protein crystallographic structures now permits to rapidly compare, with an evolutive perspective, the dynamical properties of protein families using CG ENMS [196]. Indeed, recent crystal structures of GPCRs were used to decipher, using ENMs, similarities in their intrinsic dynamics [197]. It was revealed that a contraction of the binding site is correlated with the formation of the cavity around the G protein–binding site on the intracellular side. Similar motions were observed in our study [155]. Interestingly, these motions are not so clear in the case of rhodopsin, a special GPCR being activated by light [198]. Altogether, the results illustrate that CG ENMs are well suited to study the link between conformational changes on the extra- and intracellular sides of GPCRs and their internal structural changes.

In general, CG ENMs are validated by comparisons with experimental data such as crystallographic B-factors or conformational flexibility in NMR ensembles [199, 200]. However, NMR studies of proteins are not always easy to perform, especially in the case of GPCRs, while B-factors are influenced by crystal defects and, less significantly, by thermal fluctuations [201–204]. A more recent strategy consists in the comparison of various functional forms of CG ENMs with AA MD results from which they are optimized [205–208]. For example, Leioatts et al. compared the results obtained from various ENM functional forms optimized from AA MD and applied them to three GPCRs [202]. They concluded that the choice of the spring function is not vital to evaluate the performance of an ENM. It emphasizes the robustness of CG ENM functional forms. However, the statement should not be considered as universally applicable as a recent study showed that some functional forms of ENMs have a large influence on the protein dynamics [205].

6.4.3 Elastic Network Models: The Design of Residue Interaction Networks

Several methods to construct RINs are available in literature [51, 66, 209, 210]. Let us note that the various methods described in this chapter are not necessarily used in the context of CG ENMs. Indeed, RIN descriptions of static protein structures do not require any CG ENM potential to decipher their dynamical behavior, as already reviewed in Section 6.1. Curiously, some RIN construction methods appear promising but are rarely used in the context of CG ENMs. The first way to construct a protein RIN is to take into account the primary sequence alone as the starting information [209]. The approach is however limited to linear and rigid proteins. Rigidity logically prevents any folding from taking place, and no interactions are taken into account between distant residues. For GPCRs, the approach has no biological meaning as the folding in a set of seven helices implies that distant residues along the sequence are actually spatially close. As a consequence, sequence-based RINs are seldom used, but are often completed by a notion of spatial cutoff [211]. In contrast, Cartesian coordinates of residues

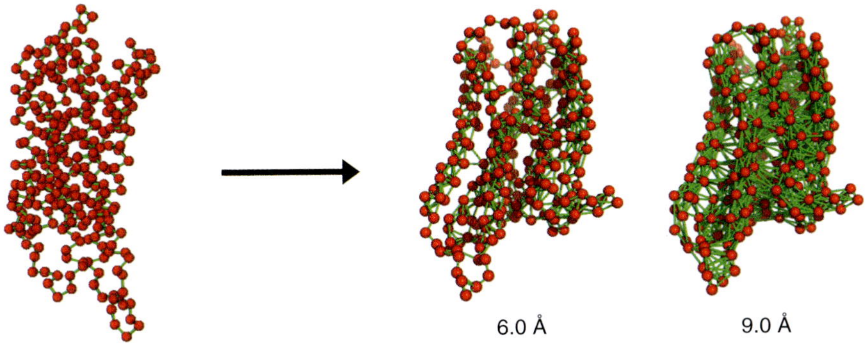

Figure 6.7 Influence of the cutoff value R_c on the connectivity of elastic network models built from Cα coordinates of μOR at the resolution level of one bead per residue (288 CG sites) [55].

are often considered to construct structure-based RINs. For instance, specific atoms such as Cα [212] or Cβ [213], COMs [86], or side-chain centroids [86] can represent the residues as RIN nodes. As already mentioned, the coordinate choice of a CG site influences the RIN topology. Once the node coordinates are obtained, the next step is to determine the links between the nodes. The most popular method consists in applying a cutoff distance, R_c. Figure 6.7 illustrates how the cutoff value R_c influences the density of edges in a RIN depicting the μOR structure. R_c strongly affects the performance of CG ENMs [214]. Indeed, too large a cutoff is not realistic as the resulting ENM might involve interactions between largely separated parts of a protein. In contrast, too weak a value may lead to ENMs missing important local interactions. Silveira et al. suggested that the value of 7 Å is an ideal lower bond cutoff to construct a RIN with one node per residue. Interestingly, the authors also observed that this threshold value is not influenced by the location of the nodes in the residues, i.e. at Cα atoms or at the side-chain geometric centers. Bartoli et al. also proposed an R_c value of 7 Å for other proteins [215]. However, such a value should certainly not be taken as universal. Indeed, Silveira et al. also showed that a decrease of R_c below 6.8 Å can be used only if specific biases are introduced in the protein structure, e.g. at the level of secondary structure elements [214]. Sun et al. proposed an even lower optimum cutoff of 5.0 Å [216]. Hence, there is still no general consensus about the choice of R_c, whose value may depend on the protein fold.

Cutoff-based RINs do not contain any explicit chemical information. Therefore, Jeong et al. proposed connection rules based on chemical interactions such as disulfide bonds, hydrogen bonds, or salt bridges [217]. Another approach is to take into account physicochemical properties, e.g. the interaction energy between residues. A purely energetic approach was, for instance, applied by Jiao et al. to elaborate RINs with weighted edges according to the contact energy between residues [218]. Energy aspects, such as solvation and energy potentials, were completed by geometric properties such as atomic occlusion by Veloso et al. [219]. Their approach allows to evaluate the importance of interatomic non-covalent contacts in the RIN construction. The concept of atomic occlusion was also considered by Noel et al. to build AA networks [220].

In 2011, Doncheva et al. introduced the software RINerator to sample contacts on the van der Waals (vdW) surface of each atom after inclusion of hydrogen atoms [221]. Different types of residue interactions are distinguished according to the overlap volume between the vdW surfaces.

Tessellation techniques belong to the family of purely geometrical RIN building approaches used to elaborate the set of residue contacts in a RIN. At that level of abstraction, tessellation of a volumetric object such as a protein, consists in partitioning the volume using one or more geometric shapes, i.e. tiles, with no overlaps and gaps between them. Tessellation of a 3D object can be assimilated to a coarse-graining process. Consequently, the connections between the tiles depict a network structure that can also be used to elaborate RINs of protein structures. There are several ways to achieve tessellation. In this section, we focus on the Voronoi tessellation, one of the most used approaches.

The Voronoi tessellation divides the Euclidean space into as many regions, called Voronoi cells, as there are nodes. Each cell is generated in such a way that all points within the cell are closer to the cell node than to any other node. Voronoi cells thus correspond to polyhedra that fill the volume with no gaps and no overlaps. Each Voronoi cell is thus seen as the set of points influencing the corresponding node. Tessellation techniques were proved successful in providing protein structure representations adapted for subsequent structural analyses, as already established more than 40 years ago by Richards [222]. Since then, the technique was largely applied to study different aspects of protein structures, as reviewed by Poupon [223]. As software, let us mention VORO3D [224] that can be used, e.g. to evaluate the burying of amino acids in protein structures [225], to assign secondary structure elements [226], or to classify the degree of importance of residue interactions in oligomer interfaces [227]. It allows generating simple or weighted tessellations starting from PDB files as well as to evaluate various Voronoi cell properties, such as volume, area, and environment accessibility. The program also provides a contact matrix between the cells that can further be used to define the edges of a RIN. In practice, for each node of a protein RIN, e.g. Cα atoms, a Voronoi cell is generated. The polygons that share a common boundary define the edges of the RIN. VORO3D is restricted to calculations at the residue scale. To consider atomic scale structures, other software are advised, such as Voronoia [228]. Another choice is the Voronoi Laguerre Delaunay Protein webserver (VLDPws) [229] that can perform a Laguerre tessellation on a given PDB protein structure embedded in a solvent. Laguerre tessellation is a weighted version of the Voronoi version [230]. In the latter, the geometry of a cell is defined by distances between points that all have the same weight, or degree of importance. In contrast, a weight is given to different points in the Laguerre tessellation approach so that some points can have a greater influence on their neighboring points to obtain *in fine* larger cells [231]. In VLDPws, the weights of the points are given according to the radius of the amino acid residues [232]. Once the protein structure is tessellated with VLDPws, the user recovers a residue–residue contact matrix that can be further used to design a RIN.

In a recent work [55], we investigated the elaboration of different ENM-like FFs by varying both their force constant schemes and connectivity patterns.

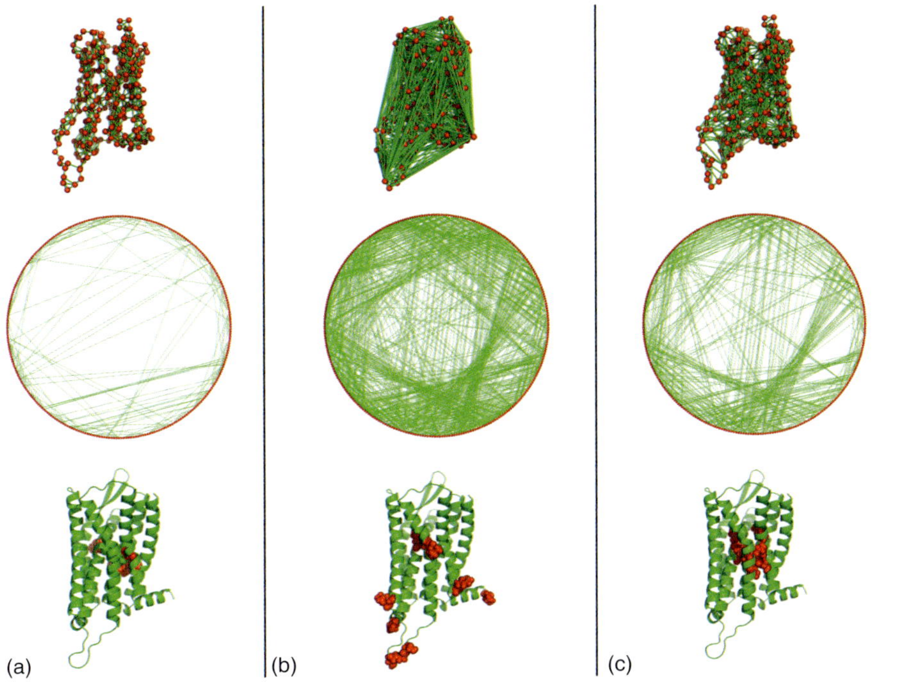

Figure 6.8 Views of the 6.0 Å cutoff (a), the VORO (b), and the VLDP (c) RINs based on the coarse-grained μOR structure along with their connectivity pattern represented in a circular view (middle). Each red dot is a residue; edges are represented by green lines. Different connectivity patterns led to various locations of the 10 most central residues in terms of their closeness centrality values (red spheres, bottom). Source: Adapted from [55].

More particularly, we compared several RINs embedding different connectivity schemes (defined from a cutoff value as well as with tessellation techniques as implemented in VORO3D and VLDPws) for the CG μOR structure. Figure 6.8 illustrates how the pattern of connectivity differs between a cutoff-, a VORO3D-, and a VLDP-based μOR RIN [55].

Apart from the cutoff- and geometric-based methods, RINs can be determined from an ensemble of structures coming from MD simulations, as emphasized in Section 6.2. Such an approach has the advantage of considering dynamical information in place of a single static structure. A cross-correlation matrix of atomic displacements is usually calculated to determine which residues have to be connected in the RIN. Therefore, MD-based RINs require not only one but two cutoff values: a value beyond which residues are not considered to be "in contact," and another one related to the frequency of the contacts. MD-based RINs can strongly differ depending on both cutoff values. Recent software useful to construct MD-based RINs ones are, e.g. Wordom [233], NetworkView [234], first introduced in [235] and used in our studies to decipher modules of μOR flexibility (cf. Section 6.5.2) [155], and PSN-Ensemble [236]. As applications of MD-based RINs to biological systems, let us mention the study of protein–RNA interactions [235] or protein allostery [237].

6.5 Network Modularization to Understand Protein Structure and Function

6.5.1 Modularization of Residue Interaction Networks

The intrinsic organization of a network is such that it can be abstracted or subdivided into modules, often called communities in the field of network science [28, 238–240]. As an example, Figure 6.9 reports the modularization of a trivial network according to the so-called INFOMap algorithm [241]. Techniques of network modularization are numerous [242–248]. Among them, the flow-based methods are the most promising ones [249, 250]. They consist in capturing the flow of information between the components of a network when a random walker goes through it to reveal its hierarchical structure [241, 251]. The trajectory of the walker is constrained by the links in the network and he should not walk from one node to another if they are not connected, this particular last case being called teleportation. In this way, the trajectory implicitly contains information about the network structure. Such an approach, however, does not guarantee that the trajectory is an ergodic solution, i.e. that the average visit rates reach a steady-state solution independent of where the random walker starts. The random walker can sometimes move to a random node irrespective of the link structure. For instance, authors of the map equation, a technique introduced in 2011 to reveal multilevel structures of a network [251], implemented that the probability that the random walker teleports to a random node is of 15% chance at every step. Such a strategy is similar to some enhanced sampling techniques in biomolecular simulations, as with replica-exchange in which conformations are exchanged between parallel copies of the same system [252, 253]. Communities thus consist in nodes among which flow persists, and the nodes that are aggregated are the ones for which information flows quickly and easily.

Multiple comparisons of clustering methods were proposed via quantitative metrics such as the historically known Rand [254], the Wallace [255], and the Fowlkes indices [256]. Such metrics allow the evaluation, at different resolutions, of clustering methods in their efficacy to characterize the similarity between community structures of different partitioning [257–261].

As reported in Section 6.2, RINs are powerful models to understand structural organization of proteins. They can also be used in the context of ENMs to simulate

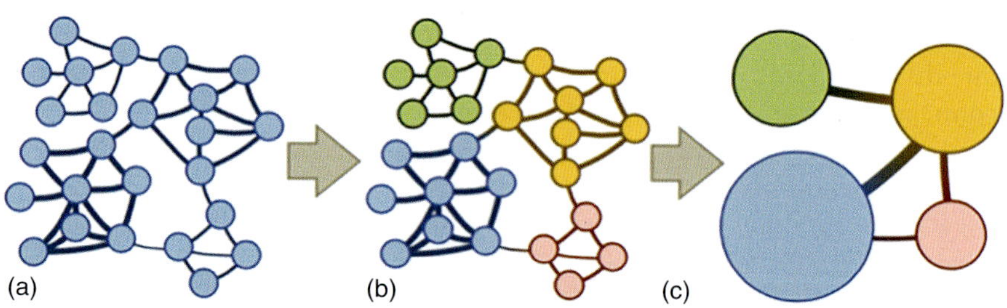

Figure 6.9 Modularization of a network with 25 nodes represented by blue circles (a) and four detected modules represented by four colors (b). The modularization leads to a modular representation of the network (c). Source: Adapted from [241].

the dynamical behavior of proteins, as described in Section 6.4. For a protein RIN, modularization will lead to subsets of amino acids that have a higher intramodular density as compared to residues associated with intermodular links [262]. Several studies already focused on how the modular organization of a protein relates to its structural properties and dynamical behavior. Modules of a RIN often correspond to protein domains, i.e. parts of the protein that move together. As such they allow highlighting the regulation of protein mechanisms and functions [263]. Identification of intermodular residues is also crucial to understand protein dynamics; these particular residues often govern conformational transitions at the domain boundaries [85, 264].

Many other biological aspects of a protein can be understood in the light of RIN modularization [209, 262]. For instance, allosteric communication in protein structures [265–267], protein–protein binding mechanisms [268, 269], and protein stability in thermophilic organisms [270] can be studied in the lens of modular properties of RIN structures. Interestingly, the comparison of RINs and modularization of several protein structures can also be helpful to understand protein evolution inside a protein family [271, 272]. Sequences of protein modules can indeed be highly different while their structure and functions are conserved, allowing new evolutionary relationships between proteins [273]. Small protein fragments or modules of a set of RINs can therefore be employed as units for comparing protein structures, in place of residues. Such a strategy helps decipher how modules are conserved in a protein family [274, 275] as well as to unveil important differences related to functional features inside a protein family [276]. Small and independent modules of a RIN have been shown to be important to explain complex biological events involving many partners such as aggregation in a crowded cellular environment [277].

Modularity of RINs at different spatial scales can also be related to biological events occurring at various timescales. It reveals how the levels of spatial organization of a protein structure can be biologically explained [278, 279]. The modularity of a protein can be obtained from AA MD simulations, as is the case for RINs. Understanding RIN modularity from AA MD is now applied to increasingly complex protein structures [143] and to biomodeling in general, as reviewed by Hospital et al. [280]. Among the most renowned biomolecules-centric modularization packages proposed so far, let us mention PiSQRD [281], a webserver dedicated to protein static structure, and GeoStaS [282], implemented as a stand-alone version or in Bio3D (cf. Section 6.2). GeoStaS can handle static structures as well as MD trajectories. More recently, we noted the release of ResiCon [283] and CoMoDo [284]. By considering covariance of atomic fluctuations, as calculated with ENMs, CoModo takes into account the dynamics of the proteins to identify their domains. The two software allow to identify domains or modules that have different degrees of compacity [283], leading therefore to a hierarchy of modular protein models. As there are many ways to partition a RIN, we expect that metrics used to compare network modularization will be increasingly helpful to rationalize modularization techniques in the context of proteins.

Up to now, the studies reported in literature were intended to detect modularity of various protein structures for the understanding of diverse biological properties. However, none of them used a dynamical modular resolution, i.e. with mutual motions of modules representing protein fragments. In the next section, we discuss the adequacy of using modularization techniques to design highly CG or mesoscale models of proteins.

6.5.2 Toward the Design of Mesoscale Protein Models with Network Modularization Techniques

As discussed in Section 6.4 in regard to the efficiency of CG ENMs to study protein dynamics, CG models often involve a regular partitioning of a protein structure, with CG sites gathering the same number of atoms. Such a representation of a protein structure actually is a particular case of modularization that is not biologically relevant. Indeed, larger domains could represent the most rigid parts, whereas the most mobile parts could be depicted with small numbers of atoms. Based on the fact that the mechanical properties of a protein are clearly related to its modular architecture [285], a description of the intrinsic dynamics of proteins in terms of the motion of a few subparts throughout RIN modularization is an interesting perspective to design dynamical modular models of proteins. Although highly CG or minimalist models of proteins exist (cf. Section 6.3), RIN modularization is, to the best of our knowledge, curiously not often used to design highly CG/mesoscale models of proteins. Yet, such models would provide insight into the hierarchical and modular organization of proteins, and into biological phenomena occurring at very long timescales, i.e. micro- to millisecond timescales. It is indeed intuitively understandable that the lower the resolution of a protein model, the more it is related to biological events occurring at long timescales [278]. Highly CG models can thus certainly help understand how modules move one versus the others. Evidence that modular models of proteins are biologically meaningful could also open the way to various levels of partitioning of the optimal RIN. As such, network modularization techniques can be applied to detect communities at various coarsening levels [286, 287] and provide a hierarchy of protein models generated at various spatial resolutions. Each of them would be adapted to a specific timescale, giving a significance between the dynamics of a protein at a given timescale and level of community detection [288]. Merging techniques of modularization and MD methods would thus lead to a rich world of truly multiscale models, a lot more versatile than the current approaches in which CG models are, in general, gathering the same number of atoms. As an example, network modularization principles were considered in our research group to show that complex MD data of the μOR can be abstracted in terms of the motions of a few modules [155]. In the study, we first used the NetworkView plugin [234] and GeoStaS [282] to build networks of residue contacts. Then, from the networks, modularization techniques were applied to partition the μOR structure. Both techniques led to different spatial levels of μOR fragmentation, as reported in Figure 6.10. Interestingly, comparisons of the resulting partitionings showed that both modularization techniques are related, i.e. one is a CG version of the other. The study thus showed that the dynamics of μOR is highly modular since it consists in the correlated motions of a few modules at different levels of spatial resolution. In this sense, the logical process would be to

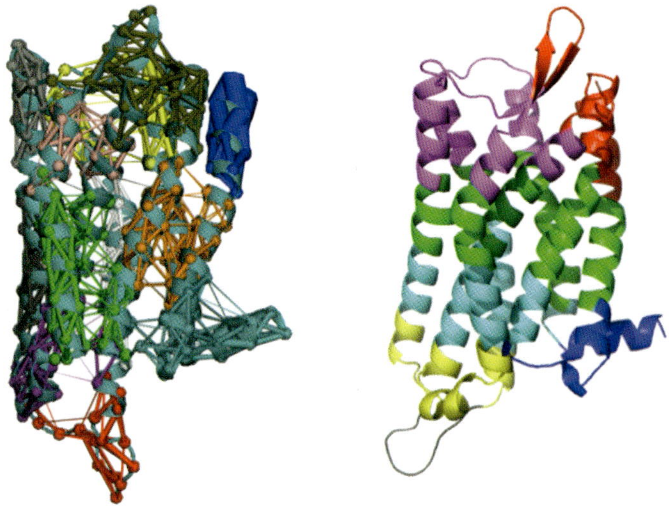

NetworkView plugin (12 modules) GeoStaS algorithm (7 modules)

Figure 6.10 Modular structures of μOR as generated from AA MD data with the NetworkView plugin and GeoStaS. Comparisons of the resulting partitionings showed that both modularization techniques are related, i.e. one is a coarse-grained version of the other. Source: Adapted from [155].

use such minimalist and modular models to apprehend the dynamics of proteins at a very low spatial resolution. A so-called multigrained (MG) model, in which a particle is not simply a CG site but rather a module, is currently in development in our group, as explained in Section 6.6.2.

The design of minimalist models that reflect a simple mechanistic description of complex system dynamics is a rising tendency [289]. The generalization of highly modular and truly multiscale models of proteins could furthermore furnish new fast techniques to generate alternative conformations of proteins. Modular models of protein could further be employed to propose new pharmaceutical compounds to target different functional states of a protein.

6.6 Laboratory Contributions in the Field of Network Science

Numerous applications of graphs to chemical systems, from small organic molecules to inorganic frameworks, DNA strands, and protein complexes, have been developed in the past about 25 years within the "Laboratoire de Physico-Chimie Informatique" at the University of Namur. More particularly, structural and dynamical properties were and are still investigated using graph representations of molecular structures and 3D molecular fields at various levels of spatial resolution.

Briefly, graph theory was first considered to represent inorganic material structures like zeolites [290]. In this field, graph nodes were defined from atoms or secondary building block centroids, which were connected to form graph edges. Symbolic array descriptions of the so-obtained graphs were determined and searched for to detect hierarchical descriptions of unit cell structural motifs, to give a human-understandable perception of complex zeolite networks [290]. Graph reduction procedures to model organic cyclic molecules were carried out through the elaboration of new structural descriptors based on graph theory [291]. In such studies, reduced graph models of small molecules were generated using a new paradigm, i.e. the presence of ring and non-ring elements, to accelerate the search for substructures in large databases. The so-obtained graph representations were helpful to enhance substructure retrieval [291].

Besides these studies, various graph representations, which now retain full 3D information, were used to study small and large systems. In the next section, we particularly describe how topological properties of smoothed 3D molecular fields such as electron density (ED) and the electrostatic potential (EP) distribution functions were analyzed to generate graphs of their CPs (cf. Figure 6.5).

6.6.1 Graph Reduction of Three-Dimensional Molecular Fields of Peptides and Proteins

CPs of a given smoothed 3D molecular field consist in locations where the gradient of the molecular field F is zero. At such locations, the curvature of the field can be determined along each of the three axes, from the eigenvalues of the Hessian matrix:

$$H = \begin{pmatrix} \frac{\partial^2 F}{\partial x^2} & \frac{\partial^2 F}{\partial x\, \partial y} & \frac{\partial^2 F}{\partial x\, \partial z} \\ \frac{\partial^2 F}{\partial y\, \partial x} & \frac{\partial^2 F}{\partial y^2} & \frac{\partial^2 F}{\partial y\, \partial z} \\ \frac{\partial^2 F}{\partial z\, \partial z} & \frac{\partial^2 F}{\partial z\, \partial x} & \frac{\partial^2 F}{\partial z^2} \end{pmatrix} \tag{6.2}$$

The eigenvalues are the three nonzero diagonal elements obtained after diagonalization of H. Four possible cases are met depending on the number of negative eigenvalues, n_E. When $n_E = 3$, the CP corresponds to a local maximum or peak; a point where $n_E = 2$ is a saddle point or pass; $n_E = 1$ corresponds to a pale, while $n_E = 0$ characterizes a local minimum or pit. In ED distributions of AA molecular structures, peaks generally correspond to atoms while passes appear at chemical bonds. Rings and cavities lead to pales and pits, respectively. The CPs can then be linked through a gradient vector field analysis to generate a graph whose vertices and edges are the CPs and gradient trajectories, respectively [292]. The change in resolution of molecular fields, especially the ED, can be achieved using various techniques: crystallography-based structure factors, multiresolution wavelet analysis, and Gaussian smoothing [293]. The last method presents the advantage to be easily used to detect local optima in molecular fields expressed as Gaussian functions, like the promolecular ED [294, 295], and CD derived from the smoothed Coulomb potential [175].

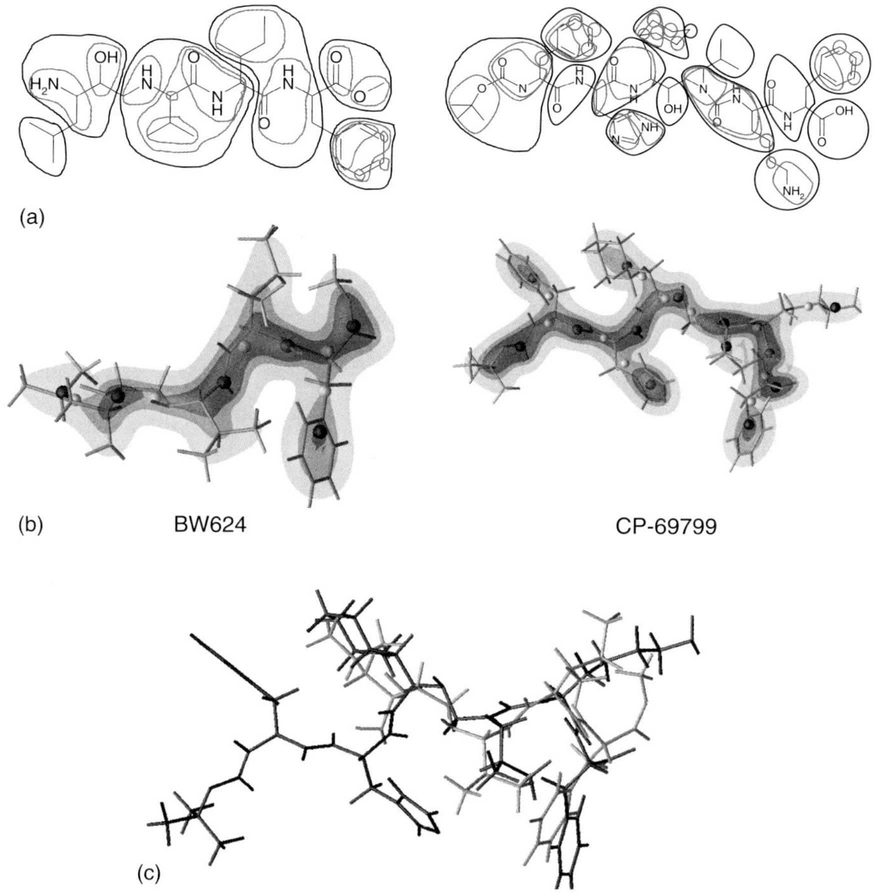

Figure 6.11 Molecular structure of endothiapepsin ligands BW624 and CP-69799 with fragments obtained from a topological analysis of the promolecular electron density distribution function smoothed through a convolution product with a Gaussian function of degree $t = 0.90$, 1.15, and 1.50 bohr^2 (a). Promolecular electron density distributions smoothed at $t = 1.5$ bohr^2 and CPs (black and gray spheres correspond to peaks and passes, respectively) (b). Superposition result obtained from the alignment of the two peak-based graphs (black, CP-69799; gray, BW624) (c).

The CP graphs were, for example, considered to perform molecular superposition as often achieved in computer-assisted drug design (cf. Figure 6.5). As an illustration, Figure 6.11 presents two small endothiapepsin ligands, BW624 and CP-69799, in terms of molecular fragments obtained from a hierarchical clustering approach applied to their progressively smoothed promolecular ED distribution function [296]. Corresponding peaks and passes, at a smoothing factor of 1.50 bohr^2, are also illustrated. The CP graphs can then be used to align the 3D structure of the two endothiapepsin ligands through an SA/MC procedure [297]. A superposition result of the two ligands is also given in Figure 6.11.

Graphs of ED maps have also been taken into account to compare the 3D structure of benzodiazepine ligands via an original genetic algorithm method. The algorithm, called GAGS for "Genetic Algorithm for Graph Similarity Search," led to the proposal of a pharmacophore of the benzodiazepine receptor [298, 299]. The graph parameters, i.e. the density values at the CPs and the CP–CP distances, can then be considered to compute the similarity distance between two CP graphs. It is evaluated as an RMSD involving differences between peak density values and peak–peak distances of two graphs labeled A and B:

$$\mathrm{RMSD}_{\mathrm{AB}} = \sqrt{\sum_{1}^{N} (\rho_{i\in \mathrm{A}} - \rho_{k\in \mathrm{B}})^2 + \sum_{1}^{M} (d_{ij\in \mathrm{A}} - d_{kl\in \mathrm{B}})^2} \tag{6.3}$$

where N and M are the number of CPs and distances to be compared. Various resolutions of the ED maps can be considered [296, 299–301], giving a multiscale method for fast and efficient comparisons of multiple ligands and molecular similarity evaluation.

Reduced graphs of ED CPs are also useful in protein map analysis, for instance, to detect secondary structure elements in protein maps [302]. As an example, Figure 6.12 illustrates the peaks and passes obtained from a topological analysis of the promolecular ED distribution of μOR (adapted from the structure reconstructed regarding the intracellular loop 3 (IL3) missing loop [155]), calculated

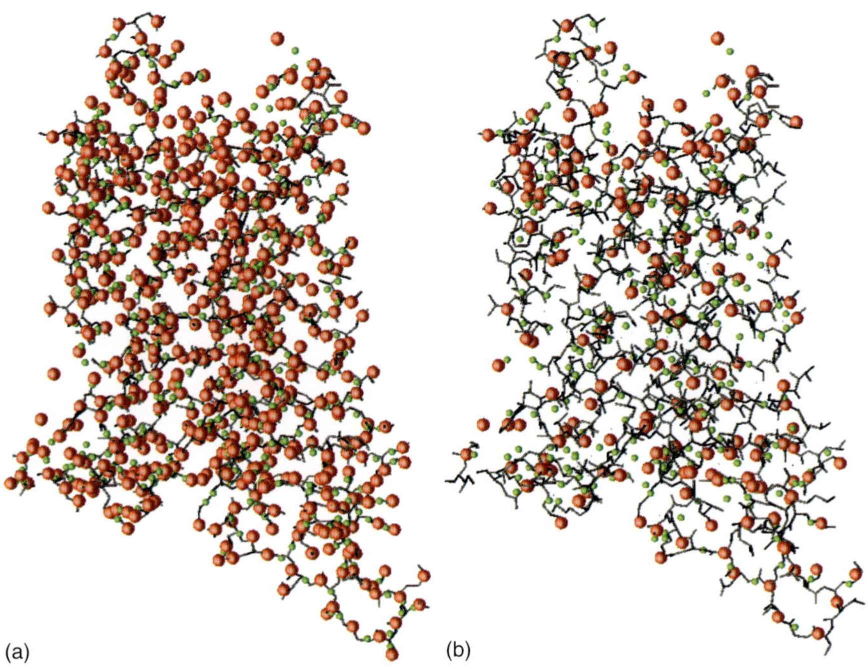

Figure 6.12 Superimposition of the atomic (black sticks) and CPs (red spheres, peaks; green spheres, passes) of the μOR receptor obtained by performing a topological analysis of the ED maps calculated at crystallographic resolution values of 2.85 (a) and 5.00 Å (b).

at crystallography resolution values of 2.85 and 5.00 Å using Xtal [303]. The CPs that appear outside the molecular structure are due to ripples originating from the Fourier transformation of the structure factors. Representations obtained at 2.85 Å allow to capture the backbone and side chain of the amino acids. They were used to study protein–protein and protein–DNA interfaces, and led to the development of original docking score functions [304, 305]. The scores proposed by Becue et al. were based on specific terms modeling the intermolecular interactions, with a geometric fit based on the knowledge extracted from a statistical survey, an electronic interaction potential using an expression of modified Coulomb type, and a penalty score based on detection of steric clashes. Particularly, the calculation of the interactions terms involved ED values and curvatures of peaks to evaluate, e.g. solvent accessible surface of the peaks [306].

ED-based graphs were also considered to study protein dynamics with ENMs [307]. Graphs were built from promolecular ED of proteins. Nodes of these graphs do not correspond to specific atoms coordinates, as it is usually the case for CG models (cf. Section 6.4), but rather to ED peaks, associated with protein fragments, leading to an original multiscale representation of proteins [307]. As illustrated in Figure 6.11, ED peaks are associated with molecular fragments whose size depends on the smoothing level. Graph edges are weighted by the magnitude of the ED overlap, or the EP values, calculated between the fragment ED distributions [308].

Studies reported in this section reported applications of networks to analyze static structures in terms of their molecular connectivity as well as their topological properties described at various spatial resolutions. In the following section, we report on how the dynamics of complex protein systems can be efficiently investigated throughout the design of ENMs at different CG resolutions.

6.6.2 Design of Multiscale Elastic Network Models to Study Protein Dynamics

The reliability of ENM-like potentials to reproduce the AA dynamics of a protein has been shown for the μOR structure [55]. The coarse-graining of the μOR structure was considered at a resolution of one particle per residue (N/1 model), corresponding therefore to 288 beads, as previously reported (cf. Figure 6.4). To select the optimum parameters of the ENM-like potentials, we compared several flexibility scores as the CD.AD score (CD.AD stands for cumulative differences of angle and dihedral angles). It consists in the cumulative difference between dynamical data, i.e. the average angle and average dihedral angle profiles, extracted from the AA and CG simulations. Such a flexibility score was proved to be sensitive enough to assess the reliability of ENMs, and thus to select their optimum parameters [55]. With this metrics, various ENM-like potentials were then investigated by varying their force constant schemes and connectivity patterns. It was shown that a μOR ENM with a fixed constant force value of 3.5 kcal/mol/Å and a connectivity pattern determined from a 9.1 Å cutoff distance can precisely reproduce the flexibility properties of μOR as obtained from AA MD simulations [155] (Figure 6.13). As the CG representation of μOR involves a reduction factor of about 300, i.e. 288 beads rather than 90 000 atoms, the

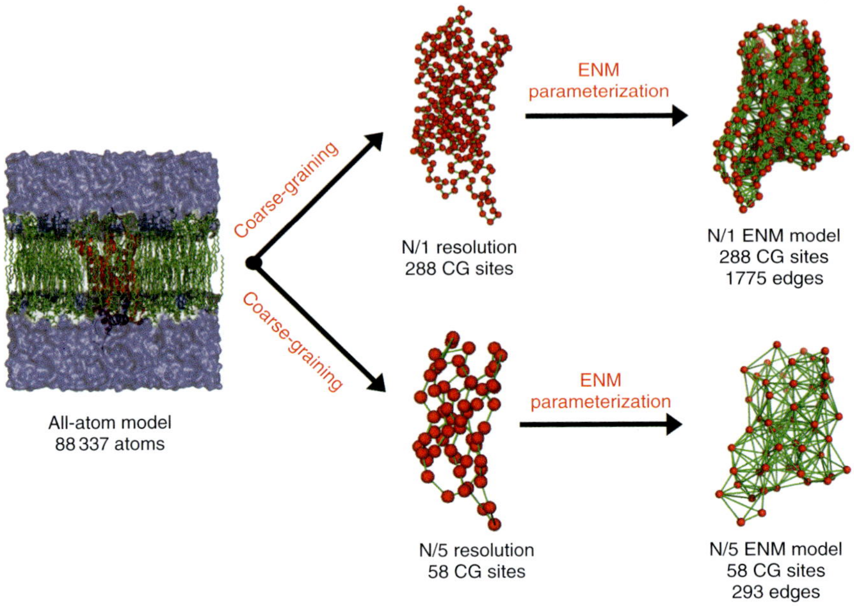

Figure 6.13 Design of the N/1 and N/5 CG ENMs of μOR. The parameterization strategy is based on data issued from AA MD simulations of a complex system, i.e. μOR embedded in a membrane environment (left), to result in a one CG site per residue system (center) or a five residues per CG site animated by a simple ENM-like potential in a solvent-free environment (right). The optimum parameters are 3.5 kcal/mol/Å and 9.1 Å for the N/1 model, and 1.5 kcal/mol/Å and 16.4 Å for the N/5 model [55].

corresponding MD simulation is 20 000 times faster than the AA MD simulation [155]. In addition, the N/1 ENM was shown to be stable up to a time step of 50 fs.

ENMs are based on the assumption that molecular motions are guided by harmonic potentials only. They allow to capture low-frequency, large-amplitude motions of proteins, i.e. highly collective modes. To generate equivalent information from AA MD trajectories, methods like essential dynamics and PCA can be applied [309]. In such methods, an MD-based covariance matrix is defined from the atom coordinates and is further diagonalized to characterize fluctuations along the eigenvectors [309]. ENMs and MD techniques thus provide similar information regarding collective protein motions [310], a consistency that depends on the cutoff parameter R_c [311]. The disadvantages of AA MD versus ENM lie in sampling inefficiencies, especially for short trajectories, and long calculation times. Therefore, the use of ENM-like potentials in the context of MD simulations is not common, but can nevertheless lead to interesting applications. The parameterization of ENM potentials from AA MD simulation data at a selected temperature is a way to, at least partly, circumvent the lack of any temperature dependency in ENMs. In addition, the parameterization may implicitly involve the influence of a solvent. There are thus two underlying reasons to consider ENM-based potentials in MD applications. The first one involves the testing of the CG ENM parameters and the evaluation of the ENM robustness under the influence of temperature and, possibly, solvent conditions. Second,

MD allows generating system states according to a probability distribution function, e.g. the Boltzmann distribution function in the NVT statistical ensemble. From the generated states, various thermodynamic averages can be calculated. Time dependency between the states that define an MD trajectory also allows determining dynamical properties, such as correlation and relaxation times, and various reorientation and vibrational spectra for possible comparison and/or interpretation of experimental results. Finally, MD allows to restore translational and rotational degrees of freedom, needed to evaluate diffusional processes useful, e.g. in oligomerization phenomena.

CG models can be enhanced with electrostatic treatments to consider interactions between monomers and hence deal with dynamical processes such as the oligomerization of protein structures. In the case of a receptor as μOR, ENM-like potentials should nevertheless not lose their ability to reproduce its important flexibility properties. In this context, we merged the optimized N/1 CG ENM model with a reduced point charge model, an original approach that was, to our knowledge, never tested at a CG resolution [312]. In that prospective work, the reduced point charge model assigned to the so-called N/1 electrostatic elastic network model (eENM), in opposition to the previous neutral elastic network model (nENM), involves charges located on the Cα atoms of the charged residues at physiological pH (Figure 6.14). Charge values can, for example, be determined using the charge fitting program QFIT [313] applied to best approximate the molecular electrostatic forces (MEFs). In our work, fittings were achieved by considering molecular electrostatic potential (MEP) grid points located at distances between 1.4 and 2.0 times the vdW radius of the atoms, and forces were calculated as detailed by Leherte [185]. The reference AA CHARMM22 MEP grid of the whole protein structure was built using CHARMM point charges [314], assigned to the amino acid atoms using the PDB2PQR software [315], with a grid step of 0.5 Å due to the large size of the system. It was calculated using the well-known Coulomb law with a permittivity $\varepsilon = 1$. During the fitting procedure, the total electric charge of the protein and the magnitude of its molecular dipole moment were constrained to be equal to the corresponding AA CHARMM values, i.e. $-13e$ and (68.4, −344.6, −1268.7) D. All charged residues could adopt a specific charge value, in contrast to usual CG models for which charges are identical for a given type of residue. Such a particular approach allows taking into account the environment of the charged residues (Figure 6.14). As the presence of oppositely charged residues may lead to an exacerbation of some charge values, the electrostatic interactions of μOR described by the eENM are either reinforced or smoothened as a function of the density of charges in the different μOR regions. For instance, the proximity of highly negative residues, i.e. E270 and D272, involves high positive charges on R263, K271, and R273. The most positively charged residue R273 is in the neighbor of the most negatively charged residue D272.

eENM potentials can then be used within an MD software [316] to perform and analyze MD simulations. The comparisons of 1-μs CG MD simulations of the μOR N/1 eENM with our nENM showed that both N/1 eENM and N/1 nENM reproduce with the same precision the AA model, the *CD.AD* score of the N/1 eENM and nENM being 16.8° and 17.4°, respectively [55]. Besides the slight

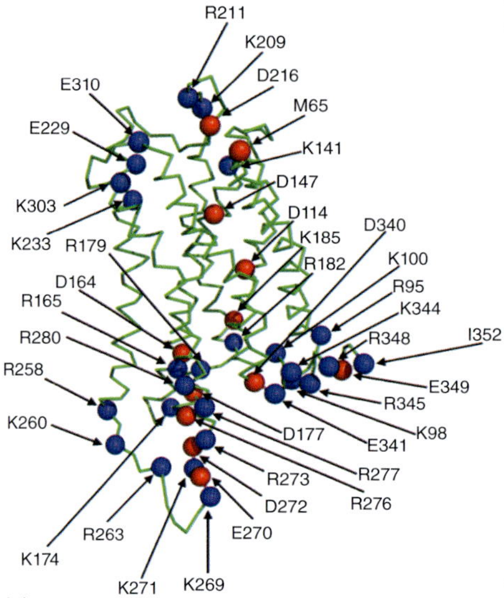

(a)

Residue type	CG partial charges
Aspartate	−2.83 (D272), −2.14 (D177), −1.72 (D216), −1.69 (D340), −1.39 (D114), −1.37 (D147), −0.65 (D164)
Glutamate	−1.98 (E270), −1.79 (E349), 0.10 (E310), 0.55 (E229), 1.26 (E341)
Arginine	−0.3 1 (R276), 0.32 (R179), 0.59 (R211), 0.63 (R280), 0.74 (R165), 1.18 (R95), 1.21 (R348), 1.24 (R277), 1.26 (R258), 1.57 (R345), 1.72 (R263), 2.38 (RI82), 3.48 (R273)
Lysine	−0.02 (K I85), 0.10 (K303), 0.18 (K233), 0.28 (K100), 0.38 (K344), 0.58 (K98), 0.62 (K269), 1.11 (K260), 1.26 (K209), 1.36 (K141), 2.25 (K271), 2.29 (K174)
Terminal	−1.17 (M65) 1.42 (I352)

(b)

Figure 6.14 (a) Coarse-grained structure of μOR at a resolution of one CG site per residue located on Cα. Positively and negatively charged particles (for aspartate, glutamate, arginine, lysine, and terminal residues) are highlighted in blue and red, respectively. (b) Charge values of the N/1 eENM CG model of μOR. Residues are listed from the most negatively to the most positively charged ones [312].

improvements brought by the eENM, a major aspect is that eENM is compatible with electrostatic treatments without disturbing its ability to reproduce the overall intrinsic flexibility properties of μOR. We are thus confident that the strategy we considered earlier to design ENMs of μOR led to robust elastic models that can be enhanced with non-bonded interactions treatments. The mixing of two CG approaches is a promising strategy introducing eENMs as a proof of concept of a new original family of ENMs that are more realistic by considering explicitly the electrostatic interactions. eENMs are ideal tools to study electrostatically driven μOR oligomerization at a CG resolution level taking into account the flexibility of monomers.

The ENM parameterization strategy of μOR proposed by Fossépré et al. [55] was also tested at lower resolutions, i.e. for one particle gathering three residues (N/3 model) (cf. Figure 6.6). Comparisons with AA MD simulations showed that simple N/3 models, for which the dynamics is again governed by the network structures only, are indeed very reliable to study the μOR intrinsic flexibility properties. Hence, the essence of the μOR flexibility is hierarchical, and networks represent an ideal framework to describe protein dynamics at various CG levels. We also showed that the μOR coarse-graining could be pushed even further considering the parameterization of the so-called N/5 model, in which one CG particle gathers five residues (cf. Figure 6.6). In Figure 6.15, one reports the comparison between the root-mean-square fluctuation (RMSF), angle, and dihedral

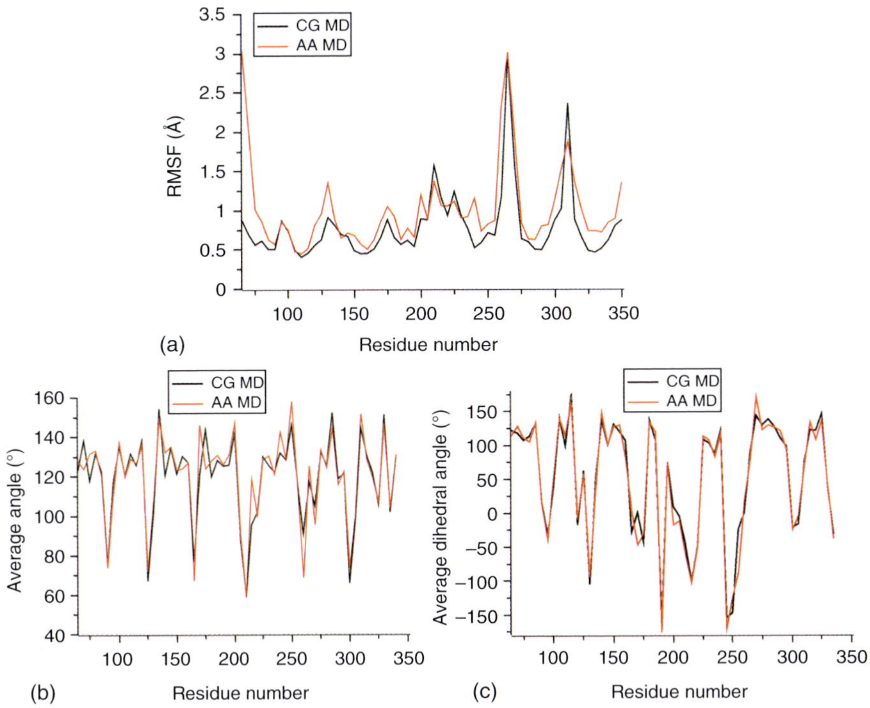

Figure 6.15 Comparisons between the AA (red) and CG (black) MD simulations based on the RMSF (a), the average angle (b), and average dihedral angle profiles (c). Comparisons were done on a 500-ns time scale. The CG FF used is the ENM-like potential at the N/5 resolution as optimized according to the minimum CD.AD score (cf. Figure 6.13).

angle profiles of AA and N/5 CG ENM MD simulations. The flexibility properties are very well conserved at the N/5 highly CG resolution. The proposed strategy to elaborate ENM-like potentials is thus relevant at different CG resolutions. The concept of highly CG ENM models opens promising perspectives to perform reliable MD simulations at various spatial resolutions. Mixing our models with highly CG models of lipids is also a possible outcome [317].

It is obvious that to push the coarse-graining of μOR further, the modularity of the μOR flexibility properties should be exploited (cf. Section 6.5.2) [155]. Very low-resolution models, called MG models, in which a particle is not simply a CG site that involves one or more residues but rather a module or MG site is under development in our group [318]. The strategy consists in exploiting the connectivity pattern of the N/1 model to modularize the μOR structure using the INFOMap algorithm [241]. The next step is then to parameterize the so-obtained ENM with specific force constants. To do so, systematic tests based on the evaluations of the CD.AD score are carried out as described earlier [55]. The process to design an MG model starting from a previously parameterized N/1 model is illustrated in Figure 6.16 for the μOR case. It results in a particular low-resolution model that is different compared to regular coarse-graining.

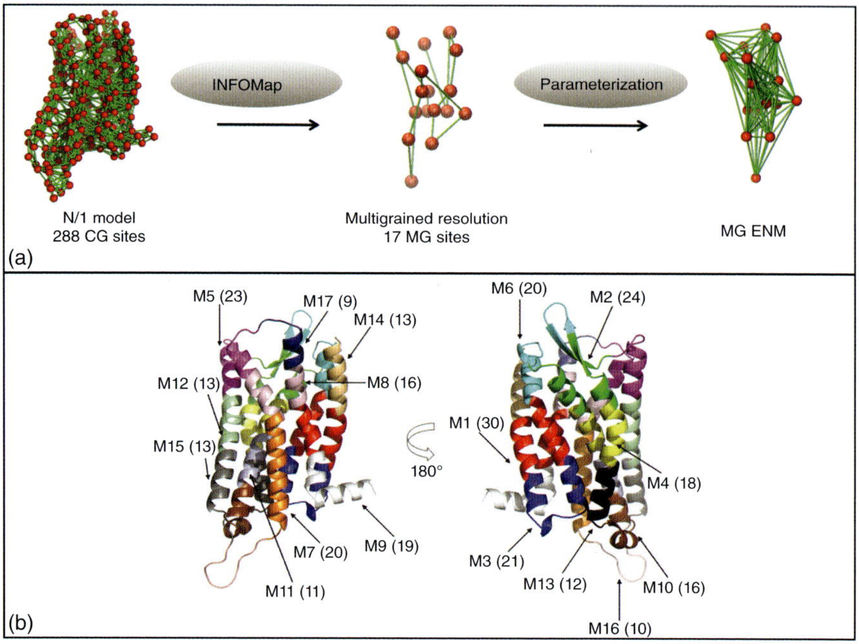

Figure 6.16 (a) Design of a multigrained (MG) model of μOR. The application of the INFOMap algorithm [241] to the N/1 RIN model yields to 17 modules (M1–M17 as depicted in (b)) in which the coordinates of each most central residue are selected as the MG interaction site. The parameterization strategy, based on the CD.AD score [55], is then used to select optimal cutoff and force constant values of the MG ENM [318]. (b) Composition of the 17 modules in terms of number of residues, indicated in parentheses, for the μOR MG model obtained from the N/1 RIN as modularized through the INFOMap algorithm. The biggest module M1 consists in 30 residues; the smallest module M17 in 9 residues [318].

In this sense, each MG interaction site gathers a different number of residues, giving a true multi-resolution nature to the MG model as reported in Figure 6.16. Such a strategy opens the way to a new category of multiscale protein models defined in a hierarchical way from existing CG models. Coupled with the numerous network modularization techniques leading to different coarsening levels (cf. Section 6.5.1), MG models could represent protein dynamics at a wide variety of spatial resolutions. MG models are truly multiscale as they mix several levels of coarsening. It is thus expected that applications of such multiscale models will be increasingly frequent in literature. In the next section, we propose some perspective works as well as challenging questions that could be investigated with network techniques.

6.7 Conclusions and Perspectives

Altogether, the numerous studies reported in this review reflect the richness of applications of networks in the domain of proteins, i.e. from the study of static structures, in terms of their molecular connectivity and topological properties described at various levels of resolution, to the dynamics of complex protein systems with ENMs at different CG resolutions, including the investigation of correlated motions from MD data.

Network parameterization must be reliable. On the whole, the wealth of the needed physicochemical parameters and coarsening techniques suggest numerous ways to parameterize molecular networks. Challenges will thus remain to determine whether particular physicochemical parameters could be universally useful or necessary to parameterize networks, i.e. whether they are transferable, or if it remains a question of good chemical sense to design molecular graphs depending on the system to be studied.

Multiscale networks are a new and promising field of research in protein modeling. Multiscale networks can be expected to be developed in order to study, e.g. protein structure and dynamics, and also multiscale materials. The design and parameterization of associated interaction potentials are still a challenging task. CG potentials can indeed be developed either directly from AA data, or in a hierarchical manner, i.e. from level to level. Lower resolutions may also require knowledge-based adjustments.

Other interesting developments, but rather more complex, could involve information transfer between graph nodes, depending, e.g. upon their neighborhood in a changing environment. A typical example is the burying of protein residues that can serve as a parameter to adapt node properties.

We also particularly emphasized how RINs can be the motor of protein dynamics in the context of CG ENM MD simulations at various spatial resolutions. Several perspectives related to the development of specific CG ENMs are conceivable. CG models are often designed in a way to be transferable to all proteins. Our approach (cf. Section 6.6.2), at purpose, took the opposite view in the sense that we designed very specific CG models. Although they are not transferable to other proteins, such models are highly precise when compared to AA MD simulations. An advantage of the approach is to be easily applicable to other proteins. By using not only a simple CG resolution but also a simple CG ENM FF, our strategy requires adjusting only two parameters. While the trend is to dream of a universal CG FF, one thinks that the lower the resolution, the higher the need of a specific FF for a given biomolecular system. As a perspective, it would be interesting to extend our strategy to generate optimized CG ENMs for other families of proteins. In this way, a series of optimized CG ENMs of diverse structures could then allow to understand how the folding and/or the particular topology of a protein influences the determination of optimum parameters of CG ENMs.

Optimized CG ENMs of a protein were built at a resolution of one bead per residue as well as at coarser resolution levels [55]. As our strategy of CG ENM parameterization proved to be valuable at different resolutions, a challenging perspective would be to design dual-resolution CG models, what we can call cross-grained (XG) models. The purpose of XG models would be to study the flexibility properties of the region of interest, whereas the rest of the system would be represented at lower resolutions. An XG model at the N/1 resolution around the binding site, whereas the rest of the system is modeled at the N/5 resolution, is presented in Figure 6.17. The design of XG models logically requires further refinements, especially to bridge both resolutions or, in other words, to find

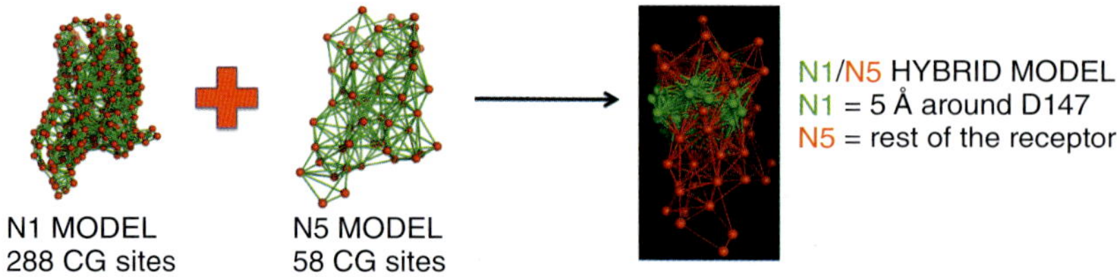

Figure 6.17 Example of an XG model. Two CG resolutions are combined. The N1 resolution is applied inside a sphere of radius 5 Å centered at a major residue of the binding site (green), whereas the rest of the receptor is described at the N5 resolution (red).

an optimal strategy for linking the particles of both N/1 and N/5 CG models. XG models would be able to describe the different degree of flexibility or rigidity along a given protein.

Finally, backmapping remains an important issue when carrying out CG calculations and/or simulations. Several strategies are already available for proteins [319–321]. For a given CG resolution, many higher resolution descriptions are indeed possible, and robustness of backmapping methods is highly necessary. Much progress is still required to automatize accurate and reliable backmapping between CG resolutions. In complement to the developments of dynamical network models of proteins, it is expected that performant and versatile backmapping methods will make very low-resolution models more workable for drug design or any task in which the atomistic resolution is needed.

Acknowledgments

All authors acknowledge the Swedish National Infrastructure for Computing (SNIC, Sweden), the Consortium des Equipements de Calcul Intensif (C.E.C.I., supported by the F.R.S.-FNRS, Belgium), and the Plateforme Technologique de Calcul Intensif (P.T.C.I., supported by the F.R.S.-FNRS, Belgium) for computing resources. MF, LL, and DPV thank Professor Renaud Lambiotte (UNamur, Belgium) for fruitful discussions. MF thanks the Belgian National Fund for Research (F.R.S.-FNRS) for his F.R.I.A. doctoral scholarship. MF, LL, and DPV also acknowledge the Interuniversity Attraction Poles Programmes No. 7/05: "Functional supramolecular systems" initiated by the Belgian Science Policy Office for partial financial support. AL thanks the Swedish Science Council (VR) for financial support. AL acknowledges a partial support by a grant of Ministry of Research and Innovation, CNCS – UEFISCDI, project number PN-III-P4-ID-PCCF-2016-0050, within PNCDI III.

References

1 Albert, R. and Barabási, A.-L. (2002). Statistical mechanics of complex networks. *Rev. Mod. Phys.* 74: 47.

2 Barabási, A.-L. (2002). *Linked: The New Science of Networks*. Perseus Book Group.
3 Prathik, A., Uma, K., and Anuradha, J. (2016). An overview of application of graph theory. *Int. J. ChemTech Res.* 9: 242.
4 Borgatti, S.P., Mehra, A., Brass, D.J., and Labianca, G. (2009). Network analysis in the social sciences. *Science* 323: 892.
5 Albert, R., Jeong, H., and Barabási, A.-L. (1999). Internet: diameter of the world-wide web. *Nature* 401: 130.
6 Riaz, F. and Ali, K.M. (2011). Applications of graph theory in computer science. *IEEE Xplore Computational Intelligence, Communication Systems and Networks (CICSyN)*, 142.
7 Galvez, J. (2012). A graph-theoretical approach to calculate vibrational energies of atomic and subatomic systems. *Open J. Phys. Chem.* 2: 204.
8 Balaban, A.T. (1985). Applications of graph theory in chemistry. *J. Chem. Inf. Comput. Sci.* 25: 334.
9 Bonchev, D. (1991). *Chemical Graph Theory: Introduction and Fundamentals (Mathematical Chemistry)*. CRC Press.
10 Heckmann, T., Schwanghart, W., and Philips, J.D. (2015). Graph theory – recent developments of its applications in geomorphology. *Geomorphology* 243: 130.
11 Barabási, A.-L. and Oltvai, Z.N. (2004). Network biology: understanding the cell's functional organization. *Nat. Rev. Genet.* 5: 101.
12 Grindrod, P. and Kibble, M. (2004). Review of uses of network and graph theory concepts within proteomics. *Expert Rev. Proteomics* 1: 229.
13 Barabási, A.-L., Gulbahce, N., and Loscaizo, J. (2011). Network medicine: a network-based approach to human disease. *Nat. Rev. Genet.* 12: 56.
14 Stam, C.J. (2014). Modern network science of neurological disorders. *Nat. Rev. Neurosci.* 15: 683.
15 Kalapala, V., Sanwalani, V., Clauset, A., and Moore, C. (2006). Scale invariance in road networks. *Phys. Rev. E.* 73: 026130.
16 Faloutsos, M., Faloutsos, P., and Faloutsos, C. (1999). On power-law relationship of the Internet topology. *SIGCOMM Comput. Commun. Rev.* 29: 251.
17 Grandjean, M. (2016). A social network analysis of Twitter: mapping the digital humanities community. *Cog. Arts Hum.* 3: 1171458.
18 Szell, M., Lambiotte, R., and Thurner, S. (2010). Multirelational organization of large-scale social networks in an online world. *Proc. Natl. Acad. Sci. U.S.A.* 107: 13636.
19 Wasserman, S. and Faust, K. (1994). *Social Network Analysis: Methods and Applications*. Cambridge University Press.
20 Dimitrios, K. and Vasileios, O. (2015). A network analysis of the Greek stock market. *Proc. Econom. Fin.* 33: 340.
21 Tse, C.K., Liu, J., and Lau, F.C.M. (2010). A network perspective of the stock market. *J. Empir. Fin.* 17: 659.
22 Hart, M.G., Ypma, R.J.F., Romero-Garcia, R. et al. (2016). Graph theory analysis of complex brain networks: new concepts in brain mapping applied to neurosurgery. *J. Neurosurg.* 124: 1665.

23 Mears, D. and Pollard, H.B. (2016). Network science and the human brain: using graph theory to understand the brain and one of its hubs, the amygdala, in health and disease. *J. Neurosci. Res.* 94: 590.
24 Telesford, Q.K., Simpson, S.L., Burdette, J.H. et al. (2011). The brain as a complex system: using network science as a tool for understanding the brain. *Brain Connect.* 1: 295.
25 Ebenhöh, O., Handorf, T., and Heinrich, R. (2005). A cross species comparison of metabolic network functions. *Genome Inf.* 16: 203.
26 Jeong, H., Tombor, B., Albert, R. et al. (2000). The large-scale organization of metabolic networks. *Nature* 407: 651.
27 Morrison, E.S. and Badyaev, A.V. (2016). Structuring evolution: biochemical networks and metabolic diversification in birds. *BMC Evol. Biol.* 16: 168.
28 Newman, M.E.J. (2003). The structure and function of complex networks. *SIAM Rev.* 45: 167.
29 Barabási, A.-L. (2012). The network takeover. *Nat. Phys.* 8: 14.
30 Fadhal, E., Gamielden, J., and Mwambene, E.C. (2014). Protein interaction networks as metric spaces: a novel perspective on distribution of hubs. *BMC Syst. Biol.* 8: 6.
31 Mulder, N.J., Akinola, R.O., Mazandu, G.K., and Rapanoel, H. (2014). Using biological networks to improve our understanding of infectious disease. *Comput. Struct. Biotechnol.* 11: 1.
32 Morrow, J.K., Tian, L., and Zhang, S. (2010). Molecular networks in drug discovery. *Crit. Rev. Biomed. Eng.* 38: 143.
33 Tang, J. and Aittokallio, T. (2014). Network pharmacology strategies toward multi-target anticancer therapies: from computational models to experimental design principles. *Curr. Pharm. Des.* 20: 23.
34 Garcia-Domenech, R., Gálvez, J., de Julián-Ortiz, J.V., and Pogliani, L. (2008). Some new trends in chemical graph theory. *Chem. Rev.* 108: 1127.
35 Csermely, P., Korcsmáros, T., Kiss, H.J.M. et al. (2013). Structure and dynamics of molecular networks: a novel paradigm of drug discovery. A comprehensive review. *Pharmacol. Ther.* 138: 333.
36 Marx, V. (2013). Biology: the big challenges of big data. *Nature* 498: 255.
37 Wu, C., Schwartz, J.-M., Brabant, G. et al. (2015). Constructing a molecular interaction network for thyroid cancer via scale text mining of gene and pathway events. *BMC Syst. Biol.* 9: S5.
38 Korcsmaros, T., Dunai, Z.A., Vellai, T., and Csermely, P. (2013). Teaching the bioinformatics of signaling networks: an integrated approach to facilitate multi-disciplinary learning. *Briefings Bioinf.* 14: 618.
39 Schlick, T., Collepardo-Guevara, R., Halvorsen, L.A. et al. (2011). Biomolecular modeling and simulation: a field coming to age. *Q. Rev. Biophys.* 44: 191.
40 Tuszynski, J.A., Winter, P., White, D. et al. (2014). Mathematical and computational modeling in biology at multiple scales. *Theor. Biol. Med. Model* 11: 52.
41 van der Kamp, M., Shaw, K.E., Woods, C.J., and Mulholland, A.J. (2008). Biomolecular simulation and modelling: status, progress and prospects. *J. R. Soc. Interface* 5: S173.

42 Karplus, M. and Lavery, R. (2014). Significance of molecular dynamics simulations for life sciences. *Isr. J. Chem.* 54: 1042.
43 Hodak, H. (2014). The Nobel prize in chemistry 2013 for the development of multiscale models of complex chemical systems: a tribute to Martin Karplus, Michael Levitt and Arieh Warshel. *J. Mol. Biol.* 426: 1.
44 Borhani, D.W. and Shaw, D.E. (2012). The future of molecular dynamics simulations in drug discovery. *J. Comput.-Aided Mol. Des.* 26: 15.
45 Buonomano, D.V. (2007). The biology of time across different scales. *Nat. Chem. Biol.* 3: 594.
46 Teilum, K., Olsen, J.G., and Kragelund, B.B. (2009). Functional aspects of protein flexibility. *Cell. Mol. Life Sci.* 66: 2231.
47 Kmiecik, S., Gront, D., Kolinski, M. et al. (2016). Coarse-grained protein models and their applications. *Chem. Rev.* 116: 7898.
48 Saunders, M.G. and Voth, G.A. (2013). Coarse-graining methods for computational biology. *Annu. Rev. Biophys.* 42: 73.
49 Nielsen, S.O., Bulo, R.E., Moore, P.B., and Ensing, B. (2010). Recent progress in adaptive multiscale molecular dynamics simulations of soft matter. *Phys. Chem. Chem. Phys.* 12: 12401.
50 Sanejouand, Y.H. (2013). Elastic network models: theoretical and empirical foundations. *Methods Mol. Biol.* 924: 601.
51 Yan, W., Zhou, J., Sun, M. et al. (2014). The construction of an amino acid network for understanding protein structure and function. *Amino Acids* 46: 1419.
52 Papaleo, E. (2015). Integrating atomistic molecular dynamics simulations, experiments, and network analysis to study protein dynamics: strength in unity. *Front. Mol. Biosci.* 2: 28.
53 Ravasz, E. and Barabási, A.-L. (2003). Hierarchical organization in complex networks. *Phys. Rev. E* 67: 026112.
54 Khan, T. and Ghosh, I. (2015). Modularity in protein structures: study on all-alpha proteins. *J. Biomol. Struct. Dyn.* 33: 2667.
55 Fossépré, M., Leherte, L., Laaksonen, A., and Vercauteren, D.P. (2016). Multiscale design of coarse-grained elastic network-based potentials for the μ opioid receptor. *J. Mol. Model.* 22: 227.
56 Dorogovtsev, S.N. and Mendes, J.F.F. (2003). *Evolution of Networks: From Biological Networks to the Internet and WWW*. Oxford University Press.
57 Pavlopoulos, G.A., Secrier, M., Moschopoulos, C.N. et al. (2011). Using graph theory to analyse biological networks. *BioData Min.* 4: 10.
58 Brinda, K.V. and Vishveshwara, S. (2005). A network representation of protein structures: implications for protein stability. *Biophys. J.* 89: 4159.
59 Hu, G., Yan, W., Zhou, J., and Shen, B. (2014). Residue interaction network analysis of Dronpa and a DNA clamp. *J. Theor. Biol.* 348: 55.
60 Jiao, X. and Chang, S. (2011). Scoring function based on weighted residue network. *Int. J. Mol. Sci.* 12: 8773.
61 Karain, W.I. and Qaraeen, N.I. (2015). Weighted protein residue networks based on joint recurrences between residues. *BMC Bioinf.* 16: 173.
62 Hinsen, K. (2009). Physical arguments for distance-weighted interactions in elastic network models for proteins. *Proc. Natl. Acad. Sci. U.S.A.* 106: E128.

63 Grewal, R.K. and Roy, S. (2015). Modeling proteins as residue interaction networks. *Protein Pept. Lett.* 22: 923.

64 O'Rourke, K.F., Gorman, S.D., and Boehr, D.D. (2016). Biophysical and computational methods to analyse amino acid interaction networks in proteins. *Comput. Struct. Biotechnol. J.* 14: 245.

65 Cheng, S., Fu, H.-L., and Cui, D.-X. (2016). Characteristics analyses and comparisons of the protein structure networks constructed by different methods. *Interdiscip. Sci.* 8: 65.

66 Di Paola, L., De Ruvo, M., Paci, P. et al. (2013). Protein contact networks: an emerging paradigm in chemistry. *Chem. Rev.* 113: 1598.

67 Amor, B.R.C., Schaub, M.T., Yaliraki, S.N., and Barahona, M. (2016). Prediction of allosteric sites and mediating interactions through bond-to-bond propensities. *Nat. Commun.* 7: 12477.

68 Chakrabarty, B. and Parekh, N. (2016). NAPS: network analysis of protein structures. *Nucleic Acids Res.* 44: W375.

69 Costa, L.D.F., Rodrigues, F.A., Travieso, G., and Villas Boas, P.R. (2007). Characterization of complex networks; a survey of measurements. *Adv. Phys.* 56: 167.

70 Lancichinetti, A., Kivelä, M., Saramäki, J., and Fortunato, S. (2010). Characterizing the community structure of complex networks. *PLoS One* 5: e11976.

71 Atilgan, A.R. and Baysal, C. (2004). Thermodynamics of residue networks in folded proteins. *WSEAS Trans. Biol. Biomed.* 1: 205.

72 Heringa, J., Argos, P., Egmond, M.R., and de Vlieg, J. (1995). Increasing thermal stability of subtilisin from mutations suggested by strongly interacting side-chain clusters. *Protein Eng.* 8: 21.

73 del Sol, A., Fujihashi, H., Amoros, D., and Nussinov, R. (2006). Residues crucial for maintaining short paths in network communication mediate signaling in proteins. *Mol. Syst. Biol.* 2: 19.

74 Süel, G.M., Lockless, S.W., Wall, M.A., and Ranganathan, R. (2003). Evolutionarily conserved networks of residues mediate allosteric communication in proteins. *Nat. Struct. Biol.* 10: 59.

75 Park, K. and Kim, D. (2011). Modeling allosteric signal propagation using protein structure networks. *BMC Bioinf.* 12: S23.

76 Zhou, J., Yan, W., Hu, G., and Shen, B. (2016). Amino acid network for prediction of catalytic residues in enzymes: a comparison survey. *Curr. Protein Pept. Sci.* 17: 41.

77 Ghasemi, M., Seidkhani, H., Tamini, F. et al. (2014). Centrality measures in biological networks. *Curr. Bioinf.* 9: 426.

78 Landherr, A., Friedl, B., and Heidemann, J. (2010). A critical review of centrality measures in social networks. *Bus. Inf. Syst. Eng.* 2: 371.

79 Jalili, M., Salehzadeh-Yazdi, A., Gupta, S. et al. (2016). Evolution of centrality measurements for the detection of essential proteins in biological networks. *Front. Physiol.* 7: 375.

80 Koschützki, D. and Schreiber, F. (2008). Centrality analysis methods for biological networks and their application to gene regulatory networks. *Gene Regul. Syst. Biol.* 2: 193.

81 Souza, V.P., Ikegami, C.M., Arantes, G.M., and Marana, S.R. (2016). Protein thermal denaturation is modulated by central residues in the protein structure network. *FEBS J.* 283: 1124.

82 Xue, W., Jiao, P., Liu, H., and Yao, X. (2014). Molecular modeling and residue interaction network studies on the mechanism of binding and resistance of the HCV NS5B polymerase mutants to VX-222 and ANA598. *Antiviral Res.* 104: 40.

83 Adamic, L. (2014). Courses on network theory. University of Michigan, Michigan, USA. www.ladamic.com.

84 Amitai, G., Shemesh, A., Sitbon, E. et al. (2004). Network analysis of protein structures identifies functional residues. *J. Mol. Biol.* 344: 1135.

85 Böde, C., Kovács, I.A., Szalay, M.S. et al. (2007). Network analysis of protein dynamics. *FEBS Lett.* 581: 2776.

86 Alves, N.A. and Martinez, A.S. (2007). Inferring topological features of proteins from amino acid residue networks. *Physica A.* 375: 336.

87 Vendruscolo, M., Dokholyan, N.V., Paci, E., and Karplus, M. (2002). Small-world view of the amino acids that play a key role in protein folding. *Phys. Rev. E* 65: 061910.

88 Vendruscolo, M., Paci, E., Karplus, M., and Dobson, C.M. (2003). Structures and relative free energies of partially folded states of proteins. *Proc. Natl. Acad. Sci. U.S.A.* 100: 14817.

89 del Sol, A., Fujihashi, H., and O'Meara, P. (2005). Topology of small-world networks of protein–protein complex structures. *Bioinformatics* 21: 1311.

90 Thibert, B., Bredesen, D.E., and del Rio, G. (2005). Improved prediction of critical residues for protein function based on network and phylogenetic analyses. *BMC Bioinf.* 6: 213.

91 Sheftel, S., Muratore, K.E., Black, M., and Costanzi, S. (2013). Graph analysis of β2 adrenergic receptor: a "social network" of GPCR residues. *In Silico Pharmacol.* 1: 16.

92 Podder, A., Jatana, N., and Latha, N. (2014). Human dopamine receptors interaction network (DRIN): a systems biology perspective on topology, stability and functionality of the network. *J. Theor. Biol.* 357: 169.

93 Liu, R. and Hu, J. (2011). Computational prediction of heme-binding residues by exploiting residue interaction network. *PLoS One* 6: e25560.

94 Paccanaro, A., Trifonov, V., Yu, H., and Gerstein, M. (2005). Inferring protein-protein interactions using interaction network topologies. *International Joint Conference on Neural Networks (IJCNN)*, 161.

95 Ye, L., Kuang, Q., Jiang, L. et al. (2014). Prediction of hot spots residues in protein–protein interface using network feature and microenvironment feature. *Chemom. Intell. Lab. Syst.* 131: 16.

96 Mathew, O.K. and Sowdhamini, R. (2016). PIMADb: a database of protein–protein interactions in huge macromolecular assemblies. *Bioinf. Biol. Insights* 10: 105.

97 Dehmer, M. and Sivakumar, L. (2016). On comparability graphs: theory and applications. In: *Advances in Mathematical Chemistry and Applications*, vol. 1, 139. Bentham Books.

98 Zhang, X., Perica, T., and Teichmann, S.A. (2013). Evolution of protein structures and interactions from the perspective of residue contact networks. *Curr. Opin. Struct. Biol.* 23: 954.

99 Pinheiro, A.V., Han, D., Shih, W.M., and Yan, H. (2011). Challenges and opportunities for structural DNA nanotechnology. *Nat. Nanotechnol.* 6: 763.

100 Rangnekar, A. and LaBean, T.H. (2014). Building DNA nanostructures for molecular computation, templated assembly, and biological applications. *Acc. Chem. Res.* 47: 1778.

101 Glover, D.J. and Clark, D.S. (2016). Protein calligraphy: a new concept begins to take shape. *ACS Cent. Sci.* 2: 438.

102 Kim, N., Petingi, L., and Schlick, T. (2013). Network theory tools for RNA modeling. *WSEAS Trans. Math.* 9: 941.

103 Benson, E., Mohammed, A., Gardell, J. et al. (2015). DNA rendering of polyhedral meshes at the nanoscale. *Nature* 523: 441.

104 Sathyapriya, R., Vijayabaskar, M.S., and Vishveshwara, S. (2008). Insights into protein–DNA interactions through structure network analysis. *PLoS Comput. Biol.* 4: e1000170.

105 Grant, B.J., Rodrigues, A.P.C., ElSawy, K.M. et al. (2006). Bio3D: an R package for the comparative analysis of protein structures. *Bioinformatics* 22: 2695.

106 Junker, B.H., Koschützki, D., and Schreiber, F. (2006). Exploration of biological network centralities with CentiBiN. *BMC Bioinf.* 7: 219.

107 Kuntal, B.K., Dutta, A., and Mande, S.S. (2016). CompNet: a GUI based tool for comparison of multiple biological interaction networks. *BMC Bioinf.* 17: 185.

108 Shannon, P., Markiel, A., Ozier, O. et al. (2003). Cytoscape: a software environment for integrated models of biomolecular interaction networks. *Genome Res.* 13: 2498.

109 Bastian, M., Heymann, S., and Jacomy, M. (2009). Gephi: an open source software for exploring and manipulating networks. *International AAAI Conference on Weblogs and Social Media.*

110 Kuchaiev, O., Stevanović, A., Hayes, W., and Pržulj, N. (2011). GraphCrunch 2: software tool for network modeling, alignment and clustering. *BMC Bioinf.* 12: 24.

111 Peixoto, T.P. (2014). The graph-tool python library. Graph-tool is downloadable at graph-tool.skewed.de.

112 Gansner, E.R. and North, S.C. (2000). An open graph visualization system and its applications to software engineering. *Software Pract. Exper.* 30: 1203.

113 Krebs, V. (2016). *InFlow*. Orgnet LLC. InFlow is downloadable at www.orgnet.com/.

114 Brohée, S., Faust, K., Lima-Mendez, G. et al. (2008). NeAT: a toolbox for the analysis of biological networks, clusters, classes and pathways. *Nucleic Acids Res.* 36: W444.

115 *NetMiner*. CYRAM. Co. Ltd. (2016). NetMiner is downloadable at www.netminer.com/.

116 Batagelj, V. and Mrvar, A. (2004). Pajek – analysis and visualization of large networks. In: *Graph Drawing Software. Mathematics and Visualization* (ed. M. Jünger and P. Mutzel), 242. Berlin, Heidelberg: Springer Verlag.

117 Martin, A.J.M., Vidotto, M., Boscariol, F. et al. (2011). RING: networking interacting residues, evolutionary information and energetics in protein structures. *Bioinformatics* 27: 2003.

118 Piovesan, D., Minervini, G., and Tosatto, S.C.E. (2016). The RING 2.0 web server for high quality residue interaction networks. *Nucleic Acids Res.* 44: W367.

119 Handcock, M.S., Hunter, D.R., Butts, C.T. et al. (2003). *Statnet: Software tools for the statistical modeling of network data*. Statnet is downloadable at http://statnet.csde.washington.edu.

120 Auber, D. (2004). Tulip – a huge graph visualization framework. In: *Graph Drawing Software* (ed. M. Jünger and P. Mutzel), 105. Berlin, Heidelberg: Springer Verlag.

121 Borgatti, S.P., Everett, M.G., and Freeman, L.C. (2002). *Ucinet for Windows: Software for Social Network Analysis*. Harvard, MA: Analytic Technologies. Ucinet is downloadable at sites.google.com/site/ucinetsoftware/.

122 Combe, C.W., Fischer, L., and Rappsilber, J. (2015). xiNET: cross-link network maps with residue resolution. *Mol. Cell Proteomics* 14: 1137.

123 Villaveces, J.M., Koti, P., and Habermann, B.H. (2015). Tools for visualization and analysis of molecular networks, pathways, and -omics data. *Adv. Appl. Bioinf. Chem.* 8: 11.

124 Su, G., Morris, J.H., Demchak, B., and Bader, G.D. (2014). Biological network exploration with Cytoscape 3. *Curr. Protoc. Bioinf.* 47: 8.13.1.

125 Doncheva, N.T., Assenov, Y., Domingues, F.S., and Albrecht, M. (2012). Topological analysis and interactive visualization of biological networks and protein structures. *Nat. Protoc.* 7: 670.

126 Pasi, M., Tiberti, M., Arrigoni, A., and Papaleo, E. (2012). XPyder: a PyMOL plugin to analyse coupled residues and their networks in protein structures. *J. Chem. Inf. Model.* 52: 1865.

127 Dror, R.O., Dirks, R.M., Grossman, J.P. et al. (2012). Biomolecular simulation: a computational microscope for molecular biology. *Annu. Rev. Biophys.* 41: 429.

128 Durrant, J.D. and McCammon, J.A. (2011). Molecular dynamics simulations and drug discovery. *BMC Biol.* 9: 71.

129 Guliaev, A.B., Cheng, S., and Hang, B. (2012). Protein dynamics via computational microscope. *World J. Methodol.* 2: 42.

130 Mortier, J., Rakers, C., Bermudez, M. et al. (2015). The impact of molecular dynamics on drug design: applications for the characterization of ligand-macromolecules complexes. *Drug Discovery Today* 20: 686.

131 Zhang, J.-L., Zheng, Q.-C., Chu, W.-T., and Zhang, H.-X. (2013). Drug design benefits from molecular dynamics: some examples. *Curr. Comput.-Aided Drug Des.* 9: 532.

132 Marco, E. and Gago, F. (2007). Overcoming the inadequacies or limitations of experimental structures as drug targets by using computational modeling tools and molecular dynamics simulations. *ChemMedChem* 2: 1388.

133 Alder, B.J. and Wainwright, T.E. (1959). Studies in molecular dynamics, I: general method. *J. Chem. Phys.* 31: 459.

134 Duan, Y. and Kollman, P.A. (1998). Pathways to a protein folding intermediate observed in a 1-microsecond simulation in aqueous solution. *Science* 282: 740.

135 Piana, S., Lindorff-Larsen, K., and Shaw, D.E. (2013). Atomic-level description of ubiquitin folding. *Proc. Natl. Acad. Sci. U.S.A.* 110: 5915.

136 Kuskin, J.S., Young, C., Grossman, J.P. et al. (2008). Incorporating flexibility in Anton, a specialized machine for molecular dynamics simulation. *Int. S. High. Perf. Comp. 4658651*, 343.

137 Vendruscolo, M. and Dobson, C.M. (2011). Protein dynamics: Moore's law in molecular biology. *Curr. Biol.* 21: R68.

138 Dodson, G.G., Lane, D.P., and Verma, C.S. (2008). Molecular simulations of protein dynamics: new windows on mechanisms in biology. *EMBO Rep.* 9: 144.

139 Zwier, M.C. and Chong, L.T. (2010). Reaching biological timescales with all-atom molecular dynamics simulations. *Curr. Opin. Pharmacol.* 10: 745.

140 Dokholyan, N.V. (2016). Controlling allosteric networks in proteins. *Chem. Rev.* 116: 6463.

141 Allain, A., Chauvot de Beauchêne, I., Langenfeld, F. et al. (2014). Allosteric pathway identification through network analysis: from molecular dynamics simulations to interactive 2D and 3D graphs. *Faraday Discuss.* 169: 303.

142 Di Paola, L. and Giuliani, A. (2015). Protein contact network topology: a natural language for allostery. *Curr. Opin. Struct. Biol.* 31: 43.

143 Stetz, G. and Verkhivker, G.M. (2015). Dancing through life: molecular dynamics simulations and network-centric modeling of allosteric mechanisms in Hsp70 and Hsp110 chaperone proteins. *PLoS One* 10: e0143752.

144 Stetz, G. and Verkhivker, G.M. (2016). Probing allosteric inhibition mechanisms of the Hsp70 chaperone proteins using molecular dynamics simulations and analysis of the residue interaction networks. *J. Chem. Inf. Model.* 56: 1490.

145 Tse, A. and Verkhivker, G.M. (2015). Molecular dynamics simulations and structural network analysis of c-Abl and c-Src kinase core proteins: capturing allosteric mechanisms and communication pathways from residue centrality. *J. Chem. Inf. Model.* 55: 1645.

146 James, K.A. and Verkhivker, G.M. (2014). Structure-based network analysis of activation mechanisms in the ErbB family of receptor tyrosine kinases: the regulatory spine residues are global mediators of structural stability and allosteric interactions. *PLoS One* 9: e113488.

147 Yao, X.-Q., Malik, R.U., Griggs, N.W. et al. (2016). Dynamic coupling and allosteric networks in the alpha subunit of heterotrimeric G proteins. *J. Biol. Chem.* 291: 4742.

148 Bhakat, S., Martin, A.J.M., and Soliman, M.E.S. (2014). An integrated molecular dynamics, principal component analysis and residue interaction network

approach reveals the impact of M184V mutation in HIV reverse transcriptase resistance to lamivudine. *Mol. Biosyst.* 10: 2215.
149 Mariani, S., Dell'Orco, D., Felline, A. et al. (2013). Network and atomistic simulations unveil the structural determinants of mutations linked to retinal diseases. *PLoS Comput. Biol.* 9: e1003207.
150 Xue, W., Jin, X., Ning, L. et al. (2013). Exploring the molecular mechanism of cross-resistance to HIV-1 integrase strand transfer inhibitors by molecular dynamics simulation and residue interaction network analysis. *J. Chem. Inf. Model.* 53: 210.
151 Blacklock, K. and Verkhivker, G.M. (2014). Computational modeling of allosteric regulation in the Hsp90 chaperones: a statistical ensemble analysis of protein structure networks and allosteric communications. *PLoS Comput. Biol.* 10: e1003679.
152 Csermely, P., Sandhu, K.S., Hazai, E. et al. (2012). Disordered proteins and network disorder in network descriptions of protein structure, dynamics and function: hypotheses and a comprehensive review. *Curr. Protein Pept. Sci.* 13: 19.
153 Maximova, T., Moffatt, R., Ma, B. et al. (2016). Principles and overview of sampling methods for modeling macromolecular structure and dynamics. *PLoS Comput. Biol.* 12: e1004619.
154 Baltoumas, F.A., Theodoropoulou, M.C., and Hamodrakas, S.J. (2016). Molecular dynamics simulations and structure-based network analysis reveal structural and functional aspects of G-protein coupled receptor dimer interactions. *J. Comput.-Aided Mol. Des.* 30: 489.
155 Fossépré, M., Leherte, L., Laaksonen, A., and Vercauteren, D.P. (2014). On the modularity of the intrinsic flexibility of the μ opioid receptor: a computational study. *PLoS One* 9: e115856.
156 Skjaerven, L., Yao, X.-Q., Scarabelli, G., and Grant, B.J. (2014). Integrating protein structural dynamics and evolutionary analysis with Bio3D. *BMC Bioinf.* 15: 399.
157 Skjaerven, L., Jariwala, S., Yao, X.-Q., and Grant, B.J. (2016). Online interactive analysis of protein structure ensembles with Bio3D-web. *Bioinformatics* 32 (22): 3510–3512, pii: btw482.
158 Miao, Y., Nichols, S.E., Gasper, P.M. et al. (2013). Activation and dynamic network of the M2 muscarinic receptor. *Proc. Natl. Acad. Sci. U.S.A.* 110: 10982.
159 Ribeiro, A.A.S.T. and Ortiz, V. (2015). MDN: a web portal for network analysis of molecular dynamics simulations. *Biophys. J.* 109: 1110.
160 Perkett, M.R., Mirijanian, D.T., and Hagan, M.F. (2016). The allosteric switching mechanism in bacteriophage MS2. *J. Chem. Phys.* 145: 035101.
161 Kurzbach, D. (2016). Network representation of protein interactions: theory of graph description and analysis. *Protein Sci.* 25: 1617.
162 Kurzbach, D., Flamm, A.G., and Sara, T. (2016). Network representation of protein interactions – experimental results. *Protein Sci.* 25: 1628.
163 Takahashi, K., Oda, T., and Naruse, K. (2014). Coarse-grained molecular dynamics simulations of biomolecules. *AIMS Biophys.* 1: 1.

164 Brini, E., Algaer, E.A., Ganguly, P. et al. (2013). Systematic coarse-graining methods for soft matter simulations – a review. *Soft Matter* 9: 2108.

165 Clementi, C. (2008). Coarse-grained models of protein folding: toy models or predictive tools ? *Curr. Opin. Struct. Biol.* 18: 10.

166 Hadley, K.R. and McCabe, C. (2012). Coarse-grained molecular models of water: a review. *Mol. Simul.* 38: 671.

167 Ingólfsson, H.I., Lopez, C.A., Uusitalo, J.J. et al. (2014). The power of coarse graining in biomolecular simulations. *WIREs Comput. Mol. Sci.* 4: 225.

168 Riniker, S., Allison, J.R., and van Gunsteren, W.F. (2012). On developing coarse-grained models for biomolecular simulation: a review. *Phys. Chem. Chem. Phys.* 14: 12423.

169 Tozzini, V. (2005). Coarse-grained models for proteins. *Curr. Opin. Chem. Biol.* 15: 144–150.

170 Tozzini, V. (2010). Minimalist models for proteins: a comparative analysis. *Q. Rev. Biophys.* 43: 333.

171 Trylska, J. (2010). Coarse-grained models to study dynamics of nanoscale biomolecules and their applications to ribosome. *J. Phys. Condens. Matter* 22: 453101.

172 Tozzini, V., Trylska, J., Chang, C.-E., and McCammon, A. (2007). Flap opening dynamics in HIV-1 protease explored with coarse-grained model. *J. Struct. Biol.* 157: 606.

173 Trylska, J., Tozzini, V., and McCammon, J.A. (2005). Exploring global motions and correlations in the ribosome. *Biophys. J.* 89: 1455–1463.

174 Gopal, S.M., Mukherjee, S., Cheng, Y.-M., and Feig, M. (2010). PRIMO/PRIMONA: a coarse-grained model for proteins and nucleic acids that preserves near-atomistic accuracy. *Proteins* 78: 1266.

175 Leherte, L. and Vercauteren, D.P. (2009). Coarse point charge models for proteins from smoothed molecular electrostatic potentials. *J. Chem. Theory Comput.* 5: 3279.

176 Leherte, L. and Vercauteren, D.P. (2014). Evaluation of reduced point charge models of proteins through molecular dynamics simulations: application to the Vps27 UIM-1–Ubiquitin complex. *J. Mol. Graphics Modell.* 47: 44.

177 Leherte, L. and Vercauteren, D.P. (2014). Comparison of reduced point charge models of proteins: molecular dynamics simulations of Ubiquitin. *Sci. China Chem.* 57: 1340.

178 Bahar, I. and Jernigan, R.L. (1997). Inter-residue potentials in globular proteins and the dominance of highly specific hydrophilic interactions at close separation. *J. Mol. Biol.* 266: 195.

179 Reith, D., Pütz, M., and Müller-Plathe, F. (2003). Deriving effective mesoscale potentials from atomistic simulations. *J. Comput. Chem.* 24: 1624.

180 Zacharias, M. (2003). Protein–protein docking with a reduced protein model accounting for side-chain flexibility. *Protein Sci.* 12: 1271.

181 Smith, A.V. and Hall, C.K. (2001). The α-helix formation: discontinuous molecular dynamics on an intermediate-resolution protein model. *Proteins* 44: 344.

182 Smith, A.V. and Hall, C.K. (2001). Assembly of a tetrameric α-helical bundle: computer simulations on an intermediate-resolution protein model. *Proteins* 44: 376.
183 Irbäck, A., Samuelsson, B., Sjunnesson, F., and Wallin, S. (2003). Thermodynamics of α- and β-structure formation in proteins. *Biophys. J.* 85: 1466.
184 Leherte, L. and Vercauteren, D.P. (2011). Implementation of a protein reduced point charge model toward molecular dynamics applications. *J. Phys. Chem. A* 115: 12531.
185 Leherte, L. (2016). Reduced point charge models of proteins: assessment based on molecular dynamics simulations. *Mol. Simul.* 42: 289.
186 Doruker, P., Jernigan, R.L., and Bahar, I. (2001). Dynamics of large proteins through hierarchical levels of coarse-grained structures. *J. Comput. Chem.* 23: 119.
187 Zhang, Z., Lu, L., Noid, W.G. et al. (2008). A systematic methodology for defining coarse-grained sites in large biomolecules. *Biophys. J.* 95: 5073.
188 Dama, J.F., Sinitskiy, A.V., McCullagh, M. et al. (2013). The theory of ultra-coarse-graining. 1. General principles. *J. Chem. Theory Comput.* 9: 2466.
189 Pfaendtner, J., Lyman, E., Pollard, T.D., and Voth, G.A. (2010). Structure and dynamics of the actin filament. *J. Mol. Biol.* 396: 252.
190 Davtyan, A., Dama, J.F., Sinitskiy, A.V., and Voth, G.A. (2014). The theory of ultra-coarse- graining. 2. Numerical implementation. *J. Chem. Theory Comput.* 10: 5265.
191 Grime, J.M.A. and Voth, G.A. (2014). Highly scalable and memory efficient ultra-coarse-grained molecular dynamics simulations. *J. Chem. Theory Comput.* 10: 423.
192 Hinsen, K. (1998). Analysis of domain motions by approximate normal mode calculations. *Proteins* 33: 417.
193 Doruker, P., Liu, Y., Yang, Z., and Bahar, I. (2012). In silico coarse-grained approaches to structural dynamics and function of proteins and their assemblies. *Compr. Biophys.* 9: 27.
194 Bahar, I., Lezon, T., Bakan, A., and Shrivastava, I.H. (2010). Normal mode analysis of biomolecular structures: functional mechanisms of membrane proteins. *Chem. Rev.* 110: 1463.
195 Isin, B., Rader, A.J., Kaur Dhiman, H. et al. (2006). Predisposition of the dark state of rhodopsin to functional changes in structure. *Proteins* 65: 970.
196 Fuglebakk, E., Tiwari, S.P., and Reuter, N. (2015). Comparing the intrinsic dynamics of multiple protein structures using elastic network models. *Biochim. Biophys. Acta.* 1850: 911.
197 Kolan, D., Fonar, G., and Samson, A.O. (2014). Elastic network normal mode dynamics reveal the GPCR activation mechanism. *Proteins* 82: 579.
198 Niv, M.Y., Skrabanek, L., Filizola, M., and Weinstein, H. (2006). Modeling activated states of GPCRs: the rhodopsin template. *J. Comput.-Aided Mol. Des.* 20: 437.
199 Kundu, S., Melton, J.S., Sorensen, D.C., and Phillips, G.N. Jr., (2002). Dynamics of proteins in crystals: comparison of experiment with simple models. *Biophys. J.* 83: 723.

200 Park, J.-K., Jernigan, R., and Wu, Z. (2013). Coarse grained normal mode analysis vs. refined gaussian network model for protein residue-level structural fluctuations. *Bull. Math. Biol.* 75: 124.

201 Hinsen, K. (2008). Structural flexibility in proteins: impact of the crystal environment. *Bioinformatics* 24: 521.

202 Leioatts, N., Romo, T.D., and Grossfield, A. (2012). Elastic network models are robust to variations in formalism. *J. Chem. Theory Comput.* 8: 2424.

203 Riccardi, D., Cui, Q., and Phillips, G.N. Jr., (2010). Evaluating elastic network models of crystalline biological molecules with temperature factors, correlated motions, and diffuse X-ray scattering. *Biophys. J.* 99: 2616.

204 Romo, T.D. and Grossfield, A. (2011). Validating and improving elastic network models with molecular dynamics simulations. *Proteins* 79: 23.

205 Fuglebakk, E., Reuter, N., and Hinsen, K. (2013). Evaluation of protein elastic network models based on an analysis of collective motions. *J. Chem. Theory Comput.* 9: 5618.

206 Micheletti, C., Carloni, P., and Maritan, A. (2004). Accurate and efficient description of protein vibrational dynamics: comparing molecular dynamics and Gaussian models. *Proteins* 55: 635.

207 Rueda, M., Chacón, P., and Orozco, M. (2007). Through validation of protein normal mode analysis: a comparative study with essential dynamics. *Structure* 15: 565.

208 Skjaerven, L., Martinez, A., and Reuter, N. (2011). Principal component and normal mode analysis of proteins: a quantitative comparison using the GroEL subunit. *Proteins* 79: 232.

209 Greene, L.H. (2012). Protein structure networks. *Brief. Funct. Genomics* 11: 469.

210 Zhou, J., Yan, W., Hu, G., and Shen, B. (2014). Amino acid network for the discrimination of native protein structures from decoys. *Curr. Protein Pept. Sci.* 15: 522.

211 Greene, L.H. and Higman, V.A. (2003). Uncovering network systems within protein structures. *J. Mol. Biol.* 334: 781.

212 Khor, S. (2012). Towards an integrated understanding of the structural characteristics of protein residue networks. *Theory Biosci.* 131: 61.

213 Estrada, E. (2010). Universality in protein residue networks. *Biophys. J.* 98: 890.

214 da Silveira, C.H., Pires, D.E.V., Minardi, R.C. et al. (2009). Protein cutoff scanning: a comparative analysis of cutoff dependent and cutoff free methods for prospecting contacts in proteins. *Proteins* 74: 727.

215 Bartoli, L., Fariselli, P., and Casadio, R. (2007). The effect of backbone on the small-world properties of protein contact maps. *Phys. Biol.* 4: L1.

216 Sun, J., Jing, R., Wu, D. et al. (2013). The effect of edge definition of complex networks on protein structure identification. *Comput. Math. Methods Med.* 2013: 365410.

217 Jeong, J.I., Jang, Y., and Kim, M.K. (2006). A connection rule for alpha-carbon coarse-grained elastic network models using chemical bond information. *J. Mol. Graph. Model.* 24: 296.

218 Jiao, X., Chang, S., Li, C.-H. et al. (2007). Construction and application of the weighted amino acid network based on energy. *Phys. Rev. E: Stat. Nonlinear Soft Matter Phys.* 75: 051903.

219 Veloso, C.J.M., Silveira, C.H., Melo, R.C. et al. (2007). On the characterization of energy networks of proteins. *Genet. Mol. Res.* 6: 799.

220 Noel, J.K., Whitford, P.C., and Onuchic, J.N. (2012). The shadow map: a general contact definition for capturing the dynamics of biomolecular folding and function. *J. Phys. Chem. B* 116: 8692.

221 Doncheva, N.T., Klein, K., Domingues, F.S., and Albrecht, M. (2011). Analyzing and visualizing residue networks of protein structures. *Trends Biochem. Sci.* 36: 179.

222 Richards, F.M. (1974). The interpration of protein structures: total volume, group volume distributions and packing density. *J. Mol. Biol.* 82: 1.

223 Poupon, A. (2004). Voronoi and Voronoi-related tessellations in studies of protein structure and interaction. *Curr. Opin. Chem. Biol.* 14: 233.

224 Dupuis, F., Sadoc, J.-F., Jullien, R. et al. (2005). VORO3D: 3D Voronoi tessellations applied to protein structures. *Bioinformatics* 21: 1715.

225 Fourty, G., Callebaut, I., and Mornon, J.-P. (2008). Characterization of non-trivial neighborhood fold constraints from protein sequences using generalized topohydrophobicity. *Bioinf. Biol. Insights* 2: 47.

226 Tyagi, M., Bornot, A., Offmann, B., and de Brevern, A.G. (2009). Analysis of loop boundaries using different local structure assignment methods. *Protein Sci.* 18: 1869.

227 Girard, E., Marchal, S., Perez, J. et al. (2010). Structure-function perturbation and dissociation of tetrameric urate oxidase by high hydrostastic pressure. *Biophys. J.* 98: 2365.

228 Rother, K., Hildebrand, P.W., Goede, A. et al. (2009). Voronoia: analyzing packing in protein stuctures. *Nucleic Acids Res.* 37: D393.

229 Esque, J., Léonard, S., de Brevern, A.G., and Oguey, C. (2013). VLDP web server: a powerful geometric tool for analyzing protein structures in their environment. *Nucleic Acid. Res.* 41: W373–W378.

230 Bourquard, T., Bernauer, J., Azé, J., and Poupon, A. (2009). Comparing Voronoi and Laguerre tessellations in the protein–protein docking context. *Sixth international symposium on Voronoi diagrams*, 225.

231 Esque, J., Oguey, C., and de Brevern, A.G. (2011). Comparative analysis of threshold and tessellation methods for determining protein contacts. *J. Chem. Inf. Model.* 51: 493.

232 Esque, J., Oguey, C., and de Brevern, A.G. (2010). A novel evaluation of residue and protein volumes by means of Laguerre tessellation. *J. Chem. Inf. Model.* 50: 947.

233 Seeber, M., Felline, A., Raimondi, F. et al. (2011). Wordom: a user-friendly program for the analysis of molecular structures, trajectories, and free energy surfaces. *J. Comput. Chem.* 32: 1183.

234 Eargle, J. and Luthey-Schulten, Z. (2012). NetworkView: 3D display and analysis of protein.RNA interaction networks. *Bioinformatics* 28: 3000.

235 Sethi, A., Eargle, J., Black, A.A., and Luthey-Schulten, Z. (2009). Dynamical networks in tRNA: protein complexes. *Proc. Natl. Acad. Sci. U.S.A.* 106: 6620.

236 Bhattacharyya, M., Bhat, C.R., and Vishveshwara, S. (2013). An automated approach to network features of protein structure ensembles. *Protein Sci.* 22: 1399.

237 Vanwart, A.T., Eargle, J., Luthey-Schulten, Z., and Amaro, R.E. (2012). Exploring residue component contributions to dynamical network models of allostery. *J. Chem. Theory Comput.* 8: 2949.

238 Girvan, M. and Newman, M.E.J. (2002). Community structure in social and biological networks. *Proc. Natl. Acad. Sci. U.S.A.* 99: 7821.

239 Lorenz, D.M., Jeng, A., and Deem, M.W. (2011). The emergence of modularity in biological systems. *Phys. Life Rev.* 8: 129.

240 Mengistu, H., Huizinga, J., Mouret, J.-B., and Clune, J. (2016). The evolutionary origins of hierarchy. *PLoS Comput. Biol.* 12: e1004829.

241 Rosvall, M. and Bergstrom, C.T. (2008). Maps of random walks on complex networks reveal community structure. *Proc. Natl. Acad. Sci. U.S.A.* 105: 1118.

242 Fortunato, S. (2010). Community detection in graphs. *Phys. Rep.* 486: 75.

243 Harenberg, S., Bello, G., Gjeltema, L. et al. (2014). Community detection in large-scale networks: a survey and empirical evaluation. *Wiley Interdiscip. Rev. Comput. Stat.* 6: 426.

244 Lambiotte, R. (2013). Multi-scale modularity and dynamics in complex networks. In: *Dynamics on and ff Complex Networks*, vol. 2 (ed. A. Mukherjee, M. Choudhury, F. Peruani, et al.), 125. Springer Science & Business Media.

245 Ravasz, E. (2009). Detecting hierarchical modularity in biological networks. *Methods Mol. Biol.* 541: 145.

246 Tripathi, S., Moutari, S., Dehmer, M., and Emmert-Streib, F. (2016). Comparison of module detection algorithms in protein networks and investigation of the biological meaning of predicted modules. *BMC Bioinf.* 17: 129.

247 Voevodski, K., Teng, S.-H., and Xia, Y. (2009). Finding local communities in protein networks. *BMC Bioinf.* 10: 297.

248 Yang, Z., Algesheimer, R., and Tessone, C.J. (2016). A comparative analysis of community detection algorithms on artificial networks. *Sci. Rep.* 6: 30750.

249 Rosvall, M., Esquivel, A.V., Lancichinetti, A. et al. (2014). Memory in network flows and its effects on spreading dynamics and community detection. *Nat. Commun.* 5: 4630.

250 Salnikov, V., Schaub, M.T., and Lambiotte, R. (2016). Using higher-order Markov models to reveal flow-based communities in networks. *Sci. Rep.* 6: 23194.

251 Rosvall, M. and Bergstrom, C.T. (2011). Multilevel compression of random walks on networks reveals hierarchical organization in large integrated systems. *PLoS One* 6: e18209.

252 Bernardi, R.C., Melo, M.C.R., and Schulten, K. (2015). Enhanced sampling techniques in molecular dynamics simulations of biological systems. *Biochim. Biophys. Acta* 1850: 872.

253 Mori, T., Miyashita, N., Im, W. et al. (2016). Molecular dynamics simulations of biological membranes and membrane proteins using enhanced conformational sampling algorithms. *Biochim. Biophys. Acta* 1858: 1635.

254 Rand, W.M. (1971). Objective criteria for the evaluation of clustering methods. *J. Am. Stat. Assoc.* 66: 846.

255 Wallace, D.L. (1983). A method for comparing two hierarchical clusterings: comment. *J. Am. Stat. Assoc.* 78: 569.

256 Fowlkes, E.B. and Mallows, C.L. (1983). A method for comparing two hierarchical clusterings. *J. Am. Stat. Assoc.* 78: 553.

257 Alexander-Bloch, A., Lambiotte, R., Roberts, B. et al. (2012). The discovery of population differences in network community structure: new methods and applications to brain functional networks in schizophrenia. *NeuroImage* 59: 3889.

258 Carriço, J.A. and Severiano, A. (2011). *Online tool for quantitative assessment of classification agreement*. Online tools for comparing partitions. Available at: http://www.comparingpartitions.info (last accessed December 2015).

259 Hubert, L. and Arabie, P. (1985). Comparing partitions. *J. Classif.* 2: 193.

260 Meilă, M. (2005). Comparing clusterings: an axiomatic view. *In ICML '05: Proc. of the 22nd Int. Conf. Mach. Learn.*, 577.

261 Meilă, M. (2007). Comparing clusterings – an information based distance. *J. Multivariate Anal.* 98: 873.

262 Hleap, J.S. and Blouin, C. (2016). The semantics of the modular architecture of protein structures. *Curr. Protein Pept. Sci.* 17: 62.

263 Kannan, N. and Vishveshwara, S. (1999). Identification of side-chain clusters in protein structures by a graph spectral method. *J. Mol. Biol.* 292: 441.

264 Vishveshwara, S., Ghosh, A., and Hansia, P. (2009). Intra and inter-molecular communications through protein structure network. *Curr. Protein Pept. Sci.* 10: 146.

265 del Sol, A., Araúzo-Bravo, M.J., Amoros, D., and Nussinov, R. (2007). Modular architecture of protein structures and allosteric communications: potential implications for signaling proteins and regulatory linkages. *Genome Biol.* 8: R92.

266 Farabella, I., Pham, T., Henderson, N.S. et al. (2014). Allosteric signalling in the outer membrane translocation domain of PapC usher. *eLife* 3: e03532.

267 Ghosh, A., Sakaguchi, R., Liu, C. et al. (2011). Allosteric communication in cysteinyl tRNA synthetase: a network of direct and indirect readout. *J. Biol. Chem.* 286: 37721.

268 del Sol, A. and Carbonell, P. (2007). The modular organization of domain structures: insights into protein-protein binding. *PLoS Comput. Biol.* 3: e239.

269 Saccà, C., Teso, S., Diligenti, M., and Passerini, A. (2014). Improved multi-level protein-protein interaction prediction with semantic-based regularization. *BMC Bioinf.* 15: 103.

270 Srivastasa, A. and Sinha, S. (2014). Thermostability of in vitro evolved *Bacillus subtilis* lipase A: a network and dynamics perspective. *PLoS One* 9: e102856.

271 Dib, L. and Carbone, A. (2012). Protein fragments: functional and structural roles of their coevolution networks. *PLoS One* 7: e48124.
272 Rorick, M.M. and Wagner, G.P. (2011). Protein structural modularity and robustness are associated with evolvability. *Genome Biol. Evol.* 3: 456.
273 Trifonov, E.N. and Frenkel, Z.M. (2009). Evolution of protein modularity. *Curr. Opin. Struct. Biol.* 19: 335.
274 Hleap, J.S., Susko, E., and Blouin, C. (2013). Defining structural and evolutionary modules in proteins: a community detection approach to explore sub-domain architecture. *BMC Struct. Biol.* 13: 20.
275 Rorick, M. (2012). Quantifying protein modularity and evolvability: a comparison of different techniques. *Biosystems* 110: 22.
276 Aguilar, D., Oliva, B., and Marino Buslje, C. (2012). Mapping the mutual information network of enzymatic families in the protein structure to unveil functional features. *PLoS One* 7: e41430.
277 Csermely, P., Palotai, R., and Nussinov, R. (2010). Induced fit, conformational selection and independent dynamic segments: an extended view of binding events. *Trends Biochem. Sci.* 35: 539.
278 Delmotte, A., Tate, E.W., Yaliraki, S.N., and Barahona, M. (2011). Protein multi-scale organization through graph partitioning and robustness analysis: application to the myosin-myosin light chain interaction. *Phys. Biol.* 8: 055010.
279 Schaub, M.T., Delvenne, J.-C., Yaliraki, S.N., and Barahona, M. (2012). Markov dynamics as a zooming lens for multiscale community detection: non clique-like communities and the field-of-view limit. *PLoS One* 7: e32210.
280 Hospital, A., Goñi, J.R., Orozco, M., and Gelpi, J.L. (2015). Molecular dynamics simulations: advances and applications. *Adv. Appl. Bioinf. Chem.* 10: 37.
281 Aleksiev, T., Potestio, R., Pontiggia, F. et al. (2009). PiSQRD: a web server for decomposing proteins into quasi-rigid dynamical domains. *Bioinformatics* 25: 2743.
282 Romanowska, J., Nowiński, K.S., and Trylska, J. (2012). Determining geometrically stable domains in molecular conformation sets. *J. Chem. Theory Comput.* 8: 2588.
283 Dziubiński, M., Daniluk, P., and Lesyng, B. (2016). ResiCon: a method for the identification of dynamics domains, hinges and interfactial regions in proteins. *Bioinformatics* 32: 25.
284 Wieninger, S.A. and Ullmann, G.M. (2015). CoMoDo: identifying dynamic protein domains based on covariances of motion. *J. Chem. Theory Comput.* 11: 2841.
285 Polles, G., Indelicato, G., Potestio, R. et al. (2013). Mechanical and assembly units of viral capsids identified via quasi-rigid domain decomposition. *PLoS Comput. Biol.* 9: e1003331.
286 Delvenne, J.-C., Yaliraki, S.N., and Barahona, M. (2010). Stability of graph communities across time scales. *Proc. Natl. Acad. Sci. U.S.A.* 107: 12755.
287 Traag, V.A., Krings, G., and Van Dooren, P. (2013). Significant scales in community structure. *Sci. Rep.* 3: 2930.

288 Lambiotte, R., Delvenne, J.-C., and Barahona, M. (2014). Random walks, Markov processes and the multiscale modular organization of complex networks. *IEEE Trans. Netw. Sci. Eng.* 1: 76.

289 Barzel, B., Liu, Y.-Y., and Barabási, A.-L. (2015). Constructing minimal models for complex system dynamics. *Nat. Commun.* 6: 7186.

290 Latour, T., Leherte, L., Derouane, E.G., and Vercauteren, D.P. (1993). Computerised structural analysis of zeolitic networks: conceptualisation of a zeolite scene through graphs comparison. *J. Comput.-Aided Mater. Des.* 1: 265.

291 Dury, L., Latour, T., Leherte, L. et al. (2001). A new graph descriptor for molecules containing cycles. Application as screening criterion for searching molecular structures within large databases of organic compounds. *J. Chem. Inf. Comput. Sci.* 41: 1437.

292 Bader, R.W. (1995). *Atoms in Molecules – A Quantum Theory*. Oxford: Clarendon Press.

293 Leherte, L. (2001). Application of multiresolution analyses to electron density maps of small molecules: critical point representations for molecular superposition. *J. Math. Chem.* 29: 47.

294 Leherte, L., Dury, L., and Vercauteren, D.P. (2003). Structural identification of local maxima in low-resolution promolecular electron density distributions. *J. Phys. Chem. A* 107: 9875.

295 Leherte, L. (2004). Hierarchical analysis of promolecular full electron-density distributions: description of protein structure fragments. *Acta Crystallogr., Sect. D: Biol. Crystallogr.* 60: 1254.

296 Leherte, L., Meurice, N., and Vercauteren, D.P. (2005). Influence of conformation on the representation of small flexible molecules at low resolution: alignment of endothiapepsin ligands. *J. Comput.-Aided Mol. Des.* 19: 525.

297 Leherte, L., Meurice, N., and Vercauteren, D.P. (2000). Critical point representations of electron density maps for the comparison of benzodiazepine-type ligands. *J. Chem. Inf. Comput. Sci.* 40: 816.

298 Meurice, N., Leherte, L., Vercauteren, D.P. et al. (1997). Development of a genetic algorithm method especially designed for the comparison of molecular models: application to the elucidation of the benzodiazepine receptor pharmacophore. In: *Computer-Assisted Lead Finding and Optimization: Current Tools for Medicinal Chemistry*, vol. 497 (ed. H. van de Waterbeemd, B. Testa and G. Folkers). Wiley-VCH.

299 Meurice, N., Leherte, L., and Vercauteren, D.P. (1998). Comparison of benzodiazepine-like compounds using topological analysis and genetic algorithms. *SAR QSAR Environ. Res.* 8: 195.

300 Binamé, J., Meurice, N., Leherte, L. et al. (2004). Use of electron density critical points as chemical function-based reduced representations of pharmacological ligands. *J. Chem. Inf. Comput. Sci.* 44: 1394.

301 Meurice, N., Maggiora, G.M., and Vercauteren, D.P. (2005). Evaluating molecular similarity using reduced representations of the electron density. *J. Mol. Mod.* 11: 237.

302 Leherte, L., Glasgow, J., Baxter, K. et al. (1997). Analysis of three-dimensional protein images. *J. Artif. Intell. Res.* 7: 125.

303 Hall, S.R., du Boulay, D.J., and Olthof-Hazekamp, R. (eds.) (2000). *Xtal3.7 System*, University of Western Australia. The source code is available at http://xtal.sourceforge.net/ (accessed 18 May 2016).

304 Becue, A., Meurice, N., Leherte, L., and Vercauteren, D.P. (2003). Description of protein–DNA complexes in terms of electron-density topological features. *Acta Crystallogr., Sect. D: Biol. Crystallogr.* 59: 2150.

305 Becue, A., Meurice, N., Leherte, L., and Vercauteren, D.P. (2004). Evaluation of the protein solvent-accessible surface using reduced representations in terms of critical points of the electron density. *J. Comput. Chem.* 25: 1117.

306 Becue, A., Meurice, N., Leherte, L., and Vercauteren, D.P. (2008). Protein–protein docking using three-dimensional reduced representations and based on a genetic algorithm. In: *Models, Mysteries and Magic of Molecules* (ed. J.C.A. Boeyens and J.F. Ogilvie), 301. Springer.

307 Leherte, L. and Vercauteren, D.P. (2008). Collective motions in protein structures: applications of elastic network models built from electron density distributions. *Comput. Phys. Commun.* 179: 171.

308 Leherte, L. and Vercauteren, D.P. (2008). Collective motions of rigid fragments in protein structures from smoothed electron density distributions. *J. Comput. Chem.* 29: 1472.

309 Hayward, S. and de Groot, B.L. (2008). Normal modes and essential dynamics. In: *Methods in Molecular Biology*, Molecular Modeling of Proteins, vol. 443 (ed. A. Kukol), 89. Humana Press.

310 Ahmed, A., Villinger, S., and Gohlke, H. (2010). Large-scale comparison of protein essential dynamics from molecular dynamics simulations and coarse-grained normal mode analyses. *Proteins* 78: 3341.

311 Pinamonti, G., Bottaro, S., Micheletti, C., and Bussi, G. (2015). Elastic network models for RNA: a comparative assessement with molecular dynamics and SHAPE experiments. *Nucleic Acids Res.* 43: 7260.

312 Fossépré, M., Leherte, L., Laaksonen, A., and Vercauteren, D.P. (2018). Combining coarse-grained elastic network and reduced point charge models: application to the μ opioid receptor (submitted for publication).

313 Borodin, O. and Smith, G.D. (2009). *Force field fitting toolkit*. The University of Utah.

314 Mackerell, A.D. Jr., Feig, M., and Brooks, C.L. (2004). Extending the treatment of backbone energetics in protein force fields: limitations of gas-phase quantum mechanics in reproducing protein conformational distributions in molecular dynamics simulations. *J. Comput. Chem.* 25: 1400.

315 Dolinsky, T.J., Nielsen, J.E., McCammon, J.A., and Baker, N.A. (2004). PDB2PQR: an automated pipeline for the setup of Poisson–Boltzmann electrostatics calculations. *Nucleic Acids Res.* 32: W665.

316 Gorecki, A., Szypowski, M., Dlugosz, M., and Trylska, J. (2009). RedMD – reduced molecular dynamics package. *J. Comput. Chem.* 30: 2364–2373.

317 Srivastasa, A. and Voth, G.A. (2014). Solvent-free, highly coarse-grained models for charged lipid systems. *J. Chem. Theory Comput.* 10: 4730.

318 Fossépré, M., Leherte, L., Laaksonen, A. and Vercauteren, D.P. (2018). Multi-graining: a modular- and multi-resolution approach for describing the flexibility of proteins – application to the μ opioid receptor (submitted for publication).
319 CGTools Plugin (2013). Theoretical and Computational Biophysics Group, University of Illinois at Urbana-Champaign, http://www.ks.uiuc.edu/Research/vmd/plugins/cgtools/ (accessed 9 October 2016).
320 Lombardi, L.E., Martí, M.A., and Capece, L. (2016). CG2AA: backmapping protein coarse-grained structures. *Bioinformatics* 32: 1235.
321 Wassenaar, T.A., Pluhackova, K., Böckmann, R.A. et al. (2014). Going backward: a flexible geometric approach to reverse transformation from coarse grained to atomistic models. *J. Chem. Theory Comput.* 10: 676.

Part III

Applications and Success Stories

7

From Computers to Bedside: Computational Chemistry Contributing to FDA Approval

Christina Athanasiou and Zoe Cournia

Biomedical Research Foundation, Academy of Athens, 4 Soranou Ephessiou, 11527 Athens, Greece

7.1 Introduction

The drug design process is unequivocally a time-consuming and expensive endeavor, with recent estimates classifying it as a \$2.6 billion expenditure [1]. From target identification and validation, to hit-and-lead discovery, as well as lead optimization, preclinical and clinical, the outlay in each consecutive stage accounts for several millions of US dollars, with the financial burden surging with every unsuccessful attempt, especially in the late phases of the development. Fortunately, the rise in validated protein targets relevant to therapeutic applications deriving from large-scale genomic sequencing adjoined with proteome analysis, held the basis of systematic efforts targeting the efficacious treatment of protein-provoking diseases [2]. In addition, the advances in high-throughput screening (HTS) experiments allowed the assessment of thousands of molecules concurrently by employing robotic automation, diminished the human labor, and dominated the area of hit identification in the past two decades [3].

Nonetheless, HTS is still time consuming and expensive, with its acquisition value and operational costs being prohibitive for most laboratories. Moreover, careful decision making to decrease attrition rates and avoid costly failures, together with the tremendous advances in computational technologies led to the advent of rational, computer-aided drug design (CADD). Molecular modeling techniques have revolutionized the conventional drug discovery processes, by enabling the reduction of time and resources allocated in the hit identification, hit-to-lead optimization and lead optimization phases of the drug discovery pipeline. Novel druglike candidates are first examined *in silico* for their expected affinity to a therapeutic target (in the case of structure-based drug design) or their similarity to previously identified active compounds (ligand-based drug design), as well as the prediction of physicochemical properties with the aid of sophisticated methods and algorithms. Subsequently, provided that desirable results have been received, the experimental part commences with molecular modeling prioritizing organic synthesis efforts [4]. Excluding drug candidates bearing no chance of demonstrating success early in the process can thus eliminate the substantial cost that derives from failures.

Biomolecular Simulations in Structure-Based Drug Discovery,
First Edition. Edited by Francesco L. Gervasio and Vojtech Spiwok.

The extensive and systematic use of computer-assisted methods became feasible only in the past two decades with the improvement in computer graphics and the development of algorithms able to simulate biomolecular systems. These efforts were intensified in the past decade due to the rapid development of faster architectures in tandem with the arrival of graphical processing unit (GPU) coding [5], the improvement of methodologies in both theoretical and application levels [6, 7], as well as better algorithms enabling more accurate atomistic description and treatment of interactions that new force fields provide [8–10]. Moreover, problems related to poor sampling and difficulty in surpassing energetic barriers have been addressed with pioneering enhanced sampling techniques [11–13]. Reviews thoroughly describing recent computer-aided methods have been published before [14–17]. To sum up, nowadays more than ever, the assistance of the methods has been recognized as a tool inextricably linked with drug design–oriented attempts.

This trend has not been unnoticed by pharmaceutical companies, which have reformed the structure of their R&D departments by incorporating CADD laboratories active in the development process. GlaxoSmithKline, one of the companies that has adopted CADD methods, contends that "design" rather than "discovery" is its primary goal, explaining that medicinal chemists exploit the maximum potential by applying true design principles [18]. On the same issue, Merck, Janssen, Vertex Pharmaceuticals, and other smaller companies discuss the involvement of CADD in their research and discovery process, highlighting its importance and cooperation with other disciplines [19–21].

All things considered, computational techniques can be a powerful tool in the discovery of new medicaments. But to what extent has a computational procedure ever successfully guided this complex procedure, leading to a safe and effective drug that is currently on the market? In the current review, we present cases of the US Food and Drug Administration (FDA)-approved drugs for the discovery of which CADD techniques played an instrumental role. This includes either strategies that were entirely dependent and guided by computational analyses results or workflows, where a computational method was selectively utilized at a specific point of the process and indicated the subsequent step of the research, which eventually led to the approved drug.

7.2 Rationalizing the Drug Discovery Process: Early Days

CADD is intrinsically based on the rational design of drugs. Rational drug design pertains to the development of drugs with favorable structural characteristics according to the three-dimensional structure of the disease target, which is usually a protein. When the structure of the target is unknown, rational drug design proceeds by examining molecules chemically similar to already known active compounds. The concept of rational drug discovery is not new and does not necessarily require the use of computers. Decades ago, medicinal chemists understood its benefits, long before the first attempts of using computer-modeling techniques in the process. Several examples of the first FDA-approved drugs, which were developed using rational design, illustrate the significant role of the latter in the discovery of potent and efficient drugs.

7.2.1 Captopril (Capoten®)

Angiotensin-converting enzyme (ACE) is a key component of the renin–angiotensin system and a pharmacological target for hypertension [22]. Captopril is the first oral ACE inhibitor and its discovery was considered a breakthrough at that time, not only in management of blood pressure but also because it was one of the first drugs developed with rational drug design [23]. At the time of the discovery, the exact structure of ACE was unknown; but previous studies had indicated the structural similarity of ACE with the pancreatic carboxypeptidase A, for which more structural data was available [24]. In 1973, Byers and Wolfenden identified a potent inhibitor of carboxypeptidase A, the D-benzylsuccinic acid [25]. This data led Ondetti and Cushman to the assumption that the active site of ACE would be similar to that of carboxypeptidase A and that a potent inhibitor of ACE would be also similar to that of D-benzylsuccinic acid. In 1977, they published the results of their study according to which they had developed a theoretical model of the active site of ACE based on that of carboxypeptidase A, concomitantly taking into consideration the nature of the ACE substrate [26, 27]. Specifically, they presumed that the active site of ACE would bear a zinc atom in accordance with the carboxypeptidase A metalloprotein, a positively charged group able to form ionic bonds with the terminal carboxyl groups of the substrates and a group capable of hydrogen bonding to interact with the COOH-terminal amide bond of the substrate. These three features were in agreement with the structure of the D-benzylsuccinic acid inhibitor of carboxypeptidase A, with the only difference that instead of a hydrogen-bonding able group, the inhibitor had a hydrophobic group as the substrate of carboxypeptidase A did. The next step was to modify appropriately this inhibitor in order to better fit to the hypothesized model of ACE. They noticed that ACE releases dipeptides rather than single amino acids, which means that the distance between the zinc atom and the cationic site should be greater than that in carboxypeptidase A. Thus, they replaced succinic acid with a longer succinyl derivative of an aminoacid, succinyl-L-proline. In addition, they replaced the zinc-interacting carboxyl group of D-benzylsuccinic acid with a mercapto group, which significantly increased the potency. Subsequent alterations in the structure of the compound led eventually to captopril. FDA approval came in 1981 and it was marketed by Bristol-Myers Squibb as an anti-hypertensive.

7.2.2 Saquinavir (Invirase®)

Human immunodeficiency virus-1 protease (HIV-1 PR) plays an important role in the replication of the virus, and inhibition of its action can lead to noninfectious HIV particles [28, 29]. Saquinavir, a drug marketed by Hoffmann-La Roche, was discovered on the basis of a rational drug design program initiated with peptide derivatives that were transition-state mimetics of a sequence found in several retroviral substrates [30]. The basic design criterion relied on the observation that HIV-1 PR does not cleave sequences containing dipeptides Tyr–Pro or Phe–Pro. Also, mammalian proteases do not cleave peptide bonds followed by a proline; thus, such inhibitors could be effective binders in the viral enzyme. Because reduced amides and hydroxyethylamine structures most readily accommodate the imino acid moiety of Phe–Pro and Tyr–Pro in retroviral substrates, they were chosen for further studies. Hydroxyethylamine compounds were eventually preferred over the reduced amides due to their higher potency, and several compounds with this characteristic were evaluated in order to determine the minimum sequence required for inhibition. Replacing a proline at the P10 subsite by (*S*,*S*,*S*)-decahydro-isoquinoline-3-carbonyl (DIQ) significantly improved the potency of the inhibitors, resulting in the development of saquinavir. Saquinavir, with a K_i of 0.12 nM against HIV-1 PR, was the first HIV-1 PR inhibitor ever discovered; it received FDA approval in 1995 and was marketed by Roche.

7.2.3 Ritonavir (Norvir®)

Ritonavir is another inhibitor of HIV-1 PR for the development of which a different strategy was followed, other than peptidomimetics. The initial design goal was to take advantage of the C_2-symmetric homodimer structure of HIV-1 PR with a single active site [31]. Starting from the tetrahedral intermediate for cleavage of an asymmetric dipeptide substrate, researchers from Abbott designed pseudosymmetric core diamines by rotating about the C_2 axis that bisects the carbon–nitrogen single bond of the substrate. A lead compound, A-80987, revealed activity against HIV-1 PR and further structure–activity relationship (SAR) studies led to the identification of ABT-538 (ritonavir) as a potent, oral HIV-1 PR inhibitor. Ritonavir was approved by the FDA in 1996 and is marketed by Abbott.

These examples demonstrate the power of rational drug design to deliver novel, potent modulators of protein function as a weapon to fight human disease as an efficient alternative to serendipitous drug discovery. With computer-aided methods, multiple calculations can be performed in a time-efficient way delivering quantitative analyses of the predictions; and advanced computational methods can probe the biological mechanisms in atomic-level detail, thus providing insights into rational design that can further boost the productivity in the drug design pipeline.

7.3 Use of Computer-Aided Methods in the Drug Discovery Process

The proper choice of the most suitable computational technique for a drug discovery program is primarily oriented by the availability of the pharmacological target three-dimensional structure and known active ligands for the target of interest. When the receptor's structure is known, either from experimental methods, e.g. X-ray crystallography, nuclear magnetic resonance (NMR) spectroscopy, or computer-aided predictions, i.e. homology modeling and *de novo* protein design, then drugs can be developed with the goal of perfectly fitting in and interacting with the receptor's binding pocket. This approach is designated as structure-based drug design. When structural data are not available, computational chemists utilize activity data for known active compounds against the protein target by applying ligand-based drug discovery.

7.3.1 Ligand-Based Methods

7.3.1.1 Overlay of Structures

In the late 1980s, and while computational techniques were still at their infancy, visualization tools, able to manipulate the translational and rotational degrees of freedom of the molecules, proved very handy. For a long time after captopril's discovery as an inhibitor of ACE, several peptide antagonists of the angiotensin II octapeptide had been discovered; however, peptides suffer from oral ineffectiveness and short plasma half-lives. This led to the development of **losartan (Cozaar®)**, the first nonpeptide, oral angiotensin II receptor antagonist to reach the market. In 1990, Duncia from DuPont turned his attention to a lead nonpeptide compound, known to be an angiotensin II antagonist [32, 33], and he assumed the compound's low potency could be due to its small structure compared to the endogenous peptide [34]. In order to enlarge its structure, Duncia used computer modeling to align a carboxyl group of the lead compound with the C-terminal carboxylic group of angiotensin II (Figure 7.1). The conformation

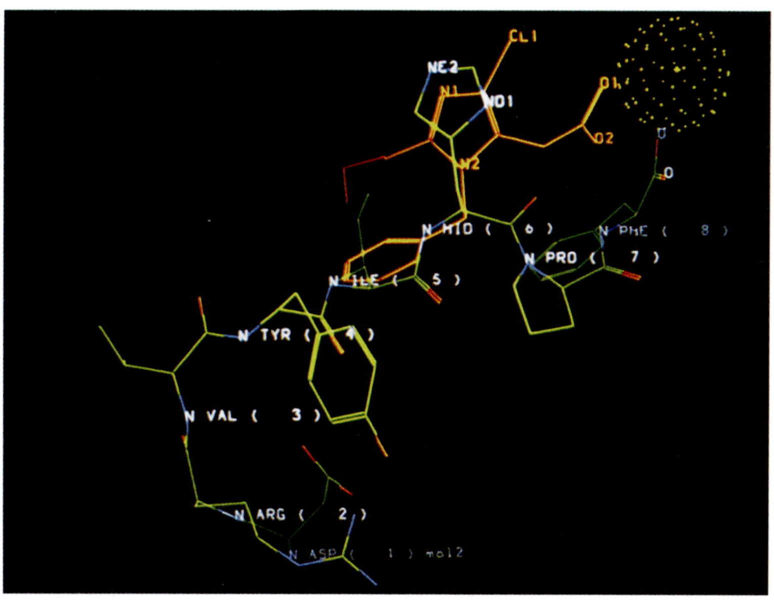

Figure 7.1 Overlap of the lead compound with the C-terminal carboxylic group of angiotensin II as it was manually performed by Duncia et al. Pending permission approval from J. Med. Chem.

of angiotensin II that was used for alignment is reported in Ref. [35] and it is in solution, since there was no available crystal structure of angiotensin II receptor at that time. The alignment was performed in 1982 with crude technology; and, as highlighted by Bhardwaj in his review [36], the overlap was completely manual without any software participating. The alignment indicated the para position of the benzyl group of the lead compound as promising for the extension of the molecule toward the N-terminus of angiotensin II. In this framework, subsequent alterations in the structure of the lead compound resulted in losartan, which was approved in 1995 and marketed by Merck.

Scaffold hopping is a medicinal chemistry technique aiming to change the molecular structure and simultaneously maintain its affinity for a given receptor [37]. This change can be achieved by first determining the molecular features that are important to activity and then searching for molecules or fragments that bear the same characteristics. An early, successful application of this approach was the angiotensin receptor II antagonist **valsartan (Diovan®)** [38]. The main goal was to modify losartan's chemical structure in order to resemble more to the substrate angiotensin II. In order to achieve this, in 1994, Bülmayer et al. created two energy-minimized conformations of losartan and angiotensin II, which were subsequently superimposed. Their initial hypothesis claimed that since the butyl group in losartan mimics the side chain of Ile5 in octapeptide angiotensin II, the imidazole ring could be a substitute for the amide bond between Ile5 and His6. The overlap of the two structures enhanced this assumption and gave them the idea of replacing the imidazole ring of losartan with an aliphatic amino acid, since the amino acid moiety proved to be crucial for activity. These efforts resulted in valsartan, a new antihypertensive drug, which was FDA approved in 2002 and is marketed by Novartis.

Another drug for the discovery of which superposition played a key role is **tirofiban (Aggrastat®)**. Aggregation of platelet-rich thrombus has been associated with arterial vaso-occlusive disorders [39, 40]. In particular, platelets aggregate through binding to fibrinogen protein via the membrane glycoprotein, integrin GPIIb/IIIa [41, 42]. Thus, inhibitors of the protein–protein interaction between fibrinogen and the platelet integrin receptor GPIIb/IIIa can have use as antithrombotic agents. The binding of fibrinogen to GPIIb/IIIa is mediated by two Arg–Gly–Asp (RGD) tripeptide sequences present in fibrinogen, and compounds that possess this sequence have been indicated as effective inhibitors of the fibrinogen–GPIIb/IIIa binding. Tirofiban is a nonpeptide inhibitor of this interaction, designed by Merck to mimic the RGD structure [43]. A previous study with the goal of designing compounds that retained several crucial characteristics of the RGD moiety, such as the amino and carboxylate functionalities separated by a distance of 10–20 Å (based on the length of the RGD sequence), led to the discovery of a lead compound with IC_{50} in the low micromolar range [44, 45]. Subsequent optimization of the lead molecule through a series of SAR studies resulted in the discovery of tirofiban ($IC_{50} = 0.009\,\mu M$). Molecular modeling enabled the overlap of tirofiban to the RGD region of peptide inhibitors, which gave significant insights about their steric and electronic similarities, thus unveiling the origins of the high potency of that compound (Figure 7.2). The molecular modeling was performed using the Merck advanced

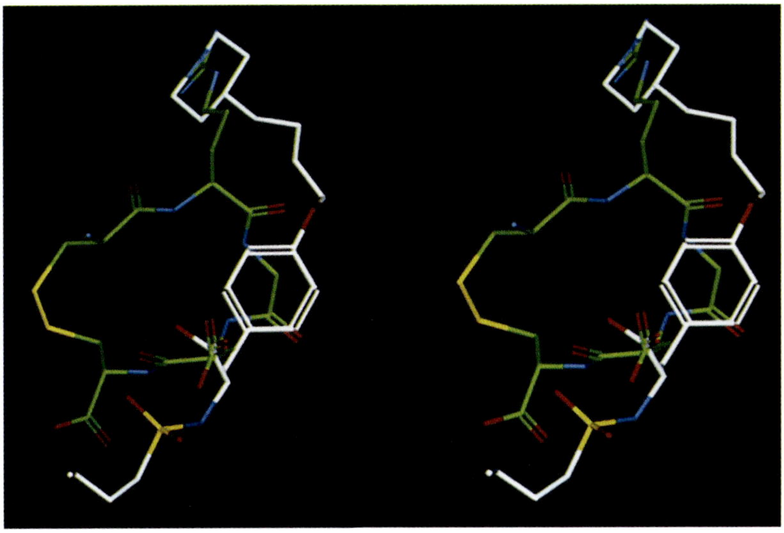

Figure 7.2 Overlap of tirofiban with the RGD region of peptide inhibitors. The piperidinyl and carboxylic acid moieties of tirofiban can substitute for the ionic groups of the Arg and Asp side chains, respectively. Pending permission approval from J. Med. Chem.

modeling facility [46], and the distance geometry algorithm JIGGLE [47] was used to produce aligned pairs of structures. Information gained from molecular modeling provided a rational explanation for the increased affinity and could be useful for the design of inhibitors for several integrin receptors that utilize the RGD sequence for their function. Tirofiban was FDA approved in 1998 and is marketed by Medicure Pharma.

7.3.1.2 Pharmacophore Modeling

The methodologies used for the discovery of the aforementioned cases of losartan, valsartan, and tirofiban can be viewed as the predecessors of pharmacophore modeling. According to Peter Gund, a pharmacophore model is "a set of structural features in a molecule that are recognized at the receptor site and is responsible for that molecule's biological activity" [48]. The main idea is the extraction of common chemical features from 3D structures of ligands known for their binding in a target, which constitute the training set. The two main steps in pharmacophore modeling include, first, performing a conformational search of the dataset ligands and then aligning the multiple conformations of the dataset to the training set in order to determine the pharmacophore features in the 3D space. Pharmacophore features can be hydrogen bond donors or acceptors, cationic, anionic, aromatic, or hydrophobic and the combinations of them. Each feature is usually represented by a sphere, the radius of which determines the tolerance of the deviation from the center of the sphere. There can also be sites of nonexistence of a feature or even excluded volumes [49].

A successful application of pharmacophore modeling is the discovery of **zolmitriptan (Zomig®)**. For years, the vasoactive hormone serotonin 5-hydroxytryptamine (5-HT) has been implicated for migraine, and thus

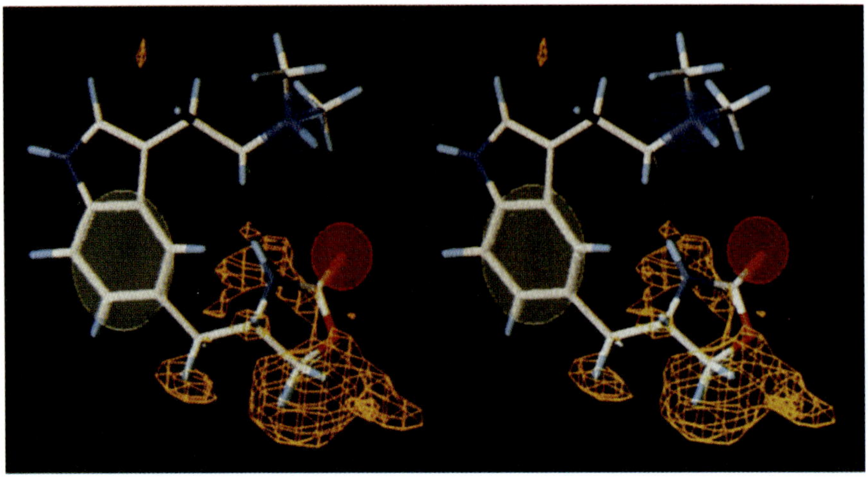

Figure 7.3 Zolmitriptan overlaid on part of the pharmacophore model and the selectivity site. Pending permission approval from J.Med. Chem.

5-HT1 receptors are pharmacological targets for the treatment of this disorder [50, 51]. Zolmitriptan is a 5-HT1 receptor agonist, indicated for the acute treatment of migraine. Zolmitriptan was first discovered by researchers at Wellcome Research Laboratories (now Glaxo Wellcome) [52], but was subsequently licensed to AstraZeneca. Zolmitriptan owes its discovery mainly in the generation of a pharmacophore model of known active molecules [53]. The conformations of the molecules were generated with molecular mechanics calculations using MOPAC-AM1-derived Mulliken charges and semiempirical quantum mechanics calculations using MOPAC-AM1 geometry optimization. The compounds were overlaid using the SYBYL 6.1 molecular modeling package [54], which indicated a pharmacophore hypothesis consistent with affinity and selectivity data. The pharmacophore model consisted of a protonated amine site, an aromatic site, a hydrophobic pocket, and two hydrogen-bonding sites (Figure 7.3). In addition, overlap of the selective and nonselective ligands of the 5-HT2A receptor was conducted in order to calculate a "selectivity site," i.e. a region of space that was occupied by the selective (but not the nonselective) compounds for 5-HT1. Furthermore, a pharmacokinetic optimization study was carried out with *C*log *P* values being calculated with Pomona89 Physico-Chemical Database & MedChem Software [55]. This procedure led to the discovery of Zolmitriptan, which was FDA approved in 2003 and is marketed by AstraZeneca.

7.3.1.3 Quantitative Structure–Activity Relationships (QSAR)

Quantitative structure–activity relationship (QSAR) methods correlate structural characteristics and motifs of compounds with their biological properties, e.g. their affinity for a receptor in a quantitative manner. A major hypothesis in QSAR studies is that the structure of the molecule is responsible for its biological activity and that chemically similar molecules will exhibit similar activities.

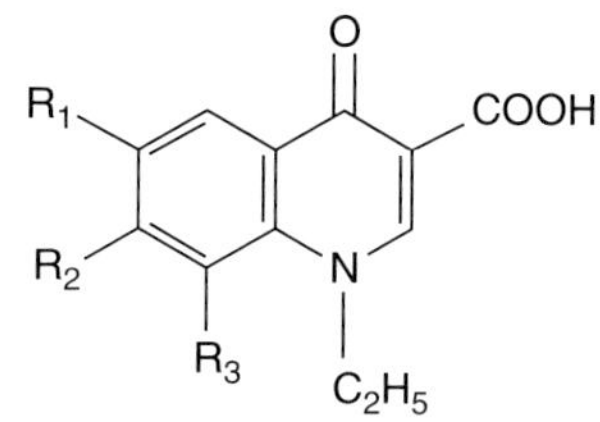

Figure 7.4 General structure of 6-,7- or 8-monosubstituted 1-ethyl-1,4-dihydro-4-oxoquinoline-3-carboxylic acids.

First, a group of ligands with the desired biological activity is determined, and then a quantitative relationship is built between the physicochemical features of the active compounds and the biological activity [56]. The steps in QSAR methodologies are as follows: Initially, compounds with experimentally known biological activity, e.g. IC_{50} values, are identified; these comprise the training set. Second, the most suitable molecular descriptors to which the biological profile can be attributed, e.g. molecular weight, number of hydrogen bond donors, etc., are determined. Then, functions that correlate the molecular descriptors with the biological activity are developed in order to describe the variations in the activity of the training set molecules. Finally, the correlation, i.e. the QSAR model, is tested for its predictive ability with molecules outside the training set.

One of the first QSAR applications in drug discovery is the development of **norfloxacin (Noroxin®)**. Norfloxacin is a fluoroquinolone antibacterial drug discovered by Kyorin Pharmaceutical in Japan in 1980 [57]. Its concept of design is partly attributed to QSARs in 6-, 7-, or 8-monosubstituted compounds of the general structure shown in Figure 7.4, relating antibacterial activity to steric parameters for the groups at position R_1 (Taft's E_s parameter) and R_3 (Verloop's B4 parameter) with a parabolic function. For substituents in position R_2, no relationship had been found, but the piperazinyl group had been shown as promising. Also, use of the Hansch equation indicated that 6,7,8-polysubstituted derivatives of the compound (Figure 7.4) could be more potent than the monosubstituted ones. This led the team to synthesize disubstituted derivatives, which proved successful. Specifically, the QSAR model predicted that a 6-fluoro-7-(1-piperazinyl) derivative would be 10 times more potent than the respective monosubstituted analog. Experimental verification was confirmed with synthesis of this derivative and its *in vitro* assessment, which showed a 16-fold increase in potency. After successful performance in clinical trials, the compound, named norfloxacin, received FDA approval in 1986 and is distributed by Merck.

7.3.2 Structure-Based Methods

The advent of the post-genomic era was accompanied by significant advances in X-ray crystallography [58, 59], NMR spectroscopy [60–62], and cryo-electron microscopy [63] that generated a wealth of three-dimensional structures of pharmacological targets in recent years. Structure-based drug discovery is rooted in the knowledge of the 3D structure of binding pockets of therapeutically relevant proteins for the generation of new chemical entities that bind in specific protein motifs through molecular recognition mechanisms. Molecular graphics tools have enabled the visualization of crystal structures since the 1970s, and

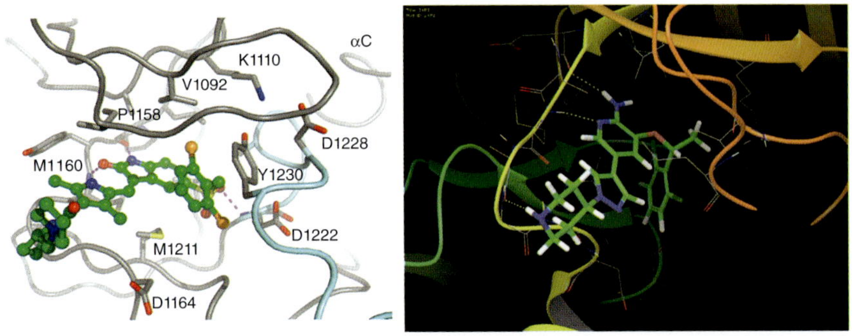

(a) (b)

Figure 7.5 (a) Co-crystal structure of PHA-665752 bound to the kinase domain of c-MET, which generated new hypotheses for optimization efforts. Pending permission approval from J. Med. Chem. (b) Crystal structure of crizotinib bound to ALK.

the design of several approved drugs has been largely influenced by structural ideas derived by visual inspection of crystallographic structures. Two examples of drugs, developed on the basis of visual inspection and detailed structural description of protein–ligand interactions, are crizotinib and nilotinib.

Crizotinib (Xalcori®) is a dual inhibitor of the receptor tyrosine kinase (RTK) c-MET and anaplastic lymphoma kinase (ALK), which was discovered through structure-based drug design [64]. Abnormal c-MET signaling, via, e.g. mutations, is implicated in many tumor processes and especially in metastasis [65]. Moreover, ALK is a drug target responsible for 5% of non-small cell lung cancer (NSCLC) cases and can be oncogenic by forming a fusion gene with several other genes, by gaining additional gene copies or by mutations [66]. Crizotinib was developed following multiple optimization steps of indolin-2-ones, a class of compounds previously described as potent kinase inhibitors. Biochemical kinase assays revealed compound SU11274 as an inhibitor of c-MET [67], which was subsequently optimized to compound PHA-665752 [68]. A co-crystal structure of the latter with c-MET brought to light a new binding conformation of c-MET (Figure 7.5), which guided the design of novel 5-aryl-3-benzyloxy-2-aminopyridine compounds, later optimized to crizotinib [69]. Cellular assays disclosed that crizotinib was also a potent inhibitor of ALK [64], which was approved in 2011 for the treatment of patients with ALK+ or ROS1+ non–small cell lung metastatic cancer.

Similar to the structure-based drug design of crizotinib, **nilotinib (Tasigna®)** was rationally designed on the basis of the crystal structure of imatinib/Bcr-Abl tyrosine kinase complexes [70–72], and therefore it is chemically similar to imatinib, a therapeutic agent for chronic myeloid leukemia (CML) [73]. Nilotinib's design answered to the need for an inhibitor of Bcr-Abl mutant forms resistant to imatinib [70]. In 2004, Manley et al. used co-crystal structures of imatinib with Abl in order to study their interactions and to propose alternative, mainly lipophilic, binding groups for the *N*-methylpiperazine of imatinib, while preserving the amide moiety, which interacted with two residues (Glu286 and Asp381). These efforts led to the discovery of nilotinib [74], which was FDA approved in 2007 for the treatment of patients with Philadelphia chromosome–positive chronic myeloid leukemia (Ph+ CML) in chronic phase.

These examples reflect the significant role of the knowledge of the protein–ligand binding interactions in the drug discovery process. Another example, where visual inspection of a crystal structure (assisted at some point by superposition algorithms) guided therapeutic development, is the case of **indinavir (Crixivan®)**. Indinavir is an inhibitor of HIV-1 protease (HIV-1 PR), which derived as a hybrid of a lead peptidomimetic compound, previously identified by Merck [75, 76], and a compound discovered by Hoffmann-La Roche displaying better solubility and bioavailability [30, 77]. In order to test the potential efficiency of the hybrid compound, Merck used computer-aided modeling to overlap the crystal structure of the Hoffmann-La Roche compound with an energy-minimized conformation of the lead compound [78]. The minimization was performed using the advanced modeling facility of Merck [46] and the OPTIMOL force field also developed by Merck (unpublished work on the development of OPTIMOL). The overlap verified the potential interactions that could be formed between the lead-derived part of the hybrid molecule and the protein, thus giving the green light for its synthesis. *In vitro* experiments indicated the high potency of the hybrid compound (IC_{50} = 7.6 nM); however, the inhibition of the spread of viral infection in MT4 human T-lymphoid cells proved to be weak. This led to further structural comparisons with the lead compound and other analogs, resulting in the development of indinavir. The co-crystallized structure of indinavir with the HIV-PR verified the modeling observations and indinavir received FDA approval in 1996 for the treatment of HIV (Merck).

While visual inspection is a necessary step in structure-based drug design, it limits the chemical space to be explored. Computer-based methods encoding the fundamentals of molecular interactions and using cheminformatics or statistical mechanics enable accessing a much larger number of possible new ligands. Next, the methods that influenced the design process of approved drugs are discussed: virtual screening, flexible molecular docking, molecular dynamics (MD) simulations, *de novo* drug design, and protein homology modeling.

7.3.2.1 Molecular Docking – Virtual Screening

Virtual screening avoids the problem of broad searches of chemical space by restricting itself to libraries of specific, accessible compounds. Virtual screening is a knowledge-driven approach for molecular design that searches compound databases to discover novel small molecule binders of a drug target [79–81]. In virtual screening, we computationally screen chemical libraries in order to predict the structure of the protein–ligand complex (docking), and rank the resulting complexes based on their predicted free energy of binding (scoring). Docking utilizes conformational search methods to explore ligand conformational space: (i) Systematic methods, which place ligands in the predicted binding site after considering all degrees of freedom; (ii) random or stochastic torsional searches about rotatable bonds, such as Monte Carlo (MC) and genetic algorithms to "evolve" new low energy conformers; and (iii) MD simulation methods and energy minimization for exploring the energy landscape of a molecule [82].

Scoring functions that predict protein–ligand energetics can be categorized in (i) force field–based functions, which approximate the binding free energy by calculating the potential energy of binding using molecular mechanics force fields as well as take into account solvation and entropy contributions; (ii) empirical scoring functions that are based on providing a score for various types of intermolecular interactions in the ligand–protein complex (e.g. hydrophobic contacts, number of hydrogen bonds, number of rotatable bonds); these are parameterized on the basis of experimental data; and (iii) knowledge-based scoring functions that use statistical observations of intermolecular contacts from receptor–ligand complexes with known conformations [83, 84].

A characteristic example of successful molecular docking application is the design of **rilpivirine (Edurant®)**. Rilpivirine is an inhibitor of HIV-1 RT, for the discovery of which molecular modeling played a key role. In 1996, molecular modeling studies suggested the replacement of the central aminotriazine ring of a diaryltriazine (DATA) compound, R106168, which had been a known inhibitor at that time [85]. This work resulted in the diarylpyrimidine (DAPY) compound, TMC120, and follow-up medicinal chemistry, crystallography, and molecular modeling studies led to the discovery of rilpivirine, a cyanovinyl DAPY compound [86]. The computational studies that led to the DAPY compounds involved docking of ligands into the non-nucleoside reverse-transcriptase inhibitor (NNRTI) binding site of reverse transcriptase (RT) and minimization of the protein–ligand complex (Figure 7.6). The initial conformations of the

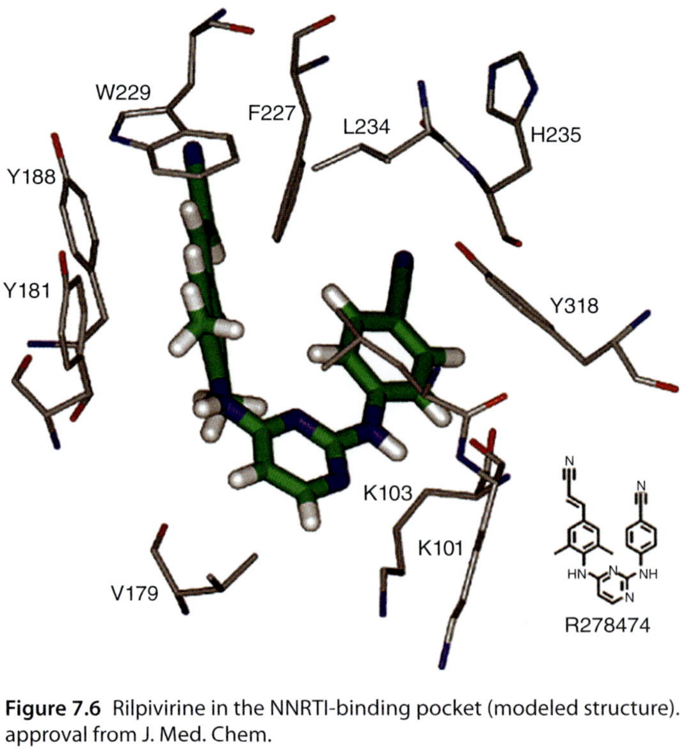

Figure 7.6 Rilpivirine in the NNRTI-binding pocket (modeled structure). Pending permission approval from J. Med. Chem.

compounds were generated using a genetic algorithm; these were subsequently docked to the HIV RT binding site, which was first kept rigid and then flexible in the next steps of the calculations. Optimization of the final docked structures was achieved with simulated annealing followed by a local minimization algorithm. The binding energy of each ligand was computed with a scoring function obtained by a molecular mechanics force field developed at the Center for Molecular Design from the MMF94 force field [87]. This computational assessment followed by the necessary experimental work resulted in the launch of rilpivirine, which was FDA approved in 2011 against HIV.

Grazoprevir (Zepatier®) is an NS3/4a protease inhibitor developed for the treatment of hepatitis C virus (HCV). Its design was mainly guided by a molecular modeling/docking-derived strategy. One of the early discovered NS3/4a protease inhibitors was BILN-2061 [88, 89]. BILN-2061 binding mode was unveiled through its co-crystallization with the 1-180 protease domain of NS3 protease [90], which, however, could not provide enough information about the interactions of the P2 thiazolyl-quinoline moiety, due to the absence of the helicase domain. Merck researchers sought to answer this by modeling the structure of BILN-2061 in complex with the full length NS3/4a protein including the helicase domain (Figure 7.7) [91]. The full-length NS3/4a modeling was achieved by merging the structure of BILN-2061, in its binding mode conformation from the previously released crystal structure [90], with a published apo-structure of the enzyme (PDB ID: 1CU1) [92], since at that time there were no full length structures with inhibitors bound. Overlapping this structure with the BILN-2061 structure, residues of the helicase domain C-terminus, which extended inside the active site, were trimmed. Examination of the resulting model led to the observation that the helicase domain offers a pocket for the binding of the P2

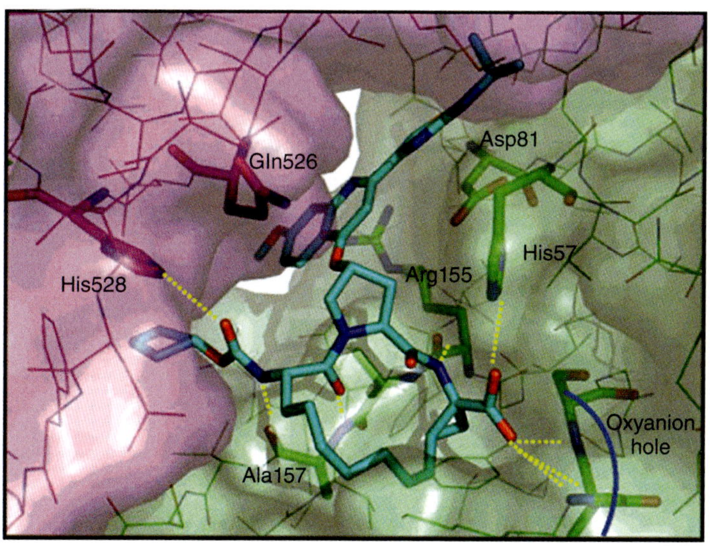

Figure 7.7 Model of BILN-2061 (cyan) bound to full length NS3/4A (protease, green; helicase, purple) with key protein–inhibitor interactions shown. Pending permission approval from J. Am. Chem. Soc.

thiazolyl-quinoline portion of BILN-2061. Also, it was observed that there is space for a linker between the P4 carbamate cyclopentane and P2 quinoline ring of BILN-2061, at the region where the truncated Glu628 residue of the helicase domain used to be, suggesting a new type of P4–P2 macrocyclization. Thus, several BILN-2061 derivatives, with different P4–P2 linker lengths, were designed and modeled accordingly to the BILN-2061 bound to the full length NS3/4a as discussed. Specifically, the derivative's common part with BILN-2061 structure adopted the same binding mode as BILN-2061, while the novel macrocyclic chain underwent a conformational search using a distance geometry algorithm [93], from which the lowest energy pose was retained. The poses were subsequently energy minimized, using the Merck molecular force fields (MMFFs) [87] keeping rigid the active site residues, and scored using X-score [94]. Scores predicted that 5- and 6-carbon linkers would be the most potent, which proved correct after organic synthesis and further assaying. Follow-up optimization steps replaced a carboxylic acid by a cyclopropylacyl-sulfonamide and an *n*-butyl with a *tert*-butyl, leading to increased potency and enhanced liver exposure, respectively. Further modeling studies took place, with the aim of generating inhibitors with activity against both 3a and 1b genotypes [95]. Thus, they focused on enlarging the P2 heterocycle with bulky substituents or fused-ring analogs. These were then docked to the aforementioned model of apo-structure, after manual adjustment of several side chains in order to accommodate the larger P2 group. Examination of the docked poses with respect to the residue mutations in the different genotypes created the hypothesis that increasing the linker flexibility could improve the activity against the 3a mutant, which was experimentally verified. Further optimization studies for improved liver exposure and enzyme activity led to compound MK-5172 (grazoprevir), which successfully passed clinical trials and was approved in 2016 against hepatitis C.

A case in which molecular docking guided the discovery of a drug is that of **betrixaban (Bevyxxa®)**. Inhibition of serine protease factor Xa (fXa) constitutes a treatment for severe cardiovascular diseases [96, 97]. Millennium Pharmaceuticals had already reported the discovery of an anthranilamide-based compound as a potent fXa inhibitor [98]. To further improve this compound's *in vitro* fXa and anticoagulant activity substitutions on the rings of the compound were added and a compound with increased potency was discovered. The binding mode of this lead compound was examined with docking calculation, using the GOLD software [99], a random search algorithm, in tandem with published and not fXa crystal structures [100]. The crucial interactions of the compound with the protein residues were monitored and an adjacent small hydrophobic pocket suggested that a small hydrophobic substituent in a specific position of the molecule could boost its affinity with the receptor. This hypothesis was confirmed by the synthesized analogs which had high affinity with fXa. Further modifications of the compound, based on observations from the docking results, as well as SAR and pharmacokinetic studies enabled the development of betrixaban which was FDA approved in 2017 for the prophylaxis of venous thromboembolism (VTE) in adult patients hospitalized for an acute medical illness, who are at risk for thromboembolic complications due to moderate or severe restricted mobility.

7.3.2.2 Flexible Receptor Molecular Docking

Conventional docking algorithms consider proteins as rigid structures, despite the fact that proteins are dynamic entities with internal motions that can undergo conformational changes upon ligand binding. More recent advances in molecular docking have led to algorithms that incorporate account receptor flexibility, either by implementing an induced fit–based method [101], or by docking in conformational ensembles, acquired via MD and MC simulations, or experimental ensembles from NMR [102]. ICM docking [103] uses induced fit and was used for the discovery of **Vaborbactam (Vabomere®)**, a β-lactamase enzyme inhibitor. Inhibitors of the β-lactamase enzymes aid the conventional β-lactam antibiotics to retrieve their initial effectiveness against resistant gram-negative bacteria. Previous reports of a potent boronic acid inhibitor [104] impelled researchers from Rempex Pharmaceuticals to design cyclic transition state mimetic compounds with a cyclic boronate ester moiety [105]. They expected that the cyclic boronates could have enhanced selectivity toward β-lactamases with respect to serine hyrdrolases, known to have linear substrates. Molecular docking studies of the designed molecules to β-lactamase enzymes from classes A, C, and D, proposed a cyclic boronate with a truncated benzo ring as the one with the higher affinity with the receptor. Specifically, the docking calculations were performed using the ICM docking module [103], which allows rotations of the ligand and the active site side chains. First, a random conformational change in the ligand takes place, followed by an energy minimization of the ligand and the side chains. Next, the surface-based solvation energy and entropy are calculated and then the next conformational change happens according to the Metropolis criterion. Additional evidence, in favor of the proposed inhibitor, was found with the examination of an available crystal structure of the Michaelis substrate with a mutant enzyme, which revealed that the putative inhibitor could reproduce important substrate–enzyme interactions. Analogs of this lead compound, designed from SAR studies, were tested *in vitro* and, eventually, the RPX-7009 (vaborbactam) compound was found. Vabomere was FDA approved on 29 August 2017 for complicated urinary tract infections (cUTI), including a type of kidney infection, and pyelonephritis, caused by specific bacteria.

7.3.2.3 Molecular Dynamics Simulations

MD simulations study the time-dependent behavior of a molecular system, by integrating Newton's law of motion. Starting from a molecular structure (derived by experimental or computational data), forces (arising from interactions between atoms) are calculated with the engagement of a force field. The latter models covalent bonds and atomic angles with springs, and dihedral angles (proper and improper) with sinusoidal functions. As far as the nonbonded interactions are concerned, the van der Waals forces are described by a Lennard-Jones potential and the electrostatic interactions by the Coulomb's law. The main advantage of MD simulations over other computational techniques, such as molecular docking, is that it allows for system flexibility by generating successive configurations of the evolving system that are combined into a trajectory. Thus, a dynamical system is utilized as opposed to the static structures usually used in docking. Moreover, the information of the microscopic system properties can be translated to macroscopic properties via statistical mechanics theory. In this way, properties such as Gibb's free energy, enthalpy, and entropy can be calculated, which can offer further energetic insights into drug binding. Today, MD simulations have become readily available due to advances in computer power with the advent of GPU cards, and by virtue of more sophisticated and efficient algorithms that properly take advantage of the new hardware, and advances in force fields and enhanced sampling methods [106–109].

Amprenavir (Agenerase®) is an HIV-1 protease inhibitor discovered by structure-based drug design by Vertex Pharmaceuticals and later co-developed and marketed by GSK and Kissei (Japan) [110]. In detail, MD calculations were carried out in order to explain the experimental observation that the P1′ amide NH of substrate sequences was not obligatory for binding and productive catalysis. The results showed that the amide formed a very weak hydrogen bond with the enzyme. The choice of the *N*,*N*-dialkyl sulfonamide moiety (supported by modeling studies) played an important role in the affinity of the ligand as it was intended to bind to a conserved water molecule, which bridges the ligand and the "flap" region of the protease, i.e. the region of the protein, which changes conformation in order for the substrate to access the active site. Also, the *N*,*N*-dialkyl sulfonamide moiety will act as a scaffold for the P1′ and P2′ groups. The Cambridge Structural Database was searched to provide likely low-energy conformations of the *N*,*N*-dialkyl sulfonamide that would bind to the enzyme. The authors report that "when representative compounds were co-crystallized with the enzyme, the bound conformation of the inhibitor backbones were substantially similar to those suggested by computational analyses." Furthermore, "good hydrogen bond distances between the conserved water molecule and the sulfonamide oxygens were observed in all cases, supporting our modeling prediction." Amprenavir was approved in 1999 as a protease inhibitor used to treat HIV infection and was marketed by GlaxoSmithKline until its discontinuation in 2004; it is now sold in its prodrug version (fosamprenavir) by ViiV Healthcare.

7.3.2.4 *De Novo* Drug Design

The concept of *de novo* drug design is based on incremental construction of a ligand in a protein-active or allosteric site. Several computer-assisted *de novo* algorithms have been developed since 1989 [111–113], with most of them relying on molecular fragments rather than atom-by-atom construction, due to the high uncertainty for a feasible synthesis of entirely novel chemical entities [114]. The fragment-based drug design process initiates with virtual screening of a library of fragments (MW < 300) with high diversity of druglike chemical space, and subsequently the linking of the most favorably positioned within the protein-binding site, with molecular spacers to create complete molecules [115]. The prediction of sites, where the new moieties can be found, varies from hydrogen bonding–only regions [116], to grids of points, where groups or fragments can be placed according to their interaction energies [117]. How strong a fragment is expected to bind is determined by the scoring function, similarly to molecular docking.

Nelfinavir (Viracept®), the first non-peptidomimetic HIV-1 PR inhibitor to receive FDA approval (1997) was developed with *de novo* drug design. Previous attempts had identified the AG-1002 and AG-1004 inhibitors, which integrated statin isosteres instead of peptide bonds [118, 119]. Moreover, the extensive analysis of peptidic inhibitors' crystal structures led to the replacement of the peptide analog amide parts with nonpeptidic substituents [120], in order to increase their oral bioavailability. Under this concept, Agouron Pharmaceuticals utilized the MCDNLG (Monte Carlo *de novo* ligand generator) to extend the structure of a nonpeptidic inhibitor in two unexplored pockets (S1–S3 and S1′–S3′ regions) [121]. During the calculations, a "super molecule" consisting of atoms that can exceed conventional chemical bond valences evolves to a low-energy molecule through Metropolis Monte Carlo and simulated annealing protocols. Simulated annealing is a technique where the system is initially heated and then gradually cooled, which enables the exploration of a large conformational space even at low temperatures [122]. The evolution is based on the interactions of the atoms with other atoms in the evolving ligand and with the protein residues. This search led to the identification of compounds with dimethylbenzyl and cyclopentyl amides as the most potent. Additional chemical synthesis attempts resulted in the discovery of nelfinavir [123]. Nelfinavir was approved by the FDA in 1997 for the treatment of HIV infection and is marketed by Hoffmann-La Roche.

Applying *de novo* inhibitor design in combination with MD simulations led to the discovery of **Zanamivir (Relenza®)**. Zanamivir was a first-in-class neuraminidase (sialidase) inhibitor targeting the influenza virus. In 1993, von Itzstein et al. published the design strategy of Zanamivir [124]. The CSIRO Division of Biomolecular Engineering had solved the crystal structure of sialidase, an influenza virus surface protein, bound with an unsaturated sialic acid analog, Neu5Ac2en, which was suggested to mimic the transition state structure of sialic acid, the product of the enzyme-catalyzed reaction [125]. Subsequently, refinement of the crystal structure using MD simulations, and specifically simulated annealing [126], verified this hypothesis [127]. The refined structure was used by von Itzstein et al. [124], who identified probable interaction sites between specific groups – referred to as probes – and the protein cavity, with the aid of GRID software [117]. The probes can be water molecules, methyl groups, amine nitrogens, carboxy oxygens, and hydroxyl groups. Each probe is successively placed in different positions inside the protein pocket and the potential energy of the probe is calculated in each of these positions, thus indicating the most favorable regions for the binding of the probe. A prediction of the most favorable substitutions in Neu5Ac2en was conducted according to this framework, and a hydroxyl group replacement by an amino group was proposed, with the aim of forming a salt bridge with the neighboring Glu119 side chain [124]. This interaction would be reinforced by replacement of the hydroxyl group with the more basic guanidino group; hence, the 4-guanidino compound (zanamivir) was synthesized and proved a potent inhibitor of the enzyme. Zanamivir was approved by the FDA in 1999 for the treatment of uncomplicated acute illness due to influenza A and B virus and is marketed by GlaxoSmithKline.

7.3.2.5 Protein Structure Prediction

Computer-aided prediction of a protein 3D structure can be invaluable for drug design projects, where protein structural information is not available. The most widely used methodology for protein structure prediction is homology modeling, which may be applied to structure-based drug design projects [128–130]. Homology modeling is based on the alignment of the amino acid sequence of the target protein to the 3D structure of an evolutionary-related, homologous protein of over 30% sequence similarity, which has been determined experimentally. The main steps followed in homology modeling, are, first, the selection of the template according to its sequence similarity with the target protein; second, the alignment of target protein sequence to the template's structure; and, finally, the construction of the model and its evaluation and refinement. Despite the highly accurate predictions that homology modeling is capable of [131], it has a basic limitation; it cannot predict new folds. This limitation can be addressed with *ab initio*, or *de novo* protein structure prediction, according to which the 3D structure of a protein is solely derived from its amino acid sequence, without any supplementary information from proteins with known structures [132–134]. This is achieved by searching for the global free energy minimum, which corresponds to the native structure of a protein using potential energy functions. For the *ab initio* protein structure prediction, a geometric representation of the protein chain, a force field, and an energy surface searching technique, such as MD and MC simulations, are required. The main limitation of *ab initio* methodologies is the hurdle of a sufficiently long simulation to sample all the conformational phase space and thus the uncertainty that the predicted conformation corresponds to the global minimum [135]. Apart from 3D structure prediction, secondary structure prediction is also possible via algorithms that examine the amino acid sequence. The Chou–Fasman method was one of the first discovered secondary structure prediction algorithms, using parameters derived from the few protein structures experimentally determined in the 1970s [136].

A drug discovery mainly attributed to homology modeling is that of **Aliskiren (Tekturna® and Rasilez®)**. Aliskiren was approved in 2007 and is a third-generation renin inhibitor, i.e. a non-peptide retaining the favorable for binding steric and electronic characteristics of the amides. At the time of the discovery, a second-generation inhibitor (CGP38560) had been found, but it, however, suffered low oral absorption and rapid biliary excretion. Researchers from Ciba-Geigy (now Novartis) used CGP38560 to generate a new molecule that would mimic its bioactive conformation. Since there were no crystal structures available, a homology model of human renin was constructed on the basis of a sequence derived from that of the gene and the 3D structures of other homologous aspartic proteinases available [137]. Subsequently, docking of CGP38560 to the homology model was performed, thus retrieving a prediction of the bioactive conformation of CGP38560 (Figure 7.8). Visualization of the conformation led the scientists to decide which features of CGP38560 should be retained and which should be removed or replaced by other scaffolds. Eventually, *in vitro* and *in vivo* assessment of the new scaffolds led to the discovery of aliskiren, used to treat hypertension, marketed by Novartis [138–140].

The RTK family of proteins has been associated with tumor progression in cases of mutations, ectopic receptors, or ligand expression [141, 142]. **Sunitinib (Sutent®)** is an inhibitor of the vascular endothelial growth factor receptor (VEGFR) from the RTK family, which was discovered with the assistance of

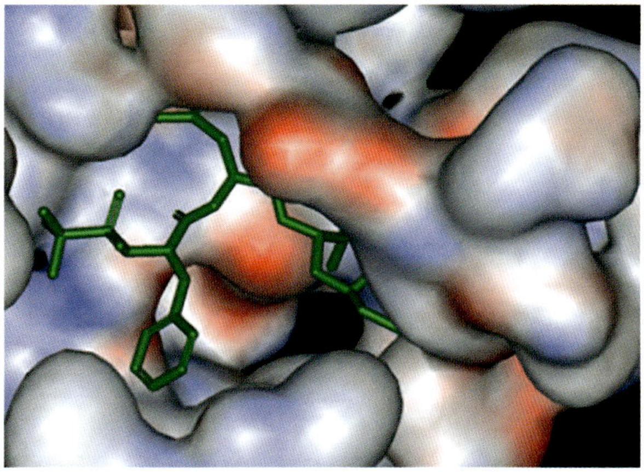

Figure 7.8 3D model of the enzyme and example of docking of CGP38560. Pending permission approval from Chem. Biol. Drug. Des.

homology modeling and docking studies. The indolin-2-one chemotype of the inhibitor was initially identified with HTS experiments [143], with three hits being promising while subsequent *in vitro* studies led to compound SU5416 [144]. However, poor pharmacokinetics and solubility were an impediment to the compound's further development. Further attempts to ensure both desirable pharmaceutical properties and an expanded RTK target profile used information deduced from crystal structures and homology models. Examination of previous analog crystal structures revealed that enhanced potency could be achieved by distorting the nucleotide-binding loop upon ligand binding, thus leading to new hypotheses concerning the ligand structures. In this framework, inhibitor SU6668 arose, which was able to inhibit both VEGFRs and PDGFRs (platelet-derived growth factor receptors) [145]. Also, a homology model built for PDGFR catalytic domain with SU6668 docked into it explained the high affinity of the compound for the receptor. Efforts to broaden the kinase selectivity spectrum resulted eventually in the discovery of SU11248 (sunitinib) [146] by SUGEN scientists, which received FDA approval in 2006 for advanced kidney cancer, a rare type of stomach cancer called gastrointestinal stromal tumor (GIST), and pancreatic neuroendocrine tumors, and is currently marketed by Pfizer.

Brigatinib (Alunbrig®) is another approved drug which was discovered using homology modeling. It is an inhibitor of ALK, an oncogenic drug target responsible for the 5% of NSCLCs as described earlier [66]. ARIAD Pharmaceuticals (now a wholly owned subsidiary of Takeda Pharmaceuticals) examined a variety of chemical scaffolds as kinase inhibitors, starting from the ceritinib inhibitor [147]. They experimentally tested a library of substituted 2-anilinopyrimidine compounds and they expected that substituents in the C2, C4, and C5 positions of the pyrimidine core would increase potency. Incorporation of a dimethylphosphine oxine (DMPO) moiety enhanced ALK activity [148]. Due to absence of an ALK structure at that time, construction of the homology model of the kinase based on the crystal structure of the activated insulin kinase (PDB ID: 1IR3), using Prime of Schrödinger [149–151], revealed the potential formation of a hydrogen bond between a substituent and the

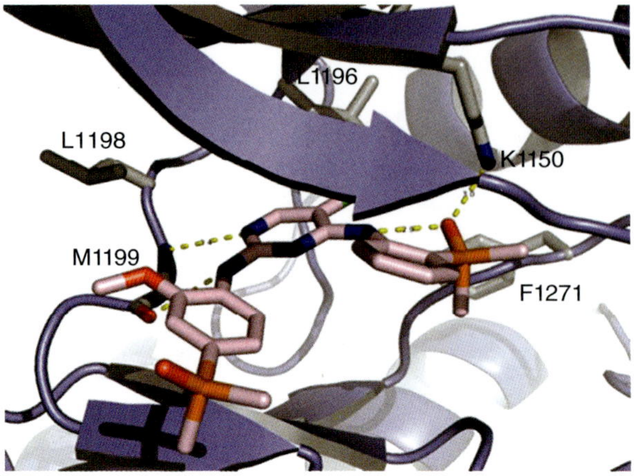

Figure 7.9 Homology model of a substituted 2-anilinopyrimidine compound bound to ALK. Dotted lines indicate hydrogen bond interactions. Pending permission approval from J. Med. Chem.

Lys-NH (K1150) on the protein (Figure 7.9). This finding urged the researchers to design a molecule with a DMPO substituent but at a different position this time, which was docked to the homology model, using Glide SP of Schrödinger [152–154], for visual inspection of the binding. The compound was synthesized and *in vitro* assays displayed improved selectivity. SAR studies followed, leading to the discovery of brigatinib and its FDA approval in 2017, for the treatment of patients with metastatic ALK-positive NSCLC who have progressed on or are intolerant to crizotinib. It is worth noting that the binding pocket of brigatinib is the same as that of crizotinib.

The discovery of **enfuvirtide (Fuzeon®)** was based on the prediction of the secondary structure of an HIV protein. The need for curative strategies for the treatment of the HIV turned the attention to inhibitors of the HIV replication. In 1987, Gallaher et al. used hydropathy plots (i.e. plots that determine the degree of hydrophobicity in a protein), sequence homology, and algorithms for the prediction of the protein structure to find similarities in the sequence of the transmembrane (TM) protein of retroviruses with other closely related branches of the virus family [155]. A highly conserved region, which was predicted to form an extended amphipathic α-helix according to Chou–Fasman algorithms, was thus discovered [136]. This prediction, along with findings which indicate that TM has a "leucine zipper" motif [156] that can be modeled by synthetic peptides [157], led Wild et al., in 1992, to the synthesis of peptides with antiviral *in vitro* activity [158]. They discovered a potent peptide fusion inhibitor, DP-107, and later the highly potent DP-178 (enfuvirtide) [159, 160], which gained FDA approval in 2003 for the treatment of HIV-1 infection in treatment-experienced patients.

7.3.2.6 Rucaparib (Zepatier®)

Poly(ADP-ribose) polymerase (PARP) constitutes a potential anticancer target because its inhibition can lead to increased cytotoxicity during radiation and treatment with monofunctional alkylating agents that are used in cancer therapy [161, 162]. Early conventional nicotinamide-based inhibitors although potent, exhibited inefficient specificity and pharmacokinetic profiles [163]. A solution to this problem was the discovery of a novel tricyclic PARP-1 inhibitor [164], further optimization of which, based on modeling studies, resulted in the structure of [5,6,7]-tricyclic indole lactam which had high affinity with the receptor. An additional 2-phenyl group able to interact with Tyr907 and Tyr896 led to increased potency, as the modeling observations suggested; however, no further information is provided by the authors as to the exact modeling techniques that were used [163]. Several medicinal chemistry optimizations of the structure led to rucaparib [165] developed by Clovis Oncology Inc., which was approved in 2016 for the treatment of patients with deleterious BRCA mutation (germline and/or somatic) associated advanced ovarian cancer who have been treated with two or more chemotherapies.

7.3.3 *Ab Initio* Quantum Chemical Methods

Ab initio quantum chemical calculations can provide the most accurate representations of molecular structure. Using quantum chemistry one attempts to solve the electronic Schrödinger equation, by calculating the electronic energy and density when the atomic nuclear coordinates and the number of electrons of the system are known. The exact solution of the electronic Schrödinger equation though is computationally intractable mainly because of the large number of the electrons; however, several approximate methodologies that converge to the exact solution have been developed [166]. For example, the Hartree–Fock approximation method [167] is used to model the N-body wave function as a single Slater determinant. Another quantum chemical approach is the density functional theory (DFT), according to which the total energy of the system is a function of the electron density [168–170]. This method is broadly used because of its advantage of depending only on three coordinates (and not on $3N$ coordinates for N electrons), making it attractive for computational implementation [166].

Dorzolamide (Trusopt®) is a carbonic anhydrase (CA) II inhibitor for the treatment of glaucoma, for the discovery of which *ab initio* calculations played a catalytic role [171]. Examination of the X-ray crystal structure of CA II revealed that the binding pocket is cone shaped, with one hydrophobic and one hydrophilic area. Thus, a general inhibitor model was built exhibiting complementarity to these structural features of the cavity. This model is exemplified by MK-927 inhibitor of CA II (Figure 7.10). Compound MK-927 has two enantiomers, with the S-enantiomer being 100-fold more potent than the R-enantiomer as determined by functional enzymatic and competition assays. In both enantiomers, the sulfonamide group is coordinated to the zinc atom of the active site. Crystal structures for both enantiomers bound to CA II were solved, indicating a difference in the binding mode that could possibly explain the difference in their potency. In order to investigate the reasons for which the conformation of the S-enantiomer is favored over that of the R-enantiomer, *ab initio* calculations at the 6-31G*, using the Gaussian 88 package, level were performed, which indicated that the preferred dihedral angle formed by atoms N-S-C-S in MK-927 structure is 72°. However, in the S-enantiomer, this angle is 150° and in the R-enantiomer 170°. This partly explains why S-enantiomer is favored over the R, although it is still far from ideal. Moreover, supplementary *ab initio*

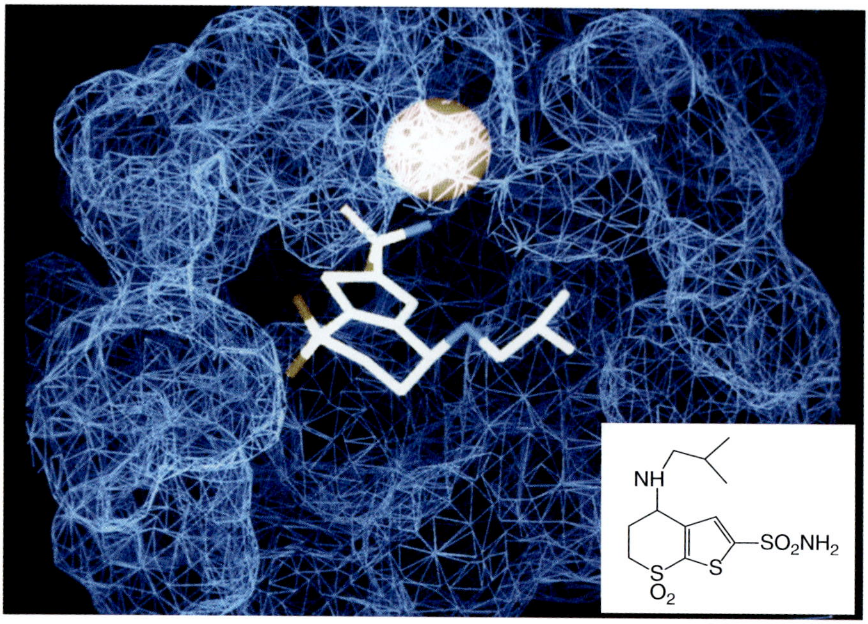

Figure 7.10 Crystal structure of MK-927 (which combines a lipophilic (propyl side chain) and hydrophilic (SO_2 and SO_2NH_2) part) in the active site of CA II. The zinc atom of the active site is shown in van der Walls representation. Inset: Molecular structure of MK-927. Pending permission approval from J. Med. Chem.

calculations suggested that the optimal conformation for the 4-isobutylamino side chain of MK-927 is the trans, as in the *S*-enantiomer. This further supports the energetically preferred structure of *S*. Subsequent *ab initio* calculations suggested that a methyl substitution in the 6-position of the initial compound could decrease the pseudoaxial conformation of the isobutylamine and consequently decrease the energy penalty paid during the binding to the enzyme. Therefore, the methyl group was introduced and, additionally, the isobutylamine was replaced by an ethylamine to compensate for the increased lipophilicity. Of the four possible diastereomers of the resulting compound, the *trans*-(*S*,*S*) form (dorzolamide) was preferred, having a K_i value of 0.37 nM. Dorzolamide was discovered at Merck and approved by the FDA in 1994 to be used in ophthalmic solutions to lower intraocular pressure (IOP) in open-angle glaucoma and ocular hypertension.

7.4 Future Outlook

The aforementioned case studies outline the substantial contribution of computational drug discovery techniques to the discovery of novel drugs in the market. In most cases, an important aspect in such projects is the estimation of the free energy of binding of small molecules into protein targets. The knowledge of this thermodynamic property can provide insight in (i) the phenomena that govern ligand binding, which is the case of protein and ligand conformational transformations occurring upon binding, (ii) the features, structural or conformational, that constitute a molecule potent or weak binder, as well as (iii) the existence of other factors affecting the binding procedure, such as water molecules. Experimental techniques able to determine the binding free energy of a compound into a protein structure of choice have been known for years, with isothermal titration calorimetry (ITC) and identification of k_D and IC_{50} values through binding assays constituting the most widespread of them. Nevertheless, the prospective knowledge of a molecule's binding free energy, prior to its synthesis, can be of even greater value to medicinal chemists by allowing them to discriminate the most potent compounds among a library, thus saving valuable time and resources that would have been otherwise allocated in the synthesis and *in vitro* testing of all library compounds. Carrying on in this framework, it can be deduced from the abovementioned statement that apart from the binding free energy of each molecule to a protein (absolute binding free energy), perhaps of more interest can be the relative binding free energies of the library molecules with respect to the target, highly helpful in lead optimization projects, where the goal is the determination of the relative binding affinity of newly designed ligands with respect to a reference one. Hence, the knowledge of the absolute binding free energy of only one compound, e.g. from experimental data, can lead to the determination of all others. The major advantage of predicting relative binding affinities, instead of absolute binding free energies, is the significant gain in the computational time. This primarily occurs by the introduction of a thermodynamic cycle in methodologies treating the former case, which enables bypassing the modeling of the whole binding process (which accounts for several microseconds), by alchemically transforming one molecule to another. This technique characterizes alchemical free energy methods, such as the free energy perturbation (FEP) method, the thermodynamic integration (TI) method, the Poisson–Boltzmann surface area (PBSA), and the generalized Born surface area (GBSA). The FEP principles are documented here as a representative example.

The FEP thermodynamic cycle in Figure 7.11 illustrates that the calculation of the binding free energy in paths 1 and 2 can be avoided by mutating the one ligand to the other in the solution and the complex phase, as shown in paths A and B, respectively. Due to the fact that free energy is a thermodynamic property, summation along the cycle equals to zero. Therefore, the difference in free energy between paths 1 and 2 can be computed by the difference in free energy between paths A and B. If this free energy difference is negative, then ligand B is expected to be more potent than A.

The transformation of one molecule to the other is performed gradually through intermediate nonphysical states, usually called lambda (λ) windows. The λ schedule is used to increase the overlap in the regions of phase space that the two states (that of potential energy U_A and the one of potential energy U_B) explore. This is important because large errors can be incurred during the estimation of the binding affinities if the potential energies U_A and U_B are significantly dissimilar. To overcome this problem, the free energy difference is divided into a series of small steps, which correspond to alchemical intermediate states, and during which the potential energy of the initial state is gradually

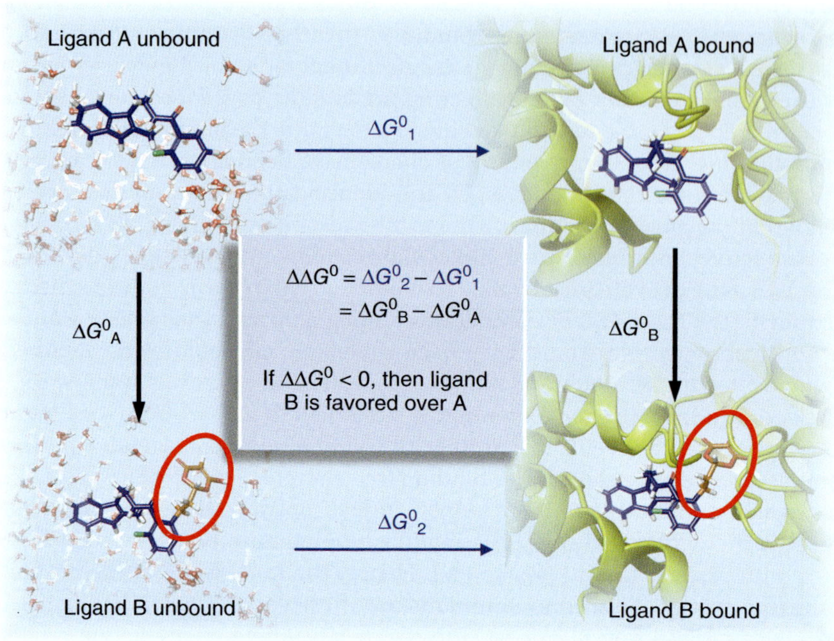

Figure 7.11 FEP thermodynamic cycle.

transformed to the potential energy of the final state. For this, a coupling parameter λ is typically used:

$$U = (1 - \lambda)U_A + \lambda U_B \tag{7.1}$$

FEP calculations are inextricably linked with the generation of configurational ensembles, either by MD or MC simulations, which allow for the microscopic properties of the systems to be interpreted to the macroscopic ones, such as the free energy, by applying the laws of statistical mechanics. Then, the free energy can be calculated by the following formula:

$$\Delta F = F_B - F_A = -kT\ln(Q_B/Q_A) = -kT\ln\langle \exp(-\beta\Delta U)\rangle_A \tag{7.2}$$

where $\beta = 1/kT, \Delta U = U_B(x) - U_A(x)$ is the difference in the potential energies calculated using the force field and the average is applied to configurations from state A. Q_i is the partition function in the Γ-phase space

$$Q_i = \int d\Gamma \exp(-\beta U_i)$$

Successful applications of FEP calculations have not reached the identification of any molecule clinical candidate yet, but hold promise for the years to come. In 2016, Janssen R&D disclosed a probe on optimization of amidine-containing spirocyclic β-secretase 1 (BACE1) inhibitors guided by FEP calculations [172]. The goal was to explore the chemical space around the lead molecules, and especially pockets P1–P3. A set of 18 molecules was submitted for FEP calculations, with some of them being synthesized on a second step. The results indicated good correlation between predicted and experimental values, with the FEP method outperforming docking. Acevedo et al. performed FEP/MC calculations on lead inhibitors of cyclophilin A (CypA), which participates in HIV-1 replication [173]. They identified nanomolar range inhibitors, with the predicted values being in close agreement with the experimental ones. In addition to these publications, the R&D department of Pfizer disclosed an application of the FEP method in tyrosine kinase (Syk) inhibitors [174]. In detail, Syk inhibitors have been associated with several therapeutic attributes, involving autoimmunity, inflammation and cancer, and thus the aforementioned study focused on the optimization of the imidazopyrazine core scaffold of previously identified Syk inhibitors [175]. The team initially examined retrospectively the predictive ability of the FEP methodology; and once encouraging results were received, they proceeded to prospective evaluation of 17 small molecules as potential inhibitors. The calculations indicated a compound with promising activity, further optimization of which culminated in compounds with cellular potency in the nanomolar range. Overall, the applications of the method extend beyond these examples, ranging from cases validating its predictive value to projects mainly directed toward the design of novel scaffolds in order for effective drugs to arise [176–179].

Apart from techniques that indicate a molecule as potent or not, there are also methods that aim to unravel the transition states and mechanisms which dominate the drug-binding process, and consequently guide future design and generate new hypotheses for compound scaffolds. These methodologies study the dynamic behavior of the systems, with MD being the most prominent of them. However, the computationally expensive atomistic models, the intrinsically complex potential energy functions, as well as the large timescales on which most phenomena of interest take place, render the detailed dynamic description of systems prohibitive. The need for a more profound understanding of the laws that govern the molecular recognition between a ligand and a target, led to the advent of more advanced techniques, which tend to introduce a bias to the system in the form of energy, in order to achieve enhanced sampling. Broadly speaking, methods such as umbrella sampling [180], steered MD [181], metadynamics [182–184], weighted histogram techniques [185], parallel tempering, and replica exchange [186, 187] can be classified in this category of enhanced sampling techniques, which rely on physical pathways. These can reconstruct the free energy landscape of the event under investigation and obtain statistics on rare events.

In 2005, Woo and Roux implemented a methodology based on umbrella sampling to calculate the equilibrium binding constant of the phosphotyrosine peptide pYEEI to the Src homology 2 domain of human Lck, acquiring close agreement with the experimental values [188]. Additional efforts by Lee and Olson, who utilized an umbrella sampling-based methodology to determine the potential of mean force by pushing the ligand to withdraw from the pocket, were capable of reproducing the experimental data, indicating a robust technique [189]. More recently, umbrella sampling, combined with weighted histogram analysis, successfully defined the binding free energy on the benzamidine–trypsin system [190]. Another approach aiming at unveiling the kinetic aspects of the binding or the dissociation of a type II inhibitor from a kinase target, employed umbrella sampling simulations to determine the channel (ATP or allosteric) that was preferred as an entrance or exit from the binding pocket [191].

Concerning applications of steered MD, in 2014, Patel et al. examined the unbinding kinetics of nine cyclin-dependent kinase 5 (CDK5) inhibitors [192], with the results revealing the aptitude of the method in distinguishing active from inactive compounds, as well as providing atomistic insights of transient and dynamical interactions. Complementary studies, from Colizzi et al., with the goal of discerning binders from nonbinders to the protein FabZ resulted in qualitatively accurate predictions, while a novel compound with enhanced biological activity was identified [193]. Steered MD was also employed for the description of the unbinding of the clinical candidate F14512 from the topoisomerase enzyme [194].

Enhanced MD simulations in the metadynamics framework have been proved valuable in deciphering the activation mechanism of p38a kinase [195], as well as in the discovery of cryptic pockets, i.e. protein cavities that reveal their nature only in the ligand binding [196]. The prediction of a molecule's true binding mode [197], the dissociation of a potent inhibitor of COX-1 and COX-2 [198], and the elucidation of the insertion mechanism of the naloxone antagonist in the pocket of the delta opioid receptor (DOR) [199], are just a few paradigmatic examples of metadynamics' possibilities.

Successful projects also based on a combination of techniques, as in the case of SSR128129E inhibitor of the fibroblast growth factor receptor (FGFR), for the discovery of which free energy, metadynamics, bias exchange metadynamics, and steered MD calculations were employed [200], may constitute the solution for highly complex systems. Similarly, utilization of steered MD and metadynamics shed light on the key interactions and transitions occurring during the detachment of the substrate of type 1 11β-hydroxysteroid dehydrogenase (11β-HSD-1) [201]. Recently, an FEP methodology coupled with the replica exchange with solute tempering protocol achieved accurate free energy predictions in a reasonable timescale framework [202].

Concluding, these examples highlight the significance of novel computational methods in pharmaceutical research, which even though they are still at their infancy, are expected to spawn considerable advances in the upcoming years. It is worth mentioning that every successful drug discovery project is the result of a multidisciplinary effort of both computational scientists and experimentalists. This observation suits well in the case of the development of an effective androgen receptor inhibitor, where various CADD methods, such as virtual screening, consensus scoring, QSAR, and pharmacophore modeling, were coordinated with experimental techniques, resulting in the largest academic licensing deal in Canadian history [203]. As the writers clearly state, it is of paramount importance to "*effectively synergize wet-lab and dry-lab efforts.*"

References

1 Mullard, A. (2014). New drugs cost US$2.6 billion to develop. *Nat. Rev. Drug Discovery* 13: 877.

2 Brown, D. and Superti-Furga, G. (2003). Rediscovering the sweet spot in drug discovery. *Drug Discovery Today* 8 (23): 1067–1077.
3 Macarron, R., Banks, M.N., Bojanic, D. et al. (2011). Impact of high-throughput screening in biomedical research. *Nat. Rev. Drug Discovery* 10 (3): 188–195.
4 Macalino, S.J., Gosu, V., Hong, S. et al. (2015). Role of computer-aided drug design in modern drug discovery. *Arch. Pharm. Res.* 38 (9): 1686–1701.
5 Kazachenko, S., Giovinazzo, M., Hall, K.W. et al. (2015). Algorithms for GPU-based molecular dynamics simulations of complex fluids: applications to water, mixtures, and liquid crystals. *J. Comput. Chem.* 36 (24): 1787–1804.
6 Mobley, D.L. and Gilson, M.K. (2017). Predicting binding free energies: frontiers and benchmarks. *Annu. Rev. Biophys.* 46 (1): 531–558.
7 De Vivo, M. (2011). Bridging quantum mechanics and structure-based drug design. *Front. Biosci. (Landmark Ed)* 16: 1619–1633.
8 Yin, J., Henriksen, N.M., Slochower, D.R. et al. (2017). Overview of the SAMPL5 host–guest challenge: are we doing better? *J. Comput.-Aided Mol. Des.* 31 (1): 1–19.
9 Lemkul, J.A., Huang, J., Roux, B. et al. (2016). An empirical polarizable force field based on the classical Drude oscillator model: development history and recent applications. *Chem. Rev.* 116 (9): 4983–5013.
10 Athanasiou, C., Vasilakaki, S., Dellis, D. et al. (2018). Using physics-based pose predictions and free energy perturbation calculations to predict binding poses and relative binding affinities for FXR ligands in the D3R Grand Challenge 2. *J. Comput.-Aided Mol. Des.* 32 (1): 21–44.
11 Bruce, N.J., Ganotra, G.K., Kokh, D.B. et al. (2017). New approaches for computing ligand-receptor binding kinetics. *Curr. Opin. Struct. Biol.* 49: 1–10.
12 Saladino, G. and Gervasio, F.L. (2016). Modeling the effect of pathogenic mutations on the conformational landscape of protein kinases. *Curr. Opin. Struct. Biol.* 37: 108–114.
13 Harpole, T.J. and Delemotte, L. (2018). Conformational landscapes of membrane proteins delineated by enhanced sampling molecular dynamics simulations. *Biochim. Biophys. Acta* 1860 (4): 909–926.
14 Leelananda, S.P. and Lindert, S. (2016). Computational methods in drug discovery. *Beilstein J. Org. Chem.* 12: 2694–2718.
15 Reynolds, C.H. (2014). Impact of computational structure-based methods on drug discovery. *Curr. Pharm. Des.* 20 (20): 3380–3386.
16 Sliwoski, G., Kothiwale, S., Meiler, J. et al. (2014). Computational methods in drug discovery. *Pharmacol. Rev.* 66 (1): 334–395.
17 Tropsha, A. and Bajorath, J. (2016). Computational methods for drug discovery and design. *J. Med. Chem.* 59 (1): 1.
18 Manas, E.S. and Green, D.V.S. (2017). CADD medicine: design is the potion that can cure my disease. *J. Comput.-Aided Mol. Des.* 31 (3): 249–253.
19 Brown, F.K., Sherer, E.C., Johnson, S.A. et al. (2017). The evolution of drug design at Merck Research Laboratories. *J. Comput.-Aided Mol. Des.* 31 (3): 255–266.

20 McGaughey, G. and Patrick Walters, W. (2017). Modeling & Informatics at Vertex Pharmaceuticals Incorporated: our philosophy for sustained impact. *J. Comput.-Aided Mol. Des.* 31 (3): 293–300.
21 van Vlijmen, H., Desjarlais, R.L., and Mirzadegan, T. (2017). Computational chemistry at Janssen. *J. Comput.-Aided Mol. Des.* 31 (3): 267–273.
22 Coates, D. (2003). The angiotensin converting enzyme (ACE). *Int. J. Biochem. Cell Biol.* 35 (6): 769–773.
23 Patlak, M. (2004). From viper's venom to drug design: treating hypertension. *FASEB J.* 18 (3): 421.
24 Das, M. and Soffer, R.L. (1975). Pulmonary angiotensin-converting enzyme. Structural and catalytic properties. *J. Biol. Chem.* 250 (17): 6762–6768.
25 Byers, L.D. and Wolfenden, R. (1973). Binding of the by-product analog benzylsuccinic acid by carboxypeptidase A. *Biochemistry* 12 (11): 2070–2078.
26 Cushman, D.W., Cheung, H.S., Sabo, E.F. et al. (1977). Design of potent competitive inhibitors of angiotensin-converting enzyme. Carboxyalkanoyl and mercaptoalkanoyl amino acids. *Biochemistry* 16 (25): 5484–5491.
27 Ondetti, M.A., Rubin, B., and Cushman, D.W. (1977). Design of specific inhibitors of angiotensin-converting enzyme: new class of orally active antihypertensive agents. *Science* 196 (4288): 441–444.
28 Kohl, N.E., Emini, E.A., Schleif, W.A. et al. (1988). Active human immunodeficiency virus protease is required for viral infectivity. *Proc. Natl. Acad. Sci. U. S. A.* 85 (13): 4686–4690.
29 Seelmeier, S., Schmidt, H., Turk, V. et al. (1988). Human immunodeficiency virus has an aspartic-type protease that can be inhibited by pepstatin A. *Proc. Natl. Acad. Sci. U. S. A.* 85 (18): 6612–6616.
30 Roberts, N.A., Martin, J.A., Kinchington, D. et al. (1990). Rational design of peptide-based HIV proteinase inhibitors. *Science* 248 (4953): 358–361.
31 Kempf, D.J., Sham, H.L., Marsh, K.C. et al. (1998). Discovery of ritonavir, a potent inhibitor of HIV protease with high oral bioavailability and clinical efficacy. *J. Med. Chem.* 41 (4): 602–617.
32 Timmermans, P.B.M.W.M., Carini, D.J., Chiu, A.T. et al. (1990). Nonpeptide angiotensin II receptor antagonists: a novel class of antihypertensive agents. *J. Vasc. Res.* 27 (2–5): 295–300.
33 Wong, P.C., Chiu, A.T., Price, W.A. et al. (1988). Nonpeptide angiotensin II receptor antagonists. I. Pharmacological characterization of 2-*n*-butyl-4-chloro-1-(2-chlorobenzyl)imidazole-5-acetic acid, sodium salt (S-8307). *J. Pharmacol. Exp. Ther.* 247 (1): 1–7.
34 Duncia, J.V., Chiu, A.T., Carini, D.J. et al. (1990). The discovery of potent nonpeptide angiotensin II receptor antagonists: a new class of potent antihypertensives. *J. Med. Chem.* 33 (5): 1312–1329.
35 Smeby, R. and Fermandjian, S. (1978). Conformation of angiotensin II. In: *Chemistry and Biochemistry of Amino Acids, Peptides and Proteins*, vol. 5 (ed. B. Weinstein), 117. New York: Marcel Dekker Inc.
36 Bhardwaj, G. (2006). How the antihypertensive losartan was discovered. *Expert Opin. Drug Discovery* 1 (6): 609–618.
37 Renner, S. and Schneider, G. (2006). Scaffold-hopping potential of ligand-based similarity concepts. *ChemMedChem* 1 (2): 181–185.

38 Bühlmayer, P., Furet, P., Criscione, L. et al. (1994). Valsartan, a potent, orally active angiotensin II antagonist developed from the structurally new amino acid series. *Bioorg. Med. Chem. Lett.* 4 (1): 29–34.
39 Anea, C.B., Lyon, M., Lee, I.A. et al. (2016). Pulmonary platelet thrombi and vascular pathology in acute chest syndrome in patients with sickle cell disease. *Am. J. Hematol.* 91 (2): 173–178.
40 Kroll, M.H. and Afshar-Kharghan, V. (2012). Platelets in pulmonary vascular physiology and pathology. *Pulm. Circ.* 2 (3): 291–308.
41 Fullard, J.F. (2004). The role of the platelet glycoprotein IIb/IIIa in thrombosis and haemostasis. *Curr. Pharm. Des.* 10 (14): 1567–1576.
42 Schneider, D.J. (2011). Anti-platelet therapy: glycoprotein IIb-IIIa antagonists. *Br. J. Clin. Pharmacol.* 72 (4): 672–682.
43 Egbertson, M.S., Chang, C.T., Duggan, M.E. et al. (1994). Non-peptide fibrinogen receptor antagonists. 2. Optimization of a tyrosine template as a mimic for Arg-Gly-Asp. *J. Med. Chem.* 37 (16): 2537–2551.
44 Chen, Y., Pitzenberger, S.M., Garsky, V.M. et al. (1991). Proton NMR assignments and secondary structure of the snake venom protein echistatin. *Biochemistry* 30 (50): 11625–11636.
45 Chen, Y., Suri, A.K., Kominos, D. et al. (1994). Three-dimensional structure of echistatin and dynamics of the active site. *J. Biomol. NMR* 4 (3): 307–324.
46 *AMF – The Advanced Modeling Facility, version 2.0.* Molecular Systems Department, Merck Research Laboratories, a division of Merck & Co. Inc., October 1992.
47 *JIGGLE – A Distance Geometry Algorithm.* Molecular Systems Department, Merck Research Laboratories, a division of Merck & Co. Inc., April 1993.
48 Gund, P. (1977). Three-dimensional pharmacophoric pattern searching. In: *Progress in Molecular and Subcellular Biology* (ed. F.E. Hahn et al.), 117–143. Berlin, Heidelberg: Springer-Verlag.
49 Yang, S.Y. (2010). Pharmacophore modeling and applications in drug discovery: challenges and recent advances. *Drug Discovery Today* 15 (11–12): 444–450.
50 Ferrari, M.D. and Saxena, P.R. (1995). 5-HT1 receptors in migraine pathophysiology and treatment. *Eur. J. Neurol.* 2 (1): 5–21.
51 Humphrey, P.P. (1991). 5-Hydroxytryptamine and the pathophysiology of migraine. *J. Neurol.* 238 (Suppl. 1): S38–S44.
52 Meng, C. (1997). Migraine: current drug discovery trend. *Curr. Med. Chem.* 4 (6): 385–404.
53 Buckingham, J., Glen, R.C., Hill, A.P. et al. (1995). Computer-aided design and synthesis of 5-substituted tryptamines and their pharmacology at the 5-HT1D receptor: discovery of compounds with potential anti-migraine properties. *J. Med. Chem.* 38 (18): 3566–3580.
54 *SYBYL 6.1 molecular modeling package.* Tripos Associates, St. Louis, MO, 1992.
55 Wang, L., Deng, Y., Wu, Y. et al. (2017). Accurate modeling of scaffold hopping transformations in drug discovery. *J. Chem. Theory Comput.* 13 (1): 42–54.

56 Aparoy, P., Reddy, K.K., and Reddanna, P. (2012). Structure and ligand based drug design strategies in the development of novel 5- LOX inhibitors. *Curr. Med. Chem.* 19 (22): 3763–3778.
57 Koga, H., Itoh, A., Murayama, S. et al. (1980). Structure-activity relationships of antibacterial 6,7- and 7,8-disubstituted 1-alkyl-1,4-dihydro-4-oxoquinoline-3-carboxylic acids. *J. Med. Chem.* 23 (12): 1358–1363.
58 Li, J. and Sun, J. (2017). Application of X-ray diffraction and electron crystallography for solving complex structure problems. *Acc. Chem. Res.* 50 (11): 2737–2745.
59 van Montfort, R.L.M. and Workman, P. (2017). Structure-based drug design: aiming for a perfect fit. *Essays Biochem.* 61 (5): 431–437.
60 Proudfoot, A., Bussiere, D.E., and Lingel, A. (2017). High-confidence protein–ligand complex modeling by NMR-guided docking enables early hit optimization. *J. Am. Chem. Soc.* 139 (49): 17824–17833.
61 Carneiro, M.G., Ab, E., Theisgen, S. et al. (2017). NMR in structure-based drug design. *Essays Biochem.* 61 (5): 485–493.
62 Nitsche, C. and Otting, G. (2017). NMR studies of ligand binding. *Curr. Opin. Struct. Biol.* 48: 16–22.
63 Boland, A., Chang, L., and Barford, D. (2017). The potential of cryo-electron microscopy for structure-based drug design. *Essays Biochem.* 61 (5): 543–560.
64 Christensen, J.G., Zou, H.Y., Arango, M.E. et al. (2007). Cytoreductive antitumor activity of PF-2341066, a novel inhibitor of anaplastic lymphoma kinase and c-Met, in experimental models of anaplastic large-cell lymphoma. *Mol. Cancer Ther.* 6 (12 Pt 1): 3314–3322.
65 Christensen, J.G., Burrows, J., and Salgia, R. (2005). c-Met as a target for human cancer and characterization of inhibitors for therapeutic intervention. *Cancer Lett.* 225 (1): 1–26.
66 Soda, M., Choi, Y.L., Enomoto, M. et al. (2007). Identification of the transforming EML4-ALK fusion gene in non-small-cell lung cancer. *Nature* 448 (7153): 561–566.
67 Wang, X., Le, P., Liang, C. et al. (2003). Potent and selective inhibitors of the Met [hepatocyte growth factor/scatter factor (HGF/SF) receptor] tyrosine kinase block HGF/SF-induced tumor cell growth and invasion. *Mol. Cancer Ther.* 2 (11): 1085–1092.
68 Christensen, J.G., Schreck, R., Burrows, J. et al. (2003). A selective small molecule inhibitor of c-Met kinase inhibits c-Met-dependent phenotypes in vitro and exhibits cytoreductive antitumor activity in vivo. *Cancer Res.* 63 (21): 7345–7355.
69 Cui, J.J., Tran-Dube, M., Shen, H. et al. (2011). Structure based drug design of crizotinib (PF-02341066), a potent and selective dual inhibitor of mesenchymal–epithelial transition factor (c-MET) kinase and anaplastic lymphoma kinase (ALK). *J. Med. Chem.* 54 (18): 6342–6363.

70 Cowan-Jacob, S.W., Guez, V., Fendrich, G. et al. (2004). Imatinib (STI571) resistance in chronic myelogenous leukemia: molecular basis of the underlying mechanisms and potential strategies for treatment. *Mini Rev. Med. Chem.* 4 (3): 285–299.

71 Manley, P.W., Cowan-Jacob, S.W., Buchdunger, E. et al. (2002). Imatinib: a selective tyrosine kinase inhibitor. *Eur. J. Cancer* 38 (Suppl 5): S19–S27.

72 Schindler, T., Bornmann, W., Pellicena, P. et al. (2000). Structural mechanism for STI-571 inhibition of abelson tyrosine kinase. *Science* 289 (5486): 1938–1942.

73 Buchdunger, E., Matter, A., and Druker, B.J. (2001). Bcr-Abl inhibition as a modality of CML therapeutics. *Biochim. Biophys. Acta* 1551 (1): M11–M18.

74 Weisberg, E., Manley, P.W., Breitenstein, W. et al. (2005). Characterization of AMN107, a selective inhibitor of native and mutant Bcr-Abl. *Cancer Cell* 7 (2): 129–141.

75 Lyle, T.A., Wiscount, C.M., Guare, J.P. et al. (1991). Benzocycloalkyl amines as novel C-termini for HIV protease inhibitors. *J. Med. Chem.* 34 (3): 1228–1230.

76 Vacca, J.P., Guare, J.P., deSolms, S.J. et al. (1991). L-687,908, a potent hydroxyethylene-containing HIV protease inhibitor. *J. Med. Chem.* 34 (3): 1225–1228.

77 Krohn, A., Redshaw, S., Ritchie, J.C. et al. (1991). Novel binding mode of highly potent HIV-proteinase inhibitors incorporating the (R)-hydroxyethylamine isostere. *J. Med. Chem.* 34 (11): 3340–3342.

78 Dorsey, B.D., Levin, R.B., McDaniel, S.L. et al. (1994). L-735,524: the design of a potent and orally bioavailable HIV protease inhibitor. *J. Med. Chem.* 37 (21): 3443–3451.

79 Shoichet, B.K. (2004). Virtual screening of chemical libraries. *Nature* 432 (7019): 862–865.

80 Lavecchia, A. and Di Giovanni, C. (2013). Virtual screening strategies in drug discovery: a critical review. *Curr. Med. Chem.* 20 (23): 2839–2860.

81 Lionta, E., Spyrou, G., Vassilatis, D.K. et al. (2014). Structure-based virtual screening for drug discovery: principles, applications and recent advances. *Curr. Top. Med. Chem.* 14 (16): 1923–1938.

82 Guido, R.V., Oliva, G., and Andricopulo, A.D. (2008). Virtual screening and its integration with modern drug design technologies. *Curr. Med. Chem.* 15 (1): 37–46.

83 Irwin, J.J. and Shoichet, B.K. (2016). Docking screens for novel ligands conferring new biology. *J. Med. Chem.* 59 (9): 4103–4120.

84 Kitchen, D.B., Decornez, H., Furr, J.R. et al. (2004). Docking and scoring in virtual screening for drug discovery: methods and applications. *Nat. Rev. Drug Discovery* 3 (11): 935–949.

85 Ludovici, D.W., Kavash, R.W., Kukla, M.J. et al. (2001). Evolution of anti-HIV drug candidates. Part 2: Diaryltriazine (DATA) analogues. *Bioorg. Med. Chem. Lett.* 11 (17): 2229–2234.

86 Janssen, P.A., Lewi, P.J., Arnold, E. et al. (2005). In search of a novel anti-HIV drug: multidisciplinary coordination in the discovery of 4-[[4-[[4-[(1E)-2-cyanoethenyl]-2,6-dimethylphenyl]amino]-2- pyrimidinyl]amino]benzonitrile (R278474, rilpivirine). *J. Med. Chem.* 48 (6): 1901–1909.

87 Halgren, T.A. (1996). Merck molecular force field. I. Basis, form, scope, parameterization, and performance of MMFF94. *J. Comput. Chem.* 17 (5–6): 490–519.

88 Lamarre, D., Anderson, P.C., Bailey, M. et al. (2003). An NS3 protease inhibitor with antiviral effects in humans infected with hepatitis C virus. *Nature* 426 (6963): 186–189.

89 Llinas-Brunet, M., Bailey, M.D., Bolger, G. et al. (2004). Structure-activity study on a novel series of macrocyclic inhibitors of the hepatitis C virus NS3 protease leading to the discovery of BILN 2061. *J. Med. Chem.* 47 (7): 1605–1608.

90 Tsantrizos, Y.S., Bolger, G., Bonneau, P. et al. (2003). Macrocyclic inhibitors of the NS3 protease as potential therapeutic agents of hepatitis C virus infection. *Angew. Chem. Int. Ed. Engl.* 42 (12): 1356–1360.

91 Liverton, N.J., Holloway, M.K., McCauley, J.A. et al. (2008). Molecular modeling based approach to potent P2–P4 macrocyclic inhibitors of hepatitis C NS3/4A protease. *J. Am. Chem. Soc.* 130 (14): 4607–4609.

92 Yao, N., Reichert, P., Taremi, S.S. et al. (1999). Molecular views of viral polyprotein processing revealed by the crystal structure of the hepatitis C virus bifunctional protease–helicase. *Structure* 7 (11): 1353–1363.

93 Kuszewski, J., Nilges, M., and Brunger, A.T. (1992). Sampling and efficiency of metric matrix distance geometry: a novel partial metrization algorithm. *J. Biomol. NMR* 2 (1): 33–56.

94 Wang, R., Lu, Y., Fang, X. et al. (2004). An extensive test of 14 scoring functions using the PDBbind refined set of 800 protein–ligand complexes. *J. Chem. Inf. Comput. Sci.* 44 (6): 2114–2125.

95 Harper, S., McCauley, J.A., Rudd, M.T. et al. (2012). Discovery of MK-5172, a macrocyclic hepatitis C virus NS3/4a protease inhibitor. *ACS Med. Chem. Lett.* 3 (4): 332–336.

96 Schaffer, L.W., Davidson, J.T., Vlasuk, G.P. et al. (1992). Selective factor Xa inhibition by recombinant antistasin prevents vascular graft thrombosis in baboons. *Arterioscler. Thromb.* 12 (8): 879–885.

97 Zhu, B.-Y. and Scarborough, R.M. (2000). Chapter 9. Factor Xa inhibitors: recent advances in anticoagulant agents. *Annu. Rep. Med. Chem.* 35: 83–102.

98 Zhang, P., Bao, L., Fan, J. et al. (2009). Anthranilamide-based *N,N*-dialkylbenzamidines as potent and orally bioavailable factor Xa inhibitors: P4 SAR. *Bioorg. Med. Chem. Lett.* 19 (8): 2186–2189.

99 Jones, G., Willett, P., Glen, R.C. et al. (1997). Development and validation of a genetic algorithm for flexible docking. *J. Mol. Biol.* 267 (3): 727–748.

100 Zhang, P., Huang, W., Wang, L. et al. (2009). Discovery of betrixaban (PRT054021), *N*-(5-chloropyridin-2-yl)-2-(4-(*N,N*-dimethylcarbamimidoyl) benzamido)-5-methoxybenz amide, a highly potent, selective, and orally efficacious factor Xa inhibitor. *Bioorg. Med. Chem. Lett.* 19 (8): 2179–2185.

101 Sherman, W., Day, T., Jacobson, M.P. et al. (2006). Novel procedure for modeling ligand/receptor induced fit effects. *J. Med. Chem.* 49 (2): 534–553.

102 Ellingson, S.R., Miao, Y., Baudry, J. et al. (2015). Multi-conformer ensemble docking to difficult protein targets. *J. Phys. Chem. B* 119 (3): 1026–1034.

103 Totrov, M. and Abagyan, R. (1997). Flexible protein–ligand docking by global energy optimization in internal coordinates. *Proteins* 29 (1): 215–220.

104 Ness, S., Martin, R., Kindler, A.M. et al. (2000). Structure-based design guides the improved efficacy of deacylation transition state analogue inhibitors of TEM-1 beta-lactamase(). *Biochemistry* 39 (18): 5312–5321.

105 Hecker, S.J., Reddy, K.R., Totrov, M. et al. (2015). Discovery of a cyclic boronic acid beta-lactamase inhibitor (RPX7009) with utility vs class A serine carbapenemases. *J. Med. Chem.* 58 (9): 3682–3692.

106 Borhani, D.W. and Shaw, D.E. (2012). The future of molecular dynamics simulations in drug discovery. *J. Comput.-Aided Mol. Des.* 26 (1): 15–26.

107 De Vivo, M., Masetti, M., Bottegoni, G. et al. (2016). Role of molecular dynamics and related methods in drug discovery. *J. Med. Chem.* 59 (9): 4035–4061.

108 Durrant, J.D. and McCammon, J.A. (2011). Molecular dynamics simulations and drug discovery. *BMC Biol.* 9: 71.

109 Ganesan, A., Coote, M.L., and Barakat, K. (2017). Molecular dynamics-driven drug discovery: leaping forward with confidence. *Drug Discovery Today* 22 (2): 249–269.

110 Tung, R., Livingston, D., Rao, B. et al. (2001). Design and synthesis of amprenavir, a novel HIV protease inhibitor. In: *Protease Inhibitors in AIDS Therapy* (ed. C. Flexner and R. Ogden). New York: Taylor & Francis.

111 Jorgensen, W.L. (2009). Efficient drug lead discovery and optimization. *Acc. Chem. Res.* 42 (6): 724–733.

112 Hartenfeller, M. and Schneider, G. (2011). De novo drug design. *Methods Mol. Biol.* 672: 299–323.

113 Devi, R.V., Sathya, S.S., and Coumar, M.S. (2015). Evolutionary algorithms for de novo drug design – a survey. *Appl. Soft Comput.* 27 (Suppl. C): 543–552.

114 Schneider, G. (2013). De novo design – hop(p)ing against hope. *Drug Discovery Today Technol.* 10 (4): e453–e460.

115 Sheng, C. and Zhang, W. (2013). Fragment informatics and computational fragment-based drug design: an overview and update. *Med. Res. Rev.* 33 (3): 554–598.

116 Danziger, D.J. and Dean, P.M. (1989). Automated site-directed drug design: a general algorithm for knowledge acquisition about hydrogen-bonding regions at protein surfaces. *Proc. R. Soc. London, Ser. B, Biol. Sci.* 236 (1283): 101–113.

117 Goodford, P.J. (1985). A computational procedure for determining energetically favorable binding sites on biologically important macromolecules. *J. Med. Chem.* 28 (7): 849–857.

118 Appelt, K. (1993). Crystal structures of HIV-1 protease–inhibitor complexes. *Perspect. Drug Discovery Des.* 1 (1): 23–48.

119 Varney, M.D., Appelt, K., Kalish, V. et al. (1994). Crystal-structure-based design and synthesis of novel C-terminal inhibitors of HIV protease. *J. Med. Chem.* 37 (15): 2274–2284.

120 Reich, S.H., Melnick, M., Davies, J.F. 2nd, et al. (1995). Protein structure-based design of potent orally bioavailable, nonpeptide inhibitors of human immunodeficiency virus protease. *Proc. Natl. Acad. Sci. U. S. A.* 92 (8): 3298–3302.

121 Gehlhaar, D.K., Moerder, K.E., Zichi, D. et al. (1995). De novo design of enzyme inhibitors by Monte Carlo ligand generation. *J. Med. Chem.* 38 (3): 466–472.

122 Bernardi, R.C., Melo, M.C., and Schulten, K. (2015). Enhanced sampling techniques in molecular dynamics simulations of biological systems. *Biochim. Biophys. Acta* 1850 (5): 872–877.

123 Kaldor, S.W., Kalish, V.J., Davies, J.F. 2nd, et al. (1997). Viracept (nelfinavir mesylate, AG1343): a potent, orally bioavailable inhibitor of HIV-1 protease. *J. Med. Chem.* 40 (24): 3979–3985.

124 von Itzstein, M., Wu, W.Y., Kok, G.B. et al. (1993). Rational design of potent sialidase-based inhibitors of influenza virus replication. *Nature* 363 (6428): 418–423.

125 Colman, P.M., Varghese, J.N., and Laver, W.G. (1983). Structure of the catalytic and antigenic sites in influenza virus neuraminidase. *Nature* 303 (5912): 41–44.

126 Brunger, A.T. (1988). Crystallographic refinement by simulated annealing. Application to a 2.8 Å resolution structure of aspartate aminotransferase. *J. Mol. Biol.* 203 (3): 803–816.

127 Varghese, J.N., McKimm-Breschkin, J.L., Caldwell, J.B. et al. (1992). The structure of the complex between influenza virus neuraminidase and sialic acid, the viral receptor. *Proteins* 14 (3): 327–332.

128 Vyas, V.K., Ukawala, R.D., Ghate, M. et al. (2012). Homology modeling a fast tool for drug discovery: current perspectives. *Indian J. Pharm. Sci.* 74 (1): 1–17.

129 Beuming, T., Lenselink, B., Pala, D. et al. (2015). Docking and virtual screening strategies for GPCR drug discovery. In: *G Protein-Coupled Receptors in Drug Discovery: Methods and Protocols*, Methods in Molecular Biology (ed. M. Filizola), 251–276. New York: Springer Science + Business Media.

130 Cappel, D., Hall, M.L., Lenselink, E.B. et al. (2016). Relative binding free energy calculations applied to protein homology models. *J. Chem. Inf. Model.* 56 (12): 2388–2400.

131 Kopp, J. and Schwede, T. (2004). Automated protein structure homology modeling: a progress report. *Pharmacogenomics* 5 (4): 405–416.

132 Chevalier, A., Silva, D.A., Rocklin, G.J. et al. (2017). Massively parallel de novo protein design for targeted therapeutics. *Nature* 550 (7674): 74–79.

133 Floudas, C.A., Fung, H.K., McAllister, S.R. et al. (2006). Advances in protein structure prediction and de novo protein design: a review. *Chem. Eng. Sci.* 61 (3): 966–988.

134 Huang, P.-S., Boyken, S.E., and Baker, D. (2016). The coming of age of de novo protein design. *Nature* 537: 320.

135 Dorn, M., MB, E.S., Buriol, L.S. et al. (2014). Three-dimensional protein structure prediction: methods and computational strategies. *Comput. Biol. Chem.* 53PB: 251–276.

136 Chou, P.Y. and Fasman, G.D. (1978). Prediction of the secondary structure of proteins from their amino acid sequence. *Adv. Enzymol. Relat. Areas Mol. Biol.* 47: 45–148.

137 Sibanda, B.L., Blundell, T., Hobart, P.M. et al. (1984). Computer graphics modelling of human renin. Specificity, catalytic activity and intron–exon junctions. *FEBS Lett.* 174 (1): 102–111.

138 Cohen, N.C. (2007). Structure-based drug design and the discovery of aliskiren (Tekturna): perseverance and creativity to overcome a R&D pipeline challenge. *Chem. Biol. Drug Des.* 70 (6): 557–565.

139 Rahuel, J., Rasetti, V., Maibaum, J. et al. (2000). Structure-based drug design: the discovery of novel nonpeptide orally active inhibitors of human renin. *Chem. Biol.* 7 (7): 493–504.

140 Wood, J.M., Maibaum, J., Rahuel, J. et al. (2003). Structure-based design of aliskiren, a novel orally effective renin inhibitor. *Biochem. Biophys. Res. Commun.* 308 (4): 698–705.

141 Blume-Jensen, P. and Hunter, T. (2001). Oncogenic kinase signalling. *Nature* 411 (6835): 355–365.

142 Sun, C. and Bernards, R. (2014). Feedback and redundancy in receptor tyrosine kinase signaling: relevance to cancer therapies. *Trends Biochem. Sci.* 39 (10): 465–474.

143 Millauer, B., Shawver, L.K., Plate, K.H. et al. (1994). Glioblastoma growth inhibited in vivo by a dominant-negative Flk-1 mutant. *Nature* 367: 576.

144 Fong, T.A.T., Shawver, L.K., Sun, L. et al. (1999). SU5416 is a potent and selective inhibitor of the vascular endothelial growth factor receptor (Flk-1/KDR) that inhibits tyrosine kinase catalysis, tumor vascularization, and growth of multiple tumor types. *Cancer Res.* 59 (1): 99.

145 Laird, A.D., Vajkoczy, P., Shawver, L.K. et al. (2000). SU6668 is a potent antiangiogenic and antitumor agent that induces regression of established tumors. *Cancer Res.* 60 (15): 4152–4160.

146 Sun, L., Liang, C., Shirazian, S. et al. (2003). Discovery of 5-[5-fluoro-2-oxo-1,2- dihydroindol-(3Z)-ylidenemethyl]-2,4- dimethyl-1H-pyrrole-3-carboxylic acid (2-diethylaminoethyl)amide, a novel tyrosine kinase inhibitor targeting vascular endothelial and platelet-derived growth factor receptor tyrosine kinase. *J. Med. Chem.* 46 (7): 1116–1119.

147 Marsilje, T.H., Pei, W., Chen, B. et al. (2013). Synthesis, structure-activity relationships, and in vivo efficacy of the novel potent and selective anaplastic lymphoma kinase (ALK) inhibitor 5-chloro-N2-(2-isopropoxy-5-methyl-4-(piperidin-4-yl)phenyl)-N4-(2-(isopropylsulf onyl)phenyl) pyrimidine-2,4-diamine (LDK378) currently in phase 1 and phase 2 clinical trials. *J. Med. Chem.* 56 (14): 5675–5690.

148 Huang, W.S., Liu, S., Zou, D. et al. (2016). Discovery of brigatinib (AP26113), a phosphine oxide-containing, potent, orally active inhibitor of anaplastic lymphoma kinase. *J. Med. Chem.* 59 (10): 4948–4964.

149 Jacobson, M.P., Friesner, R.A., Xiang, Z. et al. (2002). On the role of the crystal environment in determining protein side-chain conformations. *J. Mol. Biol.* 320 (3): 597–608.
150 Jacobson, M.P., Pincus, D.L., Rapp, C.S. et al. (2004). A hierarchical approach to all-atom protein loop prediction. *Proteins* 55 (2): 351–367.
151 Beutler, T.C., Mark, A.E., van Schaik, R.C. et al. (1994). Avoiding singularities and numerical instabilities in free energy calculations based on molecular simulations. *Chem. Phys. Lett.* 222 (6): 529–539.
152 Friesner, R.A., Banks, J.L., Murphy, R.B. et al. (2004). Glide: a new approach for rapid, accurate docking and scoring. 1. Method and assessment of docking accuracy. *J. Med. Chem.* 47 (7): 1739–1749.
153 Halgren, T.A., Murphy, R.B., Friesner, R.A. et al. (2004). Glide: a new approach for rapid, accurate docking and scoring. 2. Enrichment factors in database screening. *J. Med. Chem.* 47 (7): 1750–1759.
154 Simonson, T. (1993). Free energy of particle insertion. *Mol. Phys.* 80 (2): 441–447.
155 Gallaher, W.R., Ball, J.M., Garry, R.F. et al. (1989). A general model for the transmembrane proteins of HIV and other retroviruses. *AIDS Res. Hum. Retroviruses* 5 (4): 431–440.
156 Delwart, E.L., Mosialos, G., and Gilmore, T. (1990). Retroviral envelope glycoproteins contain a "leucine zipper"-like repeat. *AIDS Res. Hum. Retroviruses* 6 (6): 703–706.
157 O'Shea, E.K., Rutkowski, R., and Kim, P.S. (1989). Evidence that the leucine zipper is a coiled coil. *Science* 243 (4890): 538–542.
158 Wild, C., Oas, T., McDanal, C. et al. (1992). A synthetic peptide inhibitor of human immunodeficiency virus replication: correlation between solution structure and viral inhibition. *Proc. Natl. Acad. Sci. U. S. A.* 89 (21): 10537–10541.
159 Wild, C., Greenwell, T., and Matthews, T. (1993). A synthetic peptide from HIV-1 gp41 is a potent inhibitor of virus-mediated cell–cell fusion. *AIDS Res. Hum. Retroviruses* 9 (11): 1051–1053.
160 LaBonte, J., Lebbos, J., and Kirkpatrick, P. (2003). Enfuvirtide. *Nat. Rev. Drug Discovery* 2: 345.
161 Griffin, R.J., Curtin, N.J., Newell, D.R. et al. (1995). The role of inhibitors of poly(ADP-ribose) polymerase as resistance-modifying agents in cancer therapy. *Biochimie* 77 (6): 408–422.
162 Sebolt-Leopold, J.S. and Scavone, S.V. (1992). Enhancement of alkylating agent activity in vitro by PD 128763, a potent poly(ADP-ribose) synthetase inhibitor. *Int. J. Radiat. Oncol. Biol. Phys.* 22 (3): 619–621.
163 Canan Koch, S.S., Thoresen, L.H., Tikhe, J.G. et al. (2002). Novel tricyclic poly(ADP-ribose) polymerase-1 inhibitors with potent anticancer chemopotentiating activity: design, synthesis, and X-ray cocrystal structure. *J. Med. Chem.* 45 (23): 4961–4974.
164 Ruf, A., de Murcia, G., and Schulz, G.E. (1998). Inhibitor and NAD+ binding to poly(ADP-ribose) polymerase as derived from crystal structures and homology modeling. *Biochemistry* 37 (11): 3893–3900.

165 Thomas, H.D., Calabrese, C.R., Batey, M.A. et al. (2007). Preclinical selection of a novel poly(ADP-ribose) polymerase inhibitor for clinical trial. *Mol. Cancer Ther.* 6 (3): 945–956.

166 Friesner, R.A. (2005). Ab initio quantum chemistry: methodology and applications. *Proc. Natl. Acad. Sci. U. S. A.* 102 (19): 6648–6653.

167 Froese Fischer, C. (1987). General Hartree–Fock program. *Comput. Phys. Commun.* 43 (3): 355–365.

168 Hohenberg, P. and Kohn, W. (1964). Inhomogeneous electron gas. *Phys. Rev.* 136 (3B): B864–B871.

169 Kohn, W. and Sham, L.J. (1965). Self-consistent equations including exchange and correlation effects. *Phys. Rev.* 140 (4A): A1133–A1138.

170 Cole, D.J. and Hine, N.D. (2016). Applications of large-scale density functional theory in biology. *J. Phys. Condens. Matter* 28 (39): 393001.

171 Greer, J., Erickson, J.W., Baldwin, J.J. et al. (1994). Application of the three-dimensional structures of protein target molecules in structure-based drug design. *J. Med. Chem.* 37 (8): 1035–1054.

172 Ciordia, M., Perez-Benito, L., Delgado, F. et al. (2016). Application of free energy perturbation for the design of BACE1 inhibitors. *J. Chem. Inf. Model.* 56 (9): 1856–1871.

173 Acevedo, O., Ambrose, Z., Flaherty, P.T. et al. (2012). Identification of HIV inhibitors guided by free energy perturbation calculations. *Curr. Pharm. Des.* 18 (9): 1199–1216.

174 Lovering, F., Aevazelis, C., Chang, J. et al. (2016). Imidazotriazines: spleen tyrosine kinase (Syk) inhibitors identified by free-energy perturbation (FEP). *ChemMedChem* 11 (2): 217–233.

175 Currie, K.S., Kropf, J.E., Lee, T. et al. (2014). Discovery of GS-9973, a selective and orally efficacious inhibitor of spleen tyrosine kinase. *J. Med. Chem.* 57 (9): 3856–3873.

176 Cummins, P.L. and Gready, J.E. (1993). Computer-aided drug design: a free energy perturbation study on the binding of methyl-substituted pterins and N5-deazapterins to dihydrofolate reductase. *J. Comput.-Aided Mol. Des.* 7 (5): 535–555.

177 Keranen, H., Perez-Benito, L., Ciordia, M. et al. (2017). Acylguanidine beta secretase 1 inhibitors: a combined experimental and free energy perturbation study. *J. Chem. Theory Comput.* 13 (3): 1439–1453.

178 Rastelli, G., Costantino, L., Vianello, P. et al. (1998). Free energy perturbation studies on binding of the inhibitor 5,6-dihydrobenzo[h]cinnolin-3(2H)one-2-acetic acid and its methoxylated analogs to aldose reductase. *Tetrahedron* 54 (32): 9415–9428.

179 Ruiza, F.M. and Grigera, J.R. (2005). Free energy perturbation calculations on glucosidase-inhibitor complexes. *Med. Chem.* 1 (5): 455–460.

180 Torrie, G.M. and Valleau, J.P. (1977). Nonphysical sampling distributions in Monte Carlo free-energy estimation: Umbrella sampling. *J. Comput. Phys.* 23 (2): 187–199.

181 Israelewitz, B., Gao, M., and Schulten, K. (2001). Steered molecular dynamics and mechanical functions of proteins. *Curr. Opin. Struct. Biol.* 11 (2): 224–230.

182 Alessandro, L. and Francesco, L.G. (2008). Metadynamics: a method to simulate rare events and reconstruct the free energy in biophysics, chemistry and material science. *Rep. Prog. Phys.* 71 (12): 126601.
183 Laio, A. and Parrinello, M. (2002). Escaping free-energy minima. *Proc. Natl. Acad. Sci. U. S. A.* 99 (20): 12562–12566.
184 Sutto, L., Marsili, S., and Gervasio, F.L. (2012). New advances in metadynamics. *Wiley Interdiscip. Rev. Comput. Mol. Sci.* 2 (5): 771–779.
185 Kumar, S., Rosenberg, J.M., Bouzida, D. et al. (1992). THE weighted histogram analysis method for free-energy calculations on biomolecules. I. The method. *J. Comput. Chem.* 13 (8): 1011–1021.
186 Liu, P., Kim, B., Friesner, R.A. et al. (2005). Replica exchange with solute tempering: a method for sampling biological systems in explicit water. *Proc. Natl. Acad. Sci. U. S. A.* 102 (39): 13749–13754.
187 Sugita, Y. and Okamoto, Y. (1999). Replica-exchange molecular dynamics method for protein folding. *Chem. Phys. Lett.* 314 (1): 141–151.
188 Woo, H.J. and Roux, B. (2005). Calculation of absolute protein–ligand binding free energy from computer simulations. *Proc. Natl. Acad. Sci. U. S. A.* 102 (19): 6825–6830.
189 Lee, M.S. and Olson, M.A. (2006). Calculation of absolute protein–ligand binding affinity using path and endpoint approaches. *Biophys. J.* 90 (3): 864–877.
190 Doudou, S., Burton, N.A., and Henchman, R.H. (2009). Standard free energy of binding from a one-dimensional potential of mean force. *J. Chem. Theory Comput.* 5 (4): 909–918.
191 Sun, H., Tian, S., Zhou, S. et al. (2015). Revealing the favorable dissociation pathway of type II kinase inhibitors via enhanced sampling simulations and two-end-state calculations. *Sci. Rep.* 5: 8457.
192 Patel, J.S., Berteotti, A., Ronsisvalle, S. et al. (2014). Steered molecular dynamics simulations for studying protein–ligand interaction in cyclin-dependent kinase 5. *J. Chem. Inf. Model.* 54 (2): 470–480.
193 Colizzi, F., Perozzo, R., Scapozza, L. et al. (2010). Single-molecule pulling simulations can discern active from inactive enzyme inhibitors. *J. Am. Chem. Soc.* 132 (21): 7361–7371.
194 Palermo, G., Minniti, E., Greco, M.L. et al. (2015). An optimized polyamine moiety boosts the potency of human type II topoisomerase poisons as quantified by comparative analysis centered on the clinical candidate F14512. *Chem. Commun. (Cambridge)* 51 (76): 14310–14313.
195 Kuzmanic, A., Sutto, L., Saladino, G. et al. (2017). Changes in the free-energy landscape of p38alpha MAP kinase through its canonical activation and binding events as studied by enhanced molecular dynamics simulations. *Elife* 6: e22175.
196 Oleinikovas, V., Saladino, G., Cossins, B.P. et al. (2016). Understanding cryptic pocket formation in protein targets by enhanced sampling simulations. *J. Am. Chem. Soc.* 138 (43): 14257–14263.
197 Gervasio, F.L., Laio, A., and Parrinello, M. (2005). Flexible docking in solution using metadynamics. *J. Am. Chem. Soc.* 127 (8): 2600–2607.

198 Limongelli, V., Bonomi, M., Marinelli, L. et al. (2010). Molecular basis of cyclooxygenase enzymes (COXs) selective inhibition. *Proc. Natl. Acad. Sci. U. S. A.* 107 (12): 5411–5416.

199 Provasi, D., Bortolato, A., and Filizola, M. (2009). Exploring molecular mechanisms of ligand recognition by opioid receptors with metadynamics. *Biochemistry* 48 (42): 10020–10029.

200 Herbert, C., Schieborr, U., Saxena, K. et al. (2013). Molecular mechanism of SSR128129E, an extracellularly acting, small-molecule, allosteric inhibitor of FGF receptor signaling. *Cancer Cell* 23 (4): 489–501.

201 Favia, A.D., Masetti, M., Recanatini, M. et al. (2011). Substrate binding process and mechanistic functioning of type 1 11beta-hydroxysteroid dehydrogenase from enhanced sampling methods. *PLoS One* 6 (9): e25375.

202 Wang, L., Wu, Y., Deng, Y. et al. (2015). Accurate and reliable prediction of relative ligand binding potency in prospective drug discovery by way of a modern free-energy calculation protocol and force field. *J. Am. Chem. Soc.* 137 (7): 2695–2703.

203 Ban, F., Dalal, K., Li, H. et al. (2017). Best practices of computer-aided drug discovery: lessons learned from the development of a preclinical candidate for prostate cancer with a new mechanism of action. *J. Chem. Inf. Model.* 57 (5): 1018–1028.

8

Application of Biomolecular Simulations to G Protein–Coupled Receptors (GPCRs)

Mariona Torrens-Fontanals, Tomasz M. Stepniewski, Ismael Rodríguez-Espigares, and Jana Selent

Universitat Pompeu Fabra (UPF) – Hospital del Mar Medical Research Institute (IMIM), GPCR Drug Discovery Group, Research Programme on Biomedical Informatics (GRIB), Department of Experimental and Health Sciences of Pompeu Fabra University (UPF) – Hospital del Mar Medical Research Institute (IMIM), Dr. Aiguader 88, E-08003, Barcelona, Spain

8.1 Introduction

G protein–coupled receptors (GPCRs) are major targets for the pharmaceutical industry and present an immense potential for future drug development [1]. Although GPCRs have been extensively studied over the past decades, the underlying molecular and structural mechanisms responsible for many critical regulatory processes of this protein superfamily remain elusive, including signal transduction, allosteric modulation, functional selectivity, and activation [2].

GPCRs are known to be highly dynamic proteins, which can exist in multiple distinct conformational states. In fact, the signaling profile of GPCRs in response to a ligand depends critically on their ability to transition between different states [3]. Because of this, a promising approach to elucidate the molecular basis of GPCR functionality are molecular dynamics (MD) simulations, a potent computational technique capable of generating atomic-resolution simulations of the structural motions of a molecular system [4–6]. Importantly, these motions can be observed over timescales ranging from femtoseconds to milliseconds. As a consequence, MD provides, simultaneously, a temporal and structural resolution greater than what, currently, is achievable by experimental methods [7]. This is especially true for membrane proteins such as GPCRs, for which the dexperimental characterization of structural dynamics is particularly challenging [2, 8]. For this reason, and thanks to ongoing technology advances, MD simulations are increasingly being applied to the study of GPCR molecular systems, as reflected by the rapid upsurge of the number of publications per year concerning this topic (see Figure 8.1) [2, 6].

Different state-of-the-art simulation methods can be applied for the study of GPCRs, depending on the features of the receptor to be investigated. These methods include quantum chemical descriptors, classical mechanistic models with an atomistic or coarse-grain representation, and phenomenological system-level models. Each of them is characterized by the structural resolution,

Biomolecular Simulations in Structure-Based Drug Discovery,
First Edition. Edited by Francesco L. Gervasio and Vojtech Spiwok.

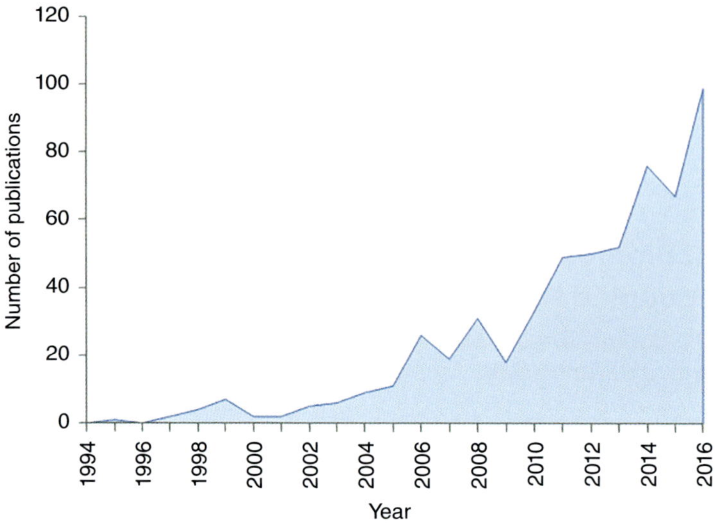

Figure 8.1 Number of publications per year indexed at Thomson Reuters' Web of Science that contain the topics "molecular dynamics" and ("GPCR" or "GPCRs"). The exponential growth of successful GPCR MD-based research is evidenced by the rapid upsurge in the number of publications per year related to this subject.

range of simulation timescales, and physicochemical accuracy they provide, which dictate the suitability of the model.

Considering the three criteria mentioned, for the study of GPCRs mechanistic models are typically the method of choice. These models are based on classical (i.e. Newtonian) mechanics, and thus the molecular system is treated as a set of classical particles. Interactions between these particles are derived from a set of empiric potential energy functions, known as force field, and quantum dynamical effects are not considered. Commonly, each system particle corresponds to a single atom (atomistic representations) or to a rigid set of atoms (coarse-grained representations) [9].

Atomistic representations capture the motion of molecular systems in full atomic detail, since each atom is represented as a point particle with its corresponding connectivity [10]. Consequently, this method leads to a better reproduction of the system's dynamics than coarse-grained representations, albeit at a higher computational cost.

Alternatively, coarse-grain representations consider a set of atoms as a single particle and each molecule as a set of particles, reducing the degrees of freedom of the system. This type of representation implies a substantial reduction of the computational requirements, and therefore can achieve longer simulation timescales. However, this is at the cost of a reduced spatial resolution [11].

In both mechanistic models, we can perform unbiased or biased simulations. Unbiased MD simulations sample the free energy landscape at equilibrium and explore the thermodynamically accessible energy landscape. This technique is limited by the inability to overcome high-energy barriers associated with large-scale motions, such as the receptor activation process. When facing

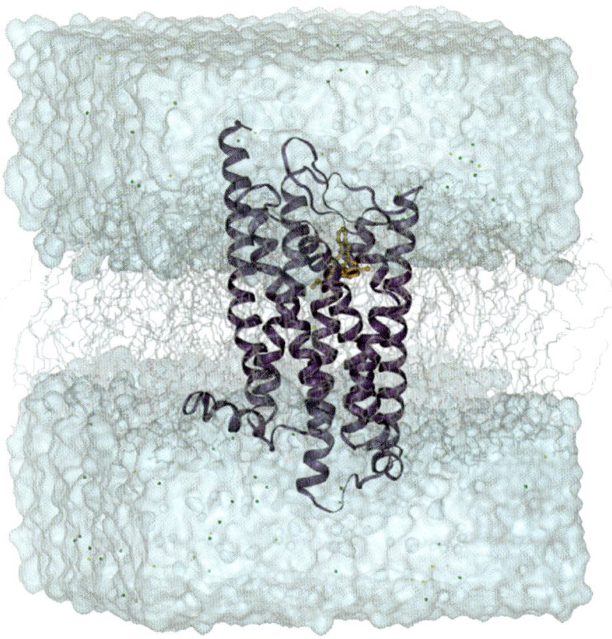

Figure 8.2 Representation of an atomistic MD simulation including a GPCR (purple) bound to a ligand (orange), the phospholipid bilayer (gray), and intercellular and intracellular water (blue).

such problems, biased techniques can be used to more efficiently overcome energetic barriers and access additional conformational states. This is achieved by including an external bias, such as a compensating force or potential. On the other hand, these external forces may drive the simulation along unrealistic deformations, producing structural artifacts [12].

In this chapter, we focus on atomistic unbiased simulations (see Figure 8.2), which provide detailed and reliable structural information but face a series of challenges, such as simulating long molecular events. With this premise, we discuss the challenges that MD-based studies of GPCRs still face and how MD can provide unique insights into GPCR research. Moreover, we review the advances of MD-based GPCR research since its origin and speculate about its future.

8.2 MD Simulations for Studying the Conformational Plasticity of GPCRs

Due to the inherent structural plasticity of GPCRs, these receptors can acquire an essentially infinite number of conformational states. In fact, several lines of evidence indicate that GPCRs do not present just one single active state and one single inactive state: they can assume multiple distinct states in each case [13–17]. MD simulations can provide a useful insight on the transitions between conformational states, such as inactive, intermediate, and active states (see Section 8.3). However, at present, MD-based studies of transition processes are limited in several respects, such as the need for simulations to extend to the millisecond timescale and the difficulty inherent in sufficiently sampling structural fluctuations to properly characterize long molecular events [18]. These constraints pose an obstacle to structure-based drug design for GPCR targets, since understanding the role of intermediate states is crucial for the comprehension of the molecular mechanisms behind receptor activation and inactivation, together with its pharmacological action. Moreover, deciphering the mechanisms by which ligands can differentially induce or stabilize different receptor populations would lead us to the design of drugs that specifically target the conformational state responsible for a desired intracellular response. This would allow, for instance, the rational design of drugs with reduced side effects [3, 19].

8.2.1 Challenges in GPCR Simulations: The Sampling Problem and Simulation Timescales

The dynamic behavior of a molecular system is intrinsically dictated by its free energy. Complex biomolecules such as GPCRs present a rugged energy landscape (see Figure 8.3), with many local minima frequently separated by high-energy barriers [20]. The conformational transition between different states requires crossing some of these high-energy barriers, which is a slow process [21]. Indeed, this shape of the free energy landscape makes it easy for systems to get trapped in one of the huge number of local minimum-energy states for a long time, which restrains the sampling process and, in turn, leads to a poor characterization of a protein's dynamic behavior [22, 23]. Therefore, most unbiased simulations explore just a small region around the energy minimum closest to the initial conformation [24]. This inadequate sampling of the conformational landscape limits the ability to analyze and reveal functional properties of the systems being examined [25]. Due to the sampling problem, complex processes such as GPCR activation or inactivation are commonly beyond the capabilities of straightforward unbiased MD simulations [25].

Furthermore, simulations are currently limited to lengths of a few microseconds on publicly available hardware, or a few milliseconds if specialized hardware [26] or distributed computing techniques [27] are used. These simulation times

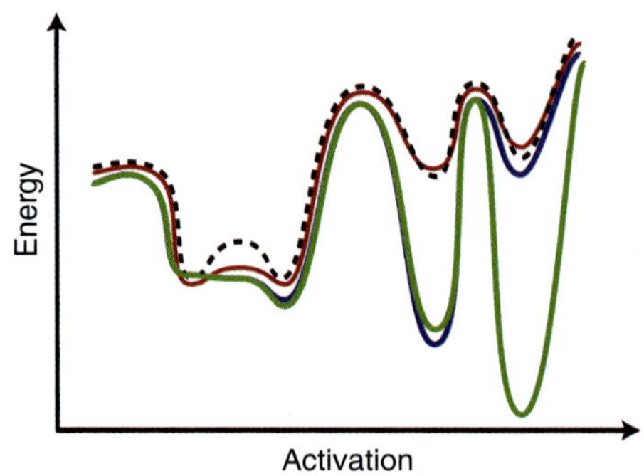

Figure 8.3 Simplified energy landscape of a GPCR in the absence of ligand (dashed line) and in the presence of an inverse agonist (red), a full agonist (blue), and a full agonist with a G protein (green). Source: Adapted with permission from Ref. [15].

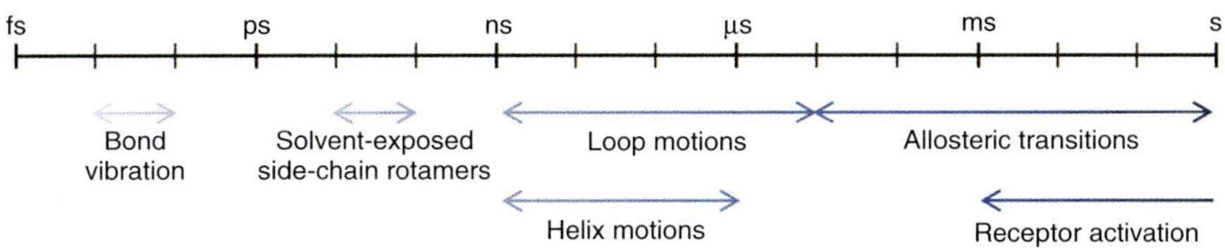

Figure 8.4 Schema of the characteristic timescales for GPCR motions (time magnitudes in log scale) [2, 9, 28, 29].

are short compared to many of the events of greatest interest in molecular physiology, which take place on longer timescales [8]. This is the case for large conformational rearrangements in GPCRs. For example, the conformational changes from the inactive to the active state of a GPCR occur at the millisecond to second timescale (see Figure 8.4) [28]. The time limitation in conventional MD simulation approaches also hampers the routine exploration of drug (un)binding, making *in silico* GPCR-related drug discovery a challenging task. Nevertheless, we experience an exciting era for GPCR research, as we expect to simulate sufficient time frames within the next five years (see Section 8.4).

8.2.2 Making Sense Out of Simulation Data

Once sufficient sampling time is obtained from unbiased all-atom MD models, one of the hardest problems is to extract information about the kinetics of relevant biological events (e.g. ligand binding) [27]. An interesting approach to overcome this problem is the application of dimensionality reduction methods. Essentially, these methods are mathematical tools that reduce the complexity of a system while minimizing the loss of information. This is possible due to the interdependence between different variables of the system, imposed by energetic constraints among atom positions [30]. Ultimately, dimensionality reduction helps visualize complex energy landscapes, improving the efficiency of simulation, analysis, and optimization. As a consequence, these methods are considered an important tool for the comprehensive analysis of MD data [31]. Among these techniques, the most popular are the ones that imply linear dimensionality reduction, especially principal component analysis (PCA) and time-dependent independent component analysis (tICA). For instance, PCA has been applied to study simulation data of the adenosine A_{2A} receptor in both the presence and absence of an inverse agonist, providing information on the convergence and reproducibility of the results. The authors found that the addition of the inverse agonist greatly improves the stability of all receptor helices. Moreover, they concluded that PCA provides a more robust assessment of convergence and sampling than other commonly used criteria such as root-mean-square deviation (RMSD), especially in comparative studies [32].

Unfortunately, the above mentioned methods (tICA and PCA) do not provide quantitative information about statistical significance or sampling quality [33]. Still, there exist other approaches that are capable of generating more robust analyses by first partitioning the conformational space into discrete conformational states and subsequently calculating the transition rates or probabilities between them, based on transitions observed in MD trajectories. The analysis of these transitions allows the reconstruction of the global behavior of the system [34, 35]. The resulting models are known as Markov state models (MSMs), where "Markovianity" means that the kinetics are modeled by a memoryless jump process between states [36]. Once the model is obtained, its reliability, convergence, and sampling quality can be tested by several methods such as Chapman–Kolmogorov tests [37], among others [38–40].

A disadvantage of applying classical MSMs is that the sampling of rare events can still be highly inefficient, which may result in improperly connected models and imprecise residence times. This can typically be solved by extending the simulation time, which is unpractical when computational resources are limited [27]. Furthermore, the sampling of the conformational space and the MSM can be improved by adaptive sampling, a type of biased sampling that identifies under-explored conformational states in MD data and generates new starting points to explore those states by resampling. Therefore, instead of requiring long simulations to model rare events, several shorter MD trajectories (~10–100 ns) can be used. This facilitates the parallelization of the computational burden of the MD among many processors [41]. A successful application of MSM adaptive sampling has recently contributed to a better understanding of functional selectivity – the phenomena explaining receptor's selectivity for certain signal transduction pathways in front of others due to the interaction with a biased ligand – in the μ-opioid receptor. The μ-opioid receptor was simulated in complex with either a balanced or a G protein–biased agonist, and MSM analysis allowed identifying a differential ligand-specific dynamic behavior of the receptor with kinetically distinct conformational states. Such information can drive the rational design of functionally selective ligands that stabilize the receptor in a specific state, which may eventually be developed into improved drugs [42].

8.3 Application of MD Simulations to GPCR Drug Design: Why Should We Use MD?

MD simulations provide unique insights that can be critical for the understanding of GPCRs. Given the fact that GPCRs are crucial drug targets, the insights yielded by MD simulations are especially relevant for the discovery of new and safer drugs [1, 43]. In this respect, drug designers are interested in understanding phenomena such as drug binding pathways, drug–receptor interaction, receptor subtype selectivity, conformational changes related to receptor (in)activation, and the role of lipids in GPCR modulation.

The entrance of a drug into the receptor binding site is a multistep process, with diverse metastable and/or intermediate binding sites. The analysis of the crystal structures of GPCRs bound to a ligand grants information on the binding mode of the ligand, but not on the pathway that it follows from the extracellular phase into the binding pocket. MD simulations generate useful insights on such binding pathways, since they enable the study of the development of the molecular system over a given timescale and provide an estimation of the rate at which a ligand associates or dissociates to the receptor. Having knowledge on the ligand binding pathways allows the identification of energetic barriers encountered along those pathways. Energetic barriers can substantially affect the binding and unbinding rates of a ligand into a receptor. Therefore, such information constitutes a foundation for the rational optimization of drug binding and unbinding kinetics [44], which play a critical role in drug efficacy, selectivity, and safety [45–47]. Moreover, the identification of metastable binding sites may be useful for the development of dimeric or bivalent compounds that can bind to different metastable binding sites, which may result in an increased affinity [48]. Some examples of ligands for which the binding pathway was described thanks to MD are (*S*)-alprenolol binding to the β_2-adrenergic receptor (β_2AR), histamine to the histamine H_4 receptor (hH_4R), and clozapine and haloperidol to the dopamine receptor D_3 (D_3R) [44, 49, 50]. Particularly relevant is the case of the β_2AR, which was the first unbiased MD simulation study capturing the full process of ligands spontaneously binding to a GPCR, achieving the final poses of the ligands without incorporation of any prior knowledge of the binding site. Results revealed not only the predominant pathway into the binding site but also the two main energetic barriers that govern drug binding and unbinding kinetics [44].

Concerning drug–receptor interaction, MD studies of this process can provide valuable information that cannot be obtained through rigid docking methods. Such insights are particularly useful for the identification or design of new ligands for a given GPCR. This is possible because MD simulations take into account the flexibility of GPCR binding pockets, which differ from one conformational state of the receptor to another and even within a single global receptor conformational state [51]. Indeed, considering multiple possible receptor structures generally increases the diversity of ligands identified. Furthermore, the goal of GPCR-related drug discovery is typically to find a ligand that not only binds to the target but also achieves a particular signaling profile. For example, the study of biased ligands, which selectively engage one signaling pathway downstream of the receptor over the others, is remarkably promising for drug design, since they can be used to inhibit the disease-associated pathways while stimulating non-pathological pathways up to their physiological levels, avoiding side effects [52]. The signaling outcome triggered by a GPCR depends on the conformational state of the receptor and thus of the binding pocket that is stabilized by the drug. In this respect, simulations can be used to compare drug–receptor interactions in different conformational states of the receptor or using different types of ligands (e.g. unbiased agonist, biased agonist, inverse agonist or antagonists) [2]. A successful story of identifying new ligands with an unprecedented level of bias was carried out for the serotonin 5-hydroxytryptamine receptor 2A (5-HT_{2A}). Martí-Solano et al. used extensive MD simulations to characterize the dynamics of ligand–receptor interactions of known biased and balanced agonists with the purpose of becoming capable of discriminating the different types of receptor agonists. Thanks to this information, they discovered new biased ligands of outstanding efficacy by tuning the structure of the balanced natural ligand serotonin. These compounds represent valuable tools to promote the design of improved, safer antipsychotic drugs [53]. Another fruitful study concerning biased agonists is the recent development of an approach for designing β-arrestin-biased ligands for the dopamine receptor D_2 (D_2R) based on structural and MD data. Particularly, specific conserved receptor–ligand contacts responsible for biased signaling were identified and modified in order to develop new biased agonists for this GPCR [54].

Another important goal of drug discovery is the creation of drugs that specifically target one receptor subtype over another. Many GPCR subtypes exhibit a highly conserved orthosteric binding site, such that a single ligand can bind to several receptors simultaneously, contributing to off-target side effects [55]. An effective strategy to achieve better selectivity is the design of drugs that target allosteric pockets, which are less conserved among GPCRs [56]. Besides their specificity, allosteric modulators have the advantage of providing a strategy to fine-tune cellular responses triggered by a ligand. Allosteric modulators can bind to a GPCR concomitantly to the orthosteric ligand, altering its binding affinity and/or cellular-signaling efficacy in a moderate manner avoiding severe effects on the cell [57–59]. For this reason, allostery has become an area of great interest for the discovery of new drugs with reduced side effects [59]. However, allosteric binding sites are not evident from crystal structures, which hamper the structure-based design of allosteric drugs. Moreover, the molecular mechanisms by which such modulators affect GPCR signaling may depend on dynamical properties that would not be evident from a single static structure. MD simulations can provide a means to detect hidden allosteric binding sites and determine the mechanistic basis of allosteric regulation [60]. One representative case of a GPCR family displaying dramatically low subtype selectivity concerning orthosteric ligands are the muscarinic acetylcholine receptors. This family comprises important drug targets for several central nervous system diseases [61]. MD-based approaches have been critical for the efficient identification of allosteric mechanisms in muscarinic receptors [58, 62], and such findings are usually paradigmatic for the entire GPCR superfamily [61]. For example, atomic-level MD simulations by Dror et al. provided a structural basis of allosteric ligand binding and elucidated mechanisms of cooperativity between the allosteric and the orthosteric ligand. These studies enabled the design of chemical modifications that substantially alter the final effect of allosteric modulators [62].

Besides GPCR subtypes, it can also be informative to study the differences between wild-type GPCRs and variants responsible for disease susceptibility or a distinct drug response. Such functional differences have been predicted to be caused by variations in ligand binding site accessibility and structure, which are determined by the dynamics of the receptor. Hence, MD simulations are a promising approach to elucidate the molecular mechanisms responsible for such interindividual differences. This remains an emerging field of study that could provide important clues on the functional single-nucleotide polymorphisms (SNPs) identified in the recent large-scale sequencing initiatives [12]. A well-defined case of SNP in GPCRs is the Arg16Gly variant of the β_2AR, which has been linked to a differential response to albuterol, a β_2AR agonist commonly used in the treatment of asthma. MD analysis of the conformational differences between the two variants revealed that divergences in the position and dynamics of the N-terminal region, where the SNP is located, lead to long-range effects at the ligand binding site, reducing the accessibility of the ligand at one of the variants [63].

Another phenomenon of interest is GPCR (in)activation, which is still not completely understood. During the process of activation, GPCRs undergo global conformational changes visiting inactive, intermediate, and active states. Currently available GPCR crystal structures provide high-resolution insights into some of these conformational states [64, 65]. However, these structures represent snapshots of GPCRs in a given conformation, and they do not reveal the mechanism by which they transition between different states. A promising approach to tackle this issue is combining high-resolution experimental data with MD simulation studies. MD simulations can provide detailed information on the transitions between conformational states, helping us explore the GPCR conformational landscape in the presence of different ligands. A notorious case of successful MD-based study of GPCR activation was performed by Dror et al. They proposed an activation mechanism for the β_2AR based on atomic-level simulations in which an agonist-bound receptor transitioned spontaneously from the active to the inactive conformation. Interestingly, their simulations highlighted the existence of an intermediate state that was suggested to represent a receptor conformation to which G protein binds during activation [14]. Yuan et al. also used MD to monitor the activation process of the β_2AR, along with the adenosine A_{2A} receptor and rhodopsin. Their simulations allowed the characterization of the formation of an intrinsic water pathway, which, in the receptors' resting state, is interrupted by a hydrophobic layer of amino acid residues. Upon agonist binding, this hydrophobic layer opens to allow the formation of a continuous intrinsic water channel [17]. Another highlight was achieved by Kohlhoff et al. when they captured the mechanism of β_2AR (in)activation in a simulation of unprecedented total length thanks to Google's Exacycle cloud computing platform. Such a study revealed multiple (in)activation pathways, and showed that agonists and inverse agonists act by modulating the receptor dynamics to prefer different pathways [66].

Finally, a lot of attention has recently been devoted to the role of the membrane environment or of specific membrane lipids in GPCR functioning. Membrane phospholipids have been found to allosterically modulate the activity [67] and oligomerization [68] of GPCRs, while membrane cholesterol can regulate its stability, ligand-binding properties and function [69–71]. Still, the precise nature of cholesterol's implication in GPCR modulation is a matter of debate [71]. Cholesterol modulation could be due to indirect effects (changes in membrane biophysical properties) [72, 73], direct cholesterol–GPCR interactions [74–77] or both. Moreover, some studies suggest a potential allosteric role of cholesterol [78], while others propose that cholesterol molecules can compete with orthosteric ligands by entering the receptor's binding site [78–80]. Understanding the influence of cholesterol on GPCR function would allow us to explore potential therapeutic uses of membrane sterols or sterol-mimetic molecules in GPCR drug discovery. Still, molecular mechanisms behind cholesterol modulation are difficult to ascertain using experimental methods alone, and thus MD simulations are increasingly being applied. For example, Manna et al. used extensive MD simulations to provide detailed insights into how the β_2AR is allosterically modulated by cholesterol. Their simulations showed a decrease in the receptor's conformational flexibility due to cholesterol binding at specific high-affinity sites located near the transmembrane helices 5–7 [81]. On the other side, Guixà-González et al. recently showed that cholesterol can spontaneously enter the adenosine A_{2A} receptor's binding pocket from the membrane milieu, unveiling a new interaction mode between cholesterol and this receptor that could potentially apply to other GPCRs [79].

8.4 Evolution of MD Timescales

It has been 40 years since the publication of the first MD simulation of a macromolecule of biological interest, which was instrumental in replacing the view of proteins as relatively rigid structures with the understanding that they are dynamic systems. That first simulation consisted of a 9.2-ps trajectory of a small protein, the bovine pancreatic trypsin inhibitor (BPTI, ~500 atoms), in vacuum [82]. From that point forward, thanks to continuous advances in both methodology and computational resources, MD simulations have gradually extended to larger systems and longer timescales. Interestingly, the first published MD simulation of a GPCR was reported in 1991, with a simulation time of 80 ps [83] (see Figure 8.5).

A noteworthy event that highly influenced the development of MD capacities was the release of the distributed volunteer computing project Folding@home in the year 2000. Thanks to its distributed computing environment, the release of

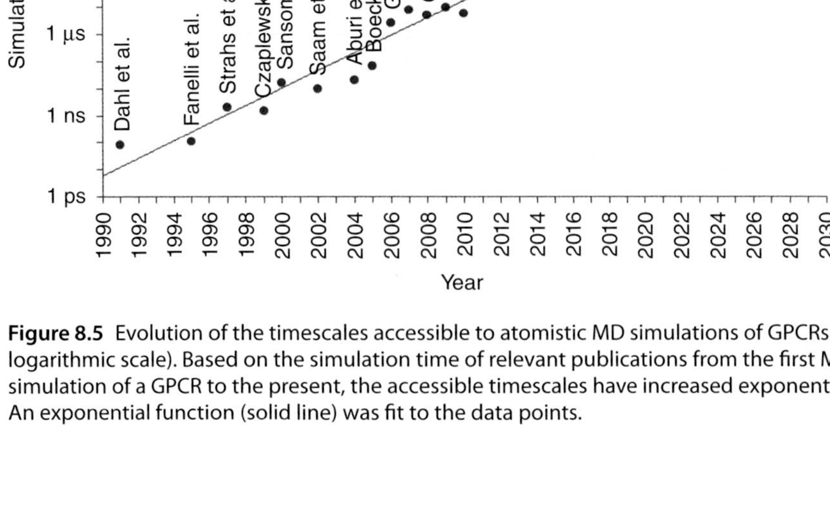

Figure 8.5 Evolution of the timescales accessible to atomistic MD simulations of GPCRs (in logarithmic scale). Based on the simulation time of relevant publications from the first MD simulation of a GPCR to the present, the accessible timescales have increased exponentially. An exponential function (solid line) was fit to the data points.

Folding@home made it possible to break previous computational barriers for MD simulations, reaching timescales of orders of magnitude longer than what had previously been achieved [84, 85]. At its release, it was already able to simulate the folding process of the 23-residue protein BBA5, which has a duration of approximately 10 μs, by producing thousands of independent, short trajectories totaling an accumulated simulation time of hundreds of microseconds [86].

The subsequent breakthrough in the MD field occurred at 2008, when the Anton supercomputer was created. What made Anton stand out from previous hardware was the fact that it is a specialized, massively parallel supercomputer particularly designed to perform MD simulations of molecular systems. This specificity makes Anton highly efficient, being able to increase the speed of MD computations and to generate simulations with longer timescales. For example, large systems such as a GPCR embedded in a lipid bilayer ($\sim 10^5$ atoms) could be simulated for hundreds of microseconds [87, 88]. This is the case of the β_2AR activation studies conducted by Dror et al., in which a total simulation time of more than 600 μs was reached thanks to Anton [14]. Moreover, Anton simulations of the muscarinic M2 and M3 receptors captured the binding of an important bronchodilator drug at an allosteric site, offering the first structural view of an orthosteric GPCR ligand binding to an allosteric site [89].

The enhancement in the performance of MD simulations has also been heavily influenced by algorithmic advances, including the fine-tuning of energy calculations, parallelization improvements and the use of graphical processing units (GPUs) [24]. Specifically, the development of algorithms exploiting GPUs has become a major breakthrough in simulation codes. During the late 2000s, such progresses in GPU exploitation allowed to use them as performance accelerators for a wide variety of scientific applications, including MD [90, 91]. For example, the use of GPUs has made the simulation of GPCRs in explicit lipid–water environments feasible within reasonable computational times [92]. In 2007, a volunteer-distributed computing project based on GPU resources named GPUGRID was launched, with the capability of simulating thousands of all-atom molecular trajectories at an average of 20 ns/day each (for systems of ~30 000–80 000 atoms) [93]. In 2010, this approach was used to reveal for the first time the molecular details of sodium ion binding to its allosteric binding site for a class A GPCR, accumulating more than 6 μs of simulation time for a system of about 60 000 atoms [94].

Despite the advances in specialized hardware and software for MD, such as supercomputers and distributed computing platforms, most of these tools are not openly available to the whole scientific community and can only be accessed by a set of research groups, which hampers the advance of the study of long-timescale processes. In this respect, cloud computing is a valuable alternative that can bring long-timescale processes, in particular those associated with GPCRs, within reach of a broader community.

All in all, current simulation timescales reach microseconds (conventional hardware [27]) to milliseconds (specialized hardware [26] / distributed computing [96, 97]). Simulations of systems having ~50 000–100 000 atoms are now routine, and simulations of approximately 500 000 atoms are common when appropriate computer facilities are available [24]. Considering the evolution of

GPCR-related MD since the publication of the first MD simulation study, it is undeniable that there has been a dramatic increase in the timescales we are able to achieve. This tendency is highly influenced by the technological development, both in terms of computational cost reduction and methodological improvement. Certainly, the evolution of GPCR-related MD is a direct consequence of Moore's law, which describes how the performance of integrated circuits has been increasing exponentially over the past half-century [98]. Based on this trend, we can extrapolate that the second timescale for GPCR simulations will be reached approximately in 2024 (see Figure 8.5). In fact, several independent authors have reached similar conclusions [92, 99].

8.5 Sharing MD Data via a Public Database

In view of the growing relevance of MD simulations for the study of GPCRs (see Figures 8.1 and 8.5), it is increasingly necessary to find efficient systems for the storage, indexing, and dissemination of all the simulation data being generated. For this reason, initiatives involving the development of MD databases are gaining importance. Regarding GPCRs, the GPCRmd database [100] is being created with the purpose of fostering GPCR MD data from all over the world. Broadly speaking, GPCRmd offers storage of MD data, citable IDs for referencing such data in publications, basic analysis tools, and online visualization of the simulations. Being the first GPCR-specific MD database, GPCRmd sets standards in archiving simulation data of this protein family in a structured and searchable manner with the aim of facilitating data retrieval for other scientists.

One important advantage of GPCRmd is that, by providing public access to the data on which scientific articles are based, it increases the transparency and credibility of MD studies. This allows other researchers to evaluate the quality of the obtained results and enhances reproducibility, which is critical for the advancement of GPCR research.

Finally, tools such as the GPCRmd database also contribute to the popularization of MD among the research community by offering freely available web resources, including user-friendly representation and analysis options as well as the display of results with easy-to-interpret and interactive graphs. These resources approach GPCR MD data to all researchers studying this protein family, including to those who are not familiarized with computational tools or complex MD analysis software, and contributes to the widespread adoption of MD.

8.6 Conclusions and Perspectives

MD simulations can provide a high-resolution view of the dynamics of biological systems relevant to human health. They can either capture atomic-level motions within conformational states or structural transitions between different conformational populations, bringing within reach information that is highly difficult to obtain by other methods [2]. For this reason, MD simulations are specially promising for the study of the functionality of complex signaling proteins such as GPCRs, which are drug targets of striking importance in the pharmaceutical industry [5].

However, there are still important drawbacks to the usability of MD simulations for the study of GPCR functionality. One main limitation to this technique is the generation of simulations long enough to allow an adequate sampling of biological events such as receptor activation upon agonist binding [12, 92].

Fortunately, with ongoing advances in both MD hardware and software, GPCR-related MD simulations are being extended to larger systems and longer timescales. If this trend continues, we will soon reach simulation times of the order of a second (see Figure 8.5), bridging the gap between the timescales of biological processes observed *in vivo* and those accessible *in silico*. This increase of the simulation time will grant us the opportunity to extend the application of MD simulations to the study of processes that were previously difficult to analyze through this method, including global conformational rearrangements, receptor dimerization, and coupling to intracellular signaling proteins [24].

Given the huge druggability of GPCRs, the predicted progress of MD simulations of this protein superfamily will have a great impact on drug discovery. Despite multiple studies underlining the importance of GPCR flexibility to molecular recognition and signaling, most drug discovery programs currently disregard MD analysis because of their computational expenses. With the forthcoming reduction of the computational costs associated to MD simulations, this technique is expected to be more commonly applied at the pharmaceutical industry and, eventually, to be commonly included in drug discovery pipelines [8, 9, 88]. Particularly, MD can be of great use through the hit discovery, hit-to-lead and lead optimization processes by identifying possible ligands for a given GPCR and assessing the stability and dynamics of the binding poses, along with receptor–ligand binding affinities and even kinetics [101].

Acknowledgments

M.T.-F. acknowledges financial support from the Spanish Ministry of Education, Culture and Sport (FPU16/01209), T.M.S acknowledges financial support from Hospital del Mar Medical Research Institute and help from the National Science Centre of Poland, project number 2017/27/N/NZ2/02571, I.R.-E. acknowledges Secretaria d'Universitats i Recerca del Departament d'Economia i Coneixement de la Generalitat de Catalunya (2015 FI_B00145) for its financial support, J.S. acknowledges financial support from Instituto de Salud Carlos III FEDER (PI15/00460).

References

1 Santos, R., Ursu, O., Gaulton, A. et al. (2017). A comprehensive map of molecular drug targets. *Nat. Rev. Drug Discov.* 16 (1): 19–34.

2 Latorraca, N.R., Venkatakrishnan, A.J., and Dror, R.O. (2017). GPCR dynamics: structures in motion. *Chem. Rev.* 117 (1): 139–155.

3 Martí-Solano, M., Schmidt, D., Kolb, P., and Selent, J. (2016). Drugging specific conformational states of GPCRs: challenges and opportunities for computational chemistry. *Drug Discov. Today* 21 (4): 625–631.

4 Rosenbaum, D.M., Rasmussen, S.G.F., and Kobilka, B.K. (2009). The structure and function of G-protein-coupled receptors. *Nature* 459 (7245): 356–363.

5 Ciancetta, A., Sabbadin, D., Federico, S. et al. (2015). Advances in computational techniques to study GPCR–ligand recognition. *Trends Pharmacol. Sci.* 36 (12): 878–890.

6 McRobb, F.M., Negri, A., Beuming, T., and Sherman, W. (2016). Molecular dynamics techniques for modeling G protein-coupled receptors. *Curr. Opin. Pharmacol.* 30: 69–75.

7 Johnston, J.M. and Filizola, M. (2011). Showcasing modern molecular dynamics simulations of membrane proteins through G protein-coupled receptors. *Curr. Opin. Struct. Biol.* 21 (4): 552–558.

8 Dror, R.O., Jensen, M.Ø., Borhani, D.W., and Shaw, D.E. (2010). Exploring atomic resolution physiology on a femtosecond to millisecond timescale using molecular dynamics simulations. *J. Gen. Physiol.* 135 (6): 555–562.

9 Adcock, S.A. and McCammon, J.A. (2006). Molecular dynamics: survey of methods for simulating the activity of proteins. *Chem. Rev.* 106 (5): 1589–1615.

10 Hug, S. (2013). Classical molecular dynamics in a nutshell. *Methods Mol. Biol. (Clifton, N.J.)* 924: 127–152.

11 Bond, P.J., Holyoake, J., Ivetac, A. et al. (2007). Coarse-grained molecular dynamics simulations of membrane proteins and peptides. *J. Struct. Biol.* 157 (3): 593–605.

12 Sengupta, D., Joshi, M., Athale, C.A., and Chattopadhyay, A. (2016). What can simulations tell us about GPCRs: integrating the scales. *Methods Cell Biol.* 132: 429–452.

13 Nygaard, R., Zou, Y., Dror, R.O. et al. (2013). The dynamic process of β2-adrenergic receptor activation. *Cell* 152 (3): 532–542.

14 Dror, R.O., Arlow, D.H., Maragakis, P. et al. (2011). Activation mechanism of the β2-adrenergic receptor. *Proc. Natl. Acad. Sci. U.S.A.* 108 (46): 18684–18689.

15 Manglik, A.I., Kim, T.H., Masureel, M. et al. (2015). Structural insights into the dynamic process of β2-adrenergic receptor signaling. *Cell* 161 (5): 1101–1111.

16 Staus, D.P., Strachan, R.T., Manglik, A. et al. (2016). Allosteric nanobodies reveal the dynamic range and diverse mechanisms of G-protein-coupled receptor activation. *Nature* 535 (7612): 448–452.

17 Yuan, S., Filipek, S., Palczewski, K., and Vogel, H. (2014). Activation of G-protein-coupled receptors correlates with the formation of a continuous internal water pathway. *Nat. Commun.* 5: 4733.

18 Freddolino, P.L., Harrison, C.B., Liu, Y., and Schulten, K. (2010). Challenges in protein folding simulations: timescale, representation, and analysis. *Nat. Phys.* 6 (10): 751–758.

19 Martí-Solano, M., Guixà-González, R., Sanz, F. et al. (2013). Novel insights into biased agonism at G protein-coupled receptors and their potential for drug design. *Curr. Pharm. Des.* 19 (28): 5156–5166.

20 Okamoto, Y. (2004). Generalized-ensemble algorithms: enhanced sampling techniques for Monte Carlo and molecular dynamics simulations. *J. Mol. Graph. Model.* 22 (5): 425–439.

21 Lei, H. and Duan, Y. (2007). Improved sampling methods for molecular simulation. *Curr. Opin. Struct. Biol.* 17 (2): 187–191.

22 Liao, C. and Zhou, J. (2014). Replica-exchange molecular dynamics simulation of basic fibroblast growth factor adsorption on hydroxyapatite. *J. Phys. Chem. B* 118 (22): 5843–5852.

23 Chen, C., Huang, Y., and Xiao, Y. (2013). Enhanced sampling of molecular dynamics simulation of peptides and proteins by double coupling to thermal bath. *J. Biomol. Struct. Dyn.* 31 (2): 206–214.

24 Hospital, A., Goñi, J.R., Orozco, M., and Gelpí, J.L. (2015). Molecular dynamics simulations: advances and applications. *Adv. Appl. Bioinform. Chem.* 8: 37–47.

25 Bernardi, R.C., Melo, M.C.R., and Schulten, K. (2015). Enhanced sampling techniques in molecular dynamics simulations of biological systems. *Biochim. Biophys. Acta, Gen. Subj.* 1850 (5): 872–877.

26 Shaw, D.E., Maragakis, P., Lindorff-Larsen, K. et al. (2010). Atomic-level characterization of the structural dynamics of proteins. *Science* 330 (6002): 341–346.

27 Paul, F., Wehmeyer, C., Abualrous, E.T. et al. (2017). Protein–peptide association kinetics beyond the seconds timescale from atomistic simulations. *Nat. Commun.* 8 (1): 1095.

28 Henzler-Wildman, K. and Kern, D. (2007). Dynamic personalities of proteins. *Nature* 450 (7172): 964–972.

29 Manglik, A. and Kobilka, B. (2014). The role of protein dynamics in GPCR function: insights from the β2AR and rhodopsin. *Curr. Opin. Cell Biol.* 27: 136–143.

30 Rodríguez-Espigares, I., Kaczor, A.A., and Selent, J. (2016). In silico exploration of the conformational universe of GPCRs. *Mol. Inform.* 35 (6–7): 227–237.

31 Brown, W.M., Martin, S., Pollock, S.N. et al. (2008). Algorithmic dimensionality reduction for molecular structure analysis. *J. Chem. Phys.* 129 (6): 64118.

32 Ng, H.W., Laughton, C.A., and Doughty, S.W. (2013). Molecular dynamics simulations of the adenosine A2a receptor: structural stability, sampling, and convergence. *J. Chem. Inf. Model.* 53 (5): 1168–1178.

33 Grossfield, A. and Zuckerman, D.M. (2009). Quantifying uncertainty and sampling quality in biomolecular simulations. *Annu. Rep. Comput. Chem.* 5: 23–48.

34 Pan, A.C. and Roux, B. (2008). Building Markov state models along pathways to determine free energies and rates of transitions. *J. Chem. Phys.* 129 (6): 64107.

35 Noé, F. and Fischer, S. (2008). Transition networks for modeling the kinetics of conformational change in macromolecules. *Curr. Opin. Struct. Biol.* 18 (2): 154–162.

36 Prinz, J.-H., Wu, H., Sarich, M. et al. (2011). Markov models of molecular kinetics: generation and validation. *J. Chem. Phys.* 134 (17): 174105.

37 Noé, F., Schütte, C., Vanden-Eijnden, E. et al. (2009). Constructing the equilibrium ensemble of folding pathways from short off-equilibrium simulations. *Proc. Natl. Acad. Sci. U.S.A.* 106 (45): 19011–19016.

38 Swope, W.C., Pitera, J.W., and Suits, F. (2004). Describing protein folding kinetics by molecular dynamics simulations. *J. Phys. Chem. B* 108 (21): 6571–6581.

39 Park, S. and Pande, V.S. (2006). Validation of Markov state models using Shannon's entropy. *J. Chem. Phys.* 124 (5): 54118.

40 Bacallado, S., Chodera, J.D., and Pande, V. (2009). Bayesian comparison of Markov models of molecular dynamics with detailed balance constraint. *J. Chem. Phys.* 131 (4): 45106.

41 Bowman, G.R., Ensign, D.L., and Pande, V.S. (2010). Enhanced modeling via network theory: adaptive sampling of Markov state models. *J. Chem. Theory Comput.* 6 (3): 787–794.

42 Kapoor, A., Martinez-Rosell, G., Provasi, D. et al. (2017). Dynamic and kinetic elements of μ-opioid receptor functional selectivity. *Sci. Rep.* 7 (1): 11255.

43 Lundstrom, K. (2006). Latest development in drug discovery on G protein-coupled receptors. *Curr. Protein Pept. Sci.* 7 (5): 465–470.

44 Dror, R.O., Pan, A.C., Arlow, D.H. et al. (2011). Pathway and mechanism of drug binding to G-protein-coupled receptors. *Proc. Natl. Acad. Sci. U.S.A.* 108 (32): 13118–13123.

45 Guo, D., Mulder-Krieger, T., IJzerman, A.P., and Heitman, L.H. (2012). Functional efficacy of adenosine A_2A receptor agonists is positively correlated to their receptor residence time. *Br. J. Pharmacol.* 166 (6): 1846–1859.

46 Swinney, D.C. (2008). Applications of binding kinetics to drug discovery. *Pharm. Med.* 22 (1): 23–34.

47 Lu, H. and Tonge, P.J. (2010). Drug-target residence time: critical information for lead optimization. *Curr. Opin. Chem. Biol.* 14 (4): 467–474.

48 Strasser, A., Wittmann, H.-J., and Seifert, R. (2017). Binding kinetics and pathways of ligands to GPCRs. *Trends Pharmacol. Sci.* 38 (8): 717–732.

49 Wittmann, H.-J. and Strasser, A. (2015). Binding pathway of histamine to the hH4R, observed by unconstrained molecular dynamics. *Bioorg. Med. Chem. Lett.* 25 (6): 1259–1268.

50 Thomas, T., Fang, Y., Yuriev, E., and Chalmers, D.K. (2016). Ligand binding pathways of clozapine and haloperidol in the dopamine D_2 and D_3 receptors. *J. Chem. Inf. Model.* 56 (2): 308–321.

51 Lebon, G., Warne, T., and Tate, C.G. (2012). Agonist-bound structures of G protein-coupled receptors. *Curr. Opin. Struct. Biol.* 22 (4): 482–490.

52 Shukla, A.K., Singh, G., and Ghosh, E. (2014). Emerging structural insights into biased GPCR signaling. *Trends Biochem. Sci.* 39 (12): 594–602.

53 Martí-Solano, M., Iglesias, A., de Fabritiis, G. et al. (2015). Detection of new biased agonists for the serotonin 5-HT2A receptor: modeling and experimental validation. *Mol. Pharmacol.* 87 (4): 740–746.

54 McCorvy, J.D., Butler, K.V., Kelly, B. et al. (2017). Structure-inspired design of β-arrestin-biased ligands for aminergic GPCRs. *Nat. Chem. Biol.* 14 (2): 126–134.

55 Allen, J.A. and Roth, B.L. (2011). Strategies to discover unexpected targets for drugs active at G protein–coupled receptors. *Annu. Rev. Pharmacol. Toxicol.* 51 (1): 117–144.

56 Christopoulos, A. (2014). Advances in G protein-coupled receptor allostery: from function to structure. *Mol. Pharmacol.* 86 (5): 463–478.

57 Jakubík, J. and El-Fakahany, E.E. (2010). Allosteric modulation of muscarinic acetylcholine receptors. *Pharmaceuticals (Basel).* 3 (9): 2838–2860.

58 Miao, Y., Goldfeld, D.A., Von Moo, E. et al. (2016). Accelerated structure-based design of chemically diverse allosteric modulators of a muscarinic G protein-coupled receptor. *Proc. Natl. Acad. Sci. U.S.A.* 113 (38): E5675–E5684.

59 Nguyen, T., Li, J.-X., Thomas, B.F. et al. (2017). Allosteric modulation: an alternate approach targeting the cannabinoid CB1 receptor. *Med. Res. Rev.* 37 (3): 441–474.

60 Hertig, S., Latorraca, N.R., and Dror, R.O. (2016). Revealing atomic-level mechanisms of protein allostery with molecular dynamics simulations. *PLoS Comput. Biol.* 12 (6): e1004746.

61 Bock, A., Schrage, R., and Mohr, K. (2017). Allosteric modulators targeting CNS muscarinic receptors. *Neuropharmacology* 136: 427–437.

62 Dror, R.O., Green, H.F., Valant, C. et al. (2013). Structural basis for modulation of a G-protein-coupled receptor by allosteric drugs. *Nature* 503 (7475): 295.

63 Shahane, G., Parsania, C., Sengupta, D., and Joshi, M. (2014). Molecular insights into the dynamics of pharmacogenetically important N-terminal variants of the human β2-adrenergic receptor. *PLoS Comput. Biol.* 10 (12): e1004006.

64 Rasmussen, S.G.F., DeVree, B.T., Zou, Y. et al. (2011). Crystal structure of the β2 adrenergic receptor–Gs protein complex. *Nature* 477 (7366): 549–555.

65 Kang, Y., Zhou, X.E., Gao, X. et al. (2015). Crystal structure of rhodopsin bound to arrestin by femtosecond X-ray laser. *Nature* 523 (7562): 561–567.

66 Kohlhoff, K.J., Shukla, D., Lawrenz, M. et al. (2014). Cloud-based simulations on Google Exacycle reveal ligand modulation of GPCR activation pathways. *Nat. Chem.* 6 (1): 15–21.

67 Dawaliby, R., Trubbia, C., Delporte, C. et al. (2016). Allosteric regulation of G protein-coupled receptor activity by phospholipids. *Nat. Chem. Biol.* 12 (1): 35–39.

68 Guixà-González, R., Javanainen, M., Gómez-Soler, M. et al. (2016). Membrane omega-3 fatty acids modulate the oligomerisation kinetics of adenosine A2A and dopamine D2 receptors. *Sci. Rep.* 6 (1): 19839.

69 Paila, Y.D. and Chattopadhyay, A. (2009). The function of G-protein coupled receptors and membrane cholesterol: specific or general interaction? *Glycoconj. J.* 26 (6): 711–720.

70 Oates, J. and Watts, A. (2011). Uncovering the intimate relationship between lipids, cholesterol and GPCR activation. *Curr. Opin. Struct. Biol.* 21 (6): 802–807.

71 Gimpl, G. (2016). Interaction of G protein coupled receptors and cholesterol. *Chem. Phys. Lipids* 199: 61–73.

72 Zocher, M., Zhang, C., Rasmussen, S.G.F. et al. (2012). Cholesterol increases kinetic, energetic, and mechanical stability of the human 2-adrenergic receptor. *Proc. Natl. Acad. Sci. U. S. A.* 109 (50): E3463–E3472.

73 Khelashvili, G., Mondal, S., Andersen, O.S., and Weinstein, H. (2010). Cholesterol modulates the membrane effects and spatial organization of membrane-penetrating ligands for G-protein coupled receptors. *J. Phys. Chem. B* 114 (37): 12046–12057.

74 Cherezov, V., Rosenbaum, D.M., Hanson, M.A. et al. (2007). High-resolution crystal structure of an engineered human 2-adrenergic G protein-coupled receptor. *Science* 318 (5854): 1258–1265.

75 Hanson, M.A., Cherezov, V., Griffith, M.T. et al. (2008). A specific cholesterol binding site is established by the 2.8 Å structure of the human β2-adrenergic receptor. *Structure* 16 (6): 897–905.

76 Liu, W., Chun, E., Thompson, A.A. et al. (2012). Structural basis for allosteric regulation of GPCRs by sodium ions. *Science* 337 (6091): 232–236.

77 Wu, H., Wang, C., Gregory, K.J. et al. (2014). Structure of a class C GPCR metabotropic glutamate receptor 1 bound to an allosteric modulator. *Science* 344 (6179): 58–64.

78 Song, Y., Kenworthy, A.K., and Sanders, C.R. (2014). Cholesterol as a co-solvent and a ligand for membrane proteins. *Protein Sci.* 23 (1): 1–22.

79 Guixà-González, R., Albasanz, J.L., Rodriguez-Espigares, I. et al. (2017). Membrane cholesterol access into a G-protein-coupled receptor. *Nat. Commun.* 8: 14505.

80 Benned-Jensen, T., Norn, C., Laurent, S. et al. (2012). Molecular characterization of oxysterol binding to the Epstein-Barr virus-induced gene 2 (GPR183). *J. Biol. Chem.* 287 (42): 35470–35483.

81 Manna, M., Niemelä, M., Tynkkynen, J. et al. (2016). Mechanism of allosteric regulation of β2-adrenergic receptor by cholesterol. *Elife* 5.

82 McCammon, J.A., Gelin, B.R., and Karplus, M. (1977). Dynamics of folded proteins. *Nature* 267 (5612): 585–590.

83 Dahl, S.G., Edvardsen, O., and Sylte, I. (1991). Molecular dynamics of dopamine at the D2 receptor. *Proc. Natl. Acad. Sci. U.S.A.* 88 (18): 8111–8115.

84 Butler, D. (1999). Computing 2010: from black holes to biology. *Nature* 402 (Suppl 6761): C67–C70.

85 Larson, S.M., Snow, C.D., Shirts, M., and Pande, V.S. (2002). Folding@Home and Genome@Home: Using distributed computing to tackle previously intractable problems in computational biology. arXiv:0901.0866 [physics.bio-ph]

86 Snow, C.D., Nguyen, H., Pande, V.S., and Gruebele, M. (2002). Absolute comparison of simulated and experimental protein-folding dynamics. *Nature* 420 (6911): 102–106.
87 Shaw, D.E., Bowers, K.J., Chow, E. et al. (2009). Millisecond-scale molecular dynamics simulations on Anton. *Proc. Conf. High Perform. Comput. Networking, Storage Anal.* (SC '09), 1.
88 Borhani, D.W. and Shaw, D.E. (2012). The future of molecular dynamics simulations in drug discovery. *J. Comput.-Aided. Mol. Des.* 26 (1): 15–26.
89 Kruse, A.C., Hu, J., Pan, A.C. et al. (2012). Structure and dynamics of the M3 muscarinic acetylcholine receptor. *Nature* 482 (7386): 552–556.
90 Stone, J.E., Phillips, J.C., Freddolino, P.L. et al. (2007). Accelerating molecular modeling applications with graphics processors. *J. Comput. Chem.* 28 (16): 2618–2640.
91 Anderson, J.A., Lorenz, C.D., and Travesset, A. (2008). General purpose molecular dynamics simulations fully implemented on graphics processing units. *J. Comput. Phys.* 227 (10): 5342–5359.
92 Martínez-Rosell, G., Giorgino, T., Harvey, M.J., and de Fabritiis, G. (2017). Drug discovery and molecular dynamics: methods, applications and perspective beyond the second timescale. *Curr. Top. Med. Chem.* 17 (23): 2617–2625.
93 Buch, I., Harvey, M.J., Giorgino, T. et al. (2010). High-throughput all-atom molecular dynamics simulations using distributed computing. *J. Chem. Inf. Model.* 50 (3): 397–403.
94 Selent, J., Sanz, F., Pastor, M., and De Fabritiis, G. (2010). Induced effects of sodium ions on dopaminergic G-protein coupled receptors. *PLoS Comput. Biol.* 6 (8): e1000884.
95 Hellerstein, J.L., Kohlhoff, K.J., and Konerding, D.E. (2012). Science in the cloud: accelerating discovery in the 21st century. *IEEE Internet Comput.* 16 (4): 64–68.
96 Plattner, N., Doerr, S., De Fabritiis, G., and Noé, F. (2017). Complete protein–protein association kinetics in atomic detail revealed by molecular dynamics simulations and Markov modelling. *Nat. Chem.* 9 (10): 1005–1011.
97 Sultan, M.M., Denny, R.A., Unwalla, R. et al. (2017). Millisecond dynamics of BTK reveal kinome-wide conformational plasticity within the apo kinase domain. *Sci. Rep.* 7 (1): 15604.
98 Moore, G.E. (1965). Cramming more components onto integrated circuits. *Electronics* 38 (8): 114–117.
99 Vendruscolo, M. and Dobson, C.M. (2011). Protein dynamics: Moore's law in molecular biology. *Curr. Biol.* 21 (2): R68–R70.
100 Rodríguez-Espigares, I., Ramírez-Anguita, J.M., Torrens-Fontanals, M. et al. (2017). GPCRmd web page. http://www.gpcrmd.org/.
101 De Vivo, M., Masetti, M., Bottegoni, G., and Cavalli, A. (2016). Role of molecular dynamics and related methods in drug discovery. *J. Med. Chem.* 59 (9): 4035–4061.

9

Molecular Dynamics Applications to GPCR Ligand Design

Andrea Bortolato[1], *Francesca Deflorian*[1], *Giuseppe Deganutti*[2], *Davide Sabbadin*[3], *Stefano Moro*[2], *and Jonathan S. Mason*[1]

[1] *Heptares Therapeutics Ltd., BioPark, Broadwater Road, Welwyn Garden City, Herts AL7 3AX, UK*
[2] *University of Padova, Molecular Modeling Section (MMS), Department of Pharmaceutical and Pharmacological Sciences, Via Marzolo 5, Padua, 35131, Italy*
[3] *Syngenta Crop Protection Muenchwilen AG, WST-810.2.18, Schaffhauserstrasse, CH-4332 Stein, Switzerland*

9.1 Introduction

G protein–coupled receptors (GPCRs) represent the largest and most diverse superfamily of proteins coded by the human genome. These fall into three major classes that share the same overall structural topology of an extracellular N-terminus, seven transmembrane-spanning domain (TMD), and an intracellular C-terminus. The largest subfamily is class A, also known as the rhodopsin family or family A, and they can be activated by small molecules including amines, purines, fatty acids, and prostaglandins, as well as peptides to large glycoproteins. Their crucial role can be easily understood by considering that GPCRs are located in the plasma membrane of all cell types and they regulate many physiologic and pathophysiologic functions. Moreover, GPCRs mediate the action of more than 30% of all marketed drugs [1].

New opportunities for GPCR ligand discovery have been unlocked by recent breakthroughs in membrane protein crystallography. In fact, the accessibility of high-resolution X-ray structures of GPCRs has provided the possibility of applying structure-based methods to the design of drug candidates [1, 2]. Among the structure-based methods available now, the most widely used are the molecular docking and molecular dynamics (MD) approaches [3]. Such methods are now well established for soluble targets, but are only recently being applied to membrane proteins such as GPCRs. In particular, characterizations of the orthosteric, and increasingly allosteric, binding site (in terms of size, shape, and physicochemical properties) are analyzed to design ligands that are optimized to achieve high-affinity binding, in particular using molecular docking methods. Although docking is a helpful method to get insights into the orthosteric or allosteric ligand–GPCR recognition, it lacks the description of three fundamental aspects that might play a significant role in ligand binding: realistic microenvironment (in particular, for membrane proteins), protein

Biomolecular Simulations in Structure-Based Drug Discovery,
First Edition. Edited by Francesco L. Gervasio and Vojtech Spiwok.

flexibility and water molecule–mediated interactions [3]. Nowadays, MD is the methodology that better accounts for these critical aspects and it represents the most promising complement to molecular docking in medicinal chemistry and drug discovery. In particular, very recently the crucial role of water molecules in ligand–receptor recognition has emerged, even if the role of these water molecules is not always evident. It has been estimated, however, that at least in 85% of all available crystal structures the ligand–protein interaction is bridged through a water molecule. Therefore, instead of thinking of proteins and ligands interacting in vacuum, we should rather address protein–water–ligand interactions as a general view of how ligands act. Another hot topic in this field is related to the awareness that the rates at which ligands associate with and dissociate from receptors (binding kinetics) directly impact ligand efficacy. The molecular aspects of ligand–receptor binding kinetics remain poorly understood, especially when compared with the better-known factors that influence binding affinity. The rational modulation of kinetic parameters during lead optimization thus remains very challenging.

In the following, we discuss the application of MD simulations in GPCR ligand design, in particular regarding the contributions arising from water molecule–mediated interactions and from the atomistic representation of the ligand–receptor binding and unbinding events, describing several approaches based on membrane MD simulations that have been recently introduced and applied to GPCRs.

9.2 The Role of Water in GPCR Structure-Based Ligand Design

Water molecules have a crucial impact on protein function, affecting protein plasticity, allosterism, mediation of ligand binding, and protein–protein interactions [4–6]. The importance of waters for GPCRs has been supported by structural information [1, 7, 8] and studies showing how ordered waters interact with residues that are important in disease states, receptor activation, and signaling [9–11]. The available high-resolution crystal structures of GPCRs [1, 12] show a high number of water molecules present in the helical bundle [10, 11, 13]. They create a complex network of hydrogen-bond interactions connecting the orthosteric site in the extracellular part of the receptor, to the G protein and β-arresting binding location in the intracellular side [14]. Several of the recently solved pairs of active and inactive GPCR structures enabled the analysis of the role of the waters during receptor activation, intrinsically related to the central hydrophobic core of GPCRs. TM3 (transmembrane helix) and TM6 are at the heart of the common activation pathway thanks to the direct contact with every other helix except TM1 [15]. GPCR activation results in the following important conformation changes: (i) the movement in of TM5; (ii) the upward movement of TM3; (iii) the rotation of TM6; and (iv) the inward movements of TM7 and TM1. The rearrangement of the hydrophobic residues between TM3 and TM6 in the core of the receptor consists primarily of L3.43a, F6.44a, X6.40a (where X is a bulky hydrophobic residue, I, L, V, or M), and, to a lesser extent, X6.41a (residues indicated using the Generic GPCR Residue Numbering [16]). The rearrangement of these contacts seen in the active state structures allows for the gross TM movements, which opens a water channel through the hydrophobic core. The active conformation is transiently stabilized by the interactions with the water molecules, allowing the creation of an H-bonding network from the extracellular side of the receptor to the intracellular side. The overall result is the opening of the cytoplasmic face of the receptor where G proteins interact. Crystallographic data of the β_2 adrenergic receptor, the M_2 muscarinic receptor, and the μ-opioid receptor support that this extensive polar network between the ligand-binding pocket and the cytoplasmic domains has a central role in signal propagation [17].

Ligands displace unfavorable waters, replacing them with groups complementary to the protein surface, a key driving force of protein–ligand binding [18]. Water molecules in protein active sites are often (i) entropically unfavorable due to the orientational and positional restraints imposed by the protein cavity and (ii) energetically unfavorable due to the water molecule's inability to form a full complement of hydrogen bonds when solvating the protein surface. Understanding the desolvation penalty of H-bonding groups in GPCR sites enables the identification of key interactions essential for ligand binding. In 1985, Goodford [19] developed the GRID software to probe a protein binding site using a range of different functional groups, including water to identify areas of attraction (hot spots). This analysis of the binding pocket completed by the evaluation *in silico* of the thermodynamic properties of water molecules provides a valuable approach to evaluate the active site druggability [7, 8]. Small molecules suitable for drug development have balanced physicochemical properties compatible with specific regions of the protein characterized by a combination of hydrophilic and lipophilic/hydrophobic features (Figure 9.1a,b). Of particular interest are subregions of the binding site with mixed hydrophobic and hydrophilic properties occupied by displaceable water molecules predicted to have a free energy higher than bulk solvent. These areas are often exploited by high-affinity ligands to create productive interactions with the receptors, resulting in significant free energy of binding gain. For example, the druggability analysis of the binding site of CCR5 (PDB ID: 4MBS) in complex with maraviroc [22] highlights how this ligand efficiently fills three distinct small lipophilic/hydrophobic subpockets occupied by "unhappy" waters (Figure 9.1a). In another chemokine receptor, CXCR4 (PDB ID: 3ODU) [23], the ligand IT1t efficiently binds to a large and less druggable peptidic orthosteric site, mainly occupied by bulk waters and creates key interactions with a hydrophobic subpocket occupied by a cluster of three high-energy waters (Figure 9.1b).

It is also possible to evaluate the full complexity of the dynamic behavior of the water network interacting with the protein–ligand complex. This alternative approach based on MD simulation evaluates the time-dependent organization of water molecules during the final stage of the ligand–receptor recognition process [25]. It provides an analysis of the variation in fluid dynamics properties of water molecules upon ligand binding based on a clear two-dimensional graph. These fluid dynamics maps enable the identification of protein "hot spots" characterized

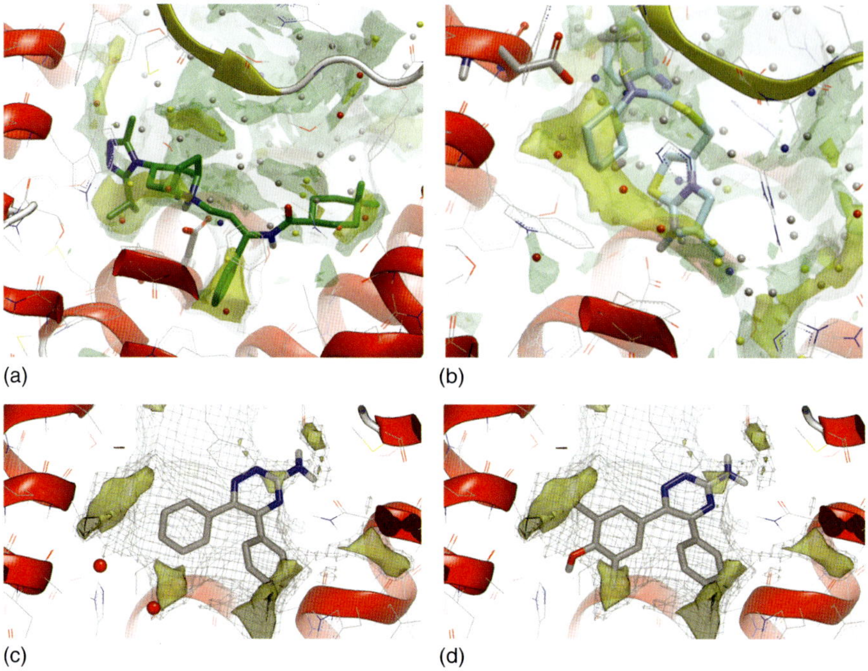

Figure 9.1 Binding site druggability analysis using GRID [19] to define the shape of the pocket and hydrophobic/hydrophilic hot spots together with analysis of locations and thermodynamic properties of water molecules using WM [20] and/or WaterFLAP [8, 21]. Lipophilic and hydrophilic hot spot regions are included as transparent surfaces, respectively, yellow (C1 = carbon sp^2 probe, contoured at −2.8 kcal mol^{-1}) and green (OH2 water probe, contoured at −5.0 kcal mol^{-1}). The shape of the pocket is shown in gray (C3 carbon sp^3 probe, contoured at used 1 kcal mol^{-1}). WM waters are shown as small spheres color coded according to their predicted relative free energy (red if $\Delta G > 3.5$ kcal mol^{-1}, yellow if ΔG between 2.5 and 3.5 kcal mol^{-1}, blue if $\Delta G < -1$ kcal mol^{-1}, and gray in the other cases). Ligands are shown in stick representation. Druggability analysis of the binding site of the chemokine receptors CCR5 (PDB ID: 4MBS) in complex with maraviroc [22] (a) and CXCR4 (PDB ID: 3ODU) in complex with IT1t [23] (b). In (c) and (d), two triazine antagonists of the A_{2A} receptor are shown [18, 24]: **4a** (c) traps 2 waters (shown as red spheres) resulting in a short ligand residence time (less than one second); these waters are displaced by **4f** (d) with a residence time of more than 12 minutes.

by peculiar shape and electrostatic properties that can play a critical role in ligand binding and SBDD.

9.2.1 WaterMap and WaterFLAP

Water molecules are a key player in ligand binding, and a number of computational approaches have been applied to evaluate them at a molecular level [26–31]. One of the most used methods in industry is WaterMap (WM) [20]. This approach consists of an all-atom explicit solvent MD simulation followed by a statistical thermodynamic analysis of water clusters (hydration sites). A restraint is applied to the protein heavy atoms and the system is relaxed with an initial minimization followed by a short MD simulation heating the system from 10 to 300 K. A final preproduction simulation of 120 ps is run at 300 K. The production simulation (2 ns at 300 K in the isothermal–isobaric ensemble (NTP)) is used for the statistical thermodynamic analysis of hydration sites in the region of interest. WM was originally tested on factor Xa to elucidate the physical properties of its active-site solvent that appeared to be missing in most continuum theories of hydration. It identified several features of the hydration of the site relevant to the structure–activity relationship (SAR) of its inhibitors [20]. In 2010, it was used to explain SAR of triazolylpurine analogs binding to the A_{2A} receptor [32]. In this series, increasing the 2-substituted aliphatic group to methyl and isopropyl results in a decrease in potency; while extending the substituent to *n*-butyl and *n*-pentyl results in a significant gain in potency. Ligand–receptor interactions, steric effects, or differences in ligand desolvation did not explain this trend. On the contrary, WM correctly predicted the rank in binding affinity for this series based on the differential water displacement patterns. Small unfavorable substituents occupied a region in the A_{2A} adenosine receptor binding site predicted to contain stable waters, while the longer favorable substituents extended to a region that contains several unstable waters. The predicted binding energies associated with displacing waters within these hydration sites correlated well with the experimental activities. WM hydration sites have been included in methods to evaluate the ligand-binding free energy like WM/molecular mechanics (MM) [33] and WScore [34]. WM/MM combines the water displacement energy with complementary terms from an implicit solvent MM description binding and it was successfully applied to factor Xa [33]. WScore incorporates WM water structure to provide an atomic-level description of ligand and protein desolvation. It also contains a detailed model for localized ligand and protein strain energy and integrates an MM-generalized Born surface area (GBSA) scoring component with these terms to assess the delocalized strain of the complex. It uses ensemble docking to take into account induced-fit effects on the receptor conformation, and protein reorganization free energies are assigned via fitting to experimental data. The performance of the method for pose prediction and enrichment in virtual screening showed significant improvement compared to classic docking approaches [34].

An alternative method to evaluate water molecules developed by molecular discovery in collaboration with Heptares is based on GRID [19]. This approach was initially tested on a series of triazine antagonists of the A_{2A} receptor [18]. The water probe was used to detect favorable locations for water molecules in a ligand–protein complex. The energy of the waters was then evaluated combing three probes: OH2 to evaluate the hydrophilic character of the pocket and DRY and C1= to evaluate its lipophilic/hydrophobic/apolar components. This approach, together with a normal force field, proved useful to rank the activity of related triazine ligands. Furthermore, it allowed the evaluation of the energy of trapped waters between the small molecules and the protein. These "unhappy" waters were shown to be important in determining the ligand residence time. In the example shown in Figure 9.1, the triazine **4a** (panel c) is predicted to trap two high-energy water molecules, resulting in its very short residence time ($<$1 s); on the contrary, triazines displacing these and interacting with the receptor in that area (like **4f**, panel d) stay in complex with A_{2A} for a longer time (>10 minutes). Since GRID is based on a united-atom force field, we introduced a short MD step to generate an all-atom system including a hydrogen-bond network useful to evaluate the SAR [8]. This is also useful to understand the water network's role in the stability of ligand docked poses. The protocol was further developed including the evaluation of the positional entropy of waters, estimating the energy landscape around its location. Trapped (low-entropy) water molecules are located in a deep and narrow energy well, while bulk-like (high-entropy) waters correspond to shallow energy basins [8]. This method became WaterFLAP and it is now an integral part of the molecular modeling suite FLAP (Fingerprints for Ligands and Proteins; molecular discovery). Recently it was applied to study the SAR of a series of novel histamine H_1 receptor ligands [35]. In this study, *in silico* prediction of water network energetics resulted to be in agreement with mutagenesis studies and SAR, helping to understand the relative importance for ligand binding of the different subpockets of the orthosteric site. This study demonstrated how X-ray structures in combination with tool compounds, assessment of water energetics, and mutagenesis studies enable SAR exploration to map GPCR–ligand binding hot spots. WaterFLAP allows also a fast estimation of the perturbation of an apo water network upon ligand binding. This approach generates first the apo water network followed by an analysis of ligand docked poses: for every pose, it predicts if the apo water is displaced, stabilized, or destabilized by the ligand binding [36].

9.3 Ligand-Binding Free Energy

The accurate estimation of protein–ligand binding free energy is another useful application of MD simulations in structure-based drug design (SBDD). Being able to predict ligand-binding affinities might be crucial at the lead generation and optimization stages of a drug discovery program. Different approaches have been used to address the binding affinity prediction, from the fast but less accurate empirical or knowledge-based scoring methods to the more rigorous free energy calculation methods. The scoring function methods, like molecular docking and virtual screening, have the advantage of speed and they have been proved successful in binding mode prediction and in enriching large libraries of compounds with new hit molecules [37–39]. These methods, however, are less successful in ranking compound by activity and to quantify ligand-binding affinities [40]. Free energy calculation methods, such as free energy perturbation (FEP), have been applied widely to investigate ligand–protein binding affinities [41, 42]. Free energy approaches are based on force field calculations, such as MM, MD, and Monte Carlo (MC) simulations. This means that, unlike the scoring function methods, free energy calculations can take into account protein flexibility and system solvation, both very important aspects in the ligand-binding process. Consequent to binding, both the protein and the ligand experience conformational changes and this flexibility should be considered when estimating free energy of binding. Moreover, upon binding the ligand would replace water molecules in the binding site; and several studies on ligand

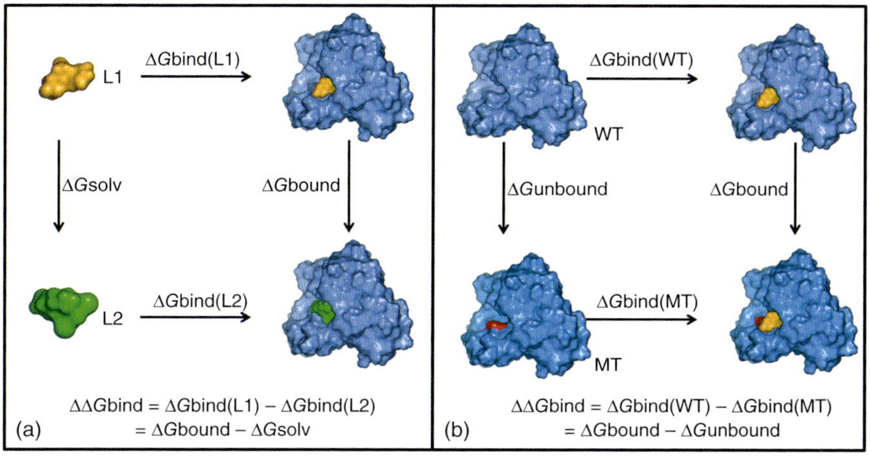

Figure 9.2 Thermodynamic cycle for relative binding free energy calculations: the alchemical transformation is applied to the ligand, going from L1 to L2 (a) or to the protein, were the wild type (WT) is mutated (MT) to evaluate the effect of residue mutations (shown in red) on binding free energy (b).

binding proved the crucial role of water-mediated interactions between ligand and protein. Considering the binding site hydration in an atomistic way is another relevant aspect during free energy calculations. A number of reviews covering free energy techniques and FEP in particular are available for a more extensive description of the methodologies [41–43].

Briefly, using a thermodynamic cycle, FEP calculates the relative binding free energies as thc difference between the binding free energies of two related ligands in both the unbound and bound states (Figure 9.2a). This is computationally less demanding compared to the simulation of the binding/unbinding process and the calculation of absolute free energies of binding, which would need impractically long MD simulations of the entire binding event. The free energy differences, both in the solvent and protein environment, are computed using a thermodynamic cycle like the one shown in Figure 9.2a and represented by the expression:

$$\Delta\Delta G\text{bind} = \Delta G\text{bind(L1)} - \Delta G\text{bind(L2)} = \Delta G\text{bound} - \Delta G\text{solv}$$

FEP is used to compute the difference of free energy between ligand 1 and ligand 2 in solvent and the difference of free energy between ligand 1 and ligand 2 bound to the protein. Eventually, the FEP calculations will tell if the alchemical transformation of ligand 1 toward ligand 2 will lead to an improved or diminished binding affinity. Despite their potentialities, until recently the use of FEP calculations has been limited, especially in industrial environment. The lack of reliability, the impractical timescale that allowed the application to only small sets of compounds, the error-prone tedious setup of the calculations, and the expensive computational costs were some of the drawbacks of initial FEP applications. The accuracy of free energy calculations is sensitive to force field parameters and inadequate conformational sampling. Recent methodological enhancements regarding force field parametrization and conformational sampling, coupled with the rapid increase in computing power have made FEP a more consolidated and attractive tool that can be applied to help guide lead optimization programs also in industrial environments; the FEP+ approach discussed subsequently is an exciting step forward to the "industrialization" of this key computational method, enabling use in timescales and numbers of compounds relevant to help drive project progression.

Successful studies applying FEP calculations for ligand–protein systems have been reported for different targets. The majority of these studies were retrospective and they showed the ability of FEP methods to reproduce experimental data on small sets of compounds. In Wang et al. [44], the FEP calculations were used in combination with the replica exchange with solute tempering (REST) method to enhance conformational sampling of the region around the ligand-binding pocket and improve the convergence of the calculations. The FEP/REST approach was applied to the L99A mutant of the T4 lysozyme and thrombin (Factor IIa) with good correlation between the predicted ligand-binding affinities and the experimental data. The FEP/REST workflow was later implemented in the automated FEP software (called FEP+) by Schrodinger [45]. The employment of the Desmond GPU MD engine, an updated OPLS force field [46] and an automatic FEP calculation setup to reduce human error, make FEP+ an interesting tool for free energy calculations to be used in drug discovery programs, including large sets of compounds.

There are cases where FEP methods were used to guide drug optimization programs, such as the optimization of non-nucleoside inhibitors of HIV-1 reverse transcriptase and inhibitors of the binding of the proinflammatory cytokine migration inhibitory factor (MIF) to its receptor, CD74 [47]. A prospective study was also conducted for the discovery of selective inhibitors of IRAK4 and TYK2 [45], where the predicted binding affinities were used to prioritize compound synthesis in order to optimize absorption, distribution, metabolism, and excretion (ADME) properties maintaining high affinity for the compounds.

FEP studies on GPCRs have been limited due to the challenges that membrane targets bring with them. MD simulations of large molecular systems formed by membrane bilayer, receptor, and explicit solvent are computationally heavier and slower to run than enzyme targets. Moreover, the starting protein structure used during the free energy calculations can affect the results. Compared to soluble targets, the number of GPCR high-resolution structures is still quite limited; nevertheless, the recent breakthroughs in GPCR structural biology have rapidly enriched structural information on this membrane receptor family, facilitating the use of free energy methodologies for binding affinity prediction. Another sensitive aspect of FEP technologies applied to GPCRs concern the problem of water molecules buried in the binding site.

Lenselink et al. [48] performed a systematic evaluation of the performances of the MD-based FEP+ tool against multiple GPCR targets. This study was applied both in a retrospective and prospective manner. The binding affinities of 45 known ligands were evaluated with FEP+ against four GPCR targets with good correlation between the predictions and the experimental data showing encouraging predictive capabilities. For the A_{2A} program, a new active molecule was discovered by the application of FEP against 46 related compounds, with a 10-fold improvement in potency. This study showed also the importance of high-quality starting structural information on the targets with an indication that the results are target dependent.

The alchemical transformations are not limited to ligands. Free energy of binding can be calculated also for one ligand binding to its point-mutated protein target with the aim of predicting the effect of protein mutations on ligand binding (Figure 9.2b). Keränen et al. applied a FEP strategy to the A_{2A} receptor in order to quantitatively estimate the effects of point mutations on the binding of the inverse agonist ZM241385 and the agonist NECA [49]. This MD-based FEP strategy employed a series of successive subperturbations during the calculations to improve the convergence and accuracy of the prediction. Moreover, the results showed the crucial role of solvent-mediated interactions in ligand binding.

The recent technological improvements in free energy calculations have made FEP methods attractive tools to calculate ligand-binding affinities in a lead optimization program. However, there are still some drawbacks that limit the applicability of FEP on a regular basis in SBDD. Small chemical transformations of the ligand are normally done in order not to compromise the convergence and the accuracy of the calculations. FEP calculations have become faster thanks to the use of graphics processing units (GPUs); nevertheless the application of FEP methods to screen very large libraries of compounds is still unpractical; access to large numbers of GPUs does, however, make the evaluation of hundreds of compounds possible in a useful timescale. Other current limitations to the use of FEP are the sensitivity of the calculations to changes of the ionization state of molecules or residues or the risk of creating trapped water or vacuum bubble by replacing water molecule during the preparation of the system. The progress in this area is however greatly accelerated compared to the past, with the significant efforts into developing FEP+ by Schrodinger yielding new and faster options, such as recent options for core ring-size changes and waters on the different lambda windows resulting in the alchemical transformation.

9.4 Ligand-Binding Kinetics

Commonly, during the drug discovery and development stages, ligand affinities toward biological targets are experimentally measured from *in vitro* assays. For binding assays, this implies that the intermolecular complex formation is detected after equilibrium conditions are reached (the concentrations of all the species in the systems remain unvaried). However, biological organisms are open systems, able to exchange energy and matter with the surrounding environment. In this scenario, data retrieved under steady-state conditions are rarely predictive for the *in vivo* behavior of a ligand, the concentration of which in proximity of the endogenous target is governed by pharmacokinetics (PK).

Over the past decade, increasing attention has been addressed to binding kinetics rate as an important parameter during lead optimization. When a ligand binds its target (through a bimolecular reaction), the formation of the complex is governed by the law of mass action and the dynamic proprieties of the equilibrium allow relating the thermodynamics dissociation constant K_D to the binding kinetics rate constants k_{on} and k_{off} (Eq. (9.1)):

$$K_D = \frac{k_{off}}{k_{on}} = \frac{[L][R]}{[LR]} \tag{9.1}$$

k_{off} is expressed in s^{-1}, while k_{on} in $M^{-1}\ s^{-1}$. This difference is due to the fact that ligand binding is a second-order reaction (depends on the local concentration of both ligand and receptor), while the intermolecular complex dissociation is a first-order reaction (is a function only of the complex concentration). The kinetics rate constant k_{on} is a crucial aspect to be considered during the development of experimental protocols for ligand-affinity determination, while dissociation k_{off} is related to the ligand residence time (t_r) and contributes to driving the biological effects triggered by the ligand. The residence time concept was first introduced in 2006 [50] and is defined as the reciprocal of the k_{off} value (Eq. (9.2)):

$$t_r = \frac{1}{k_{off}} \tag{9.2}$$

Besides the classic SAR approach, increasing efforts are addressed to the determination of structure–kinetics relationships (SKR), with the aim of identifying components in ligand structures able to influence on- and off-rates. Computational chemists are trying to develop MD tools capable of predicting binding kinetics parameters, with the purpose of facilitating and supporting SKR studies. Considering that classic MD sampling currently reaches the millisecond timescale [51] and that drug dissociation may require up to several hours, it follows that enhanced sampling MD methods are required for the evaluation of entire unbinding events. From this point of view, metadynamics [52] and scaled molecular dynamics (SMD) approaches have shown promising results when applied to both cytosolic systems [53, 54] and GPCRs [54, 55]. Moreover, statistical methods such as Markov State Models (MSM) have been employed to calculate the on-rate of ligands that recognize trypsin [56] and β-adrenoreceptors [57], respectively.

Theoretically, the prediction of a binding kinetics rate should involve the computation of the free energy profile that characterizes the ligand binding and unbinding processes. A simplified binding energy profile is shown in Figure 9.3.

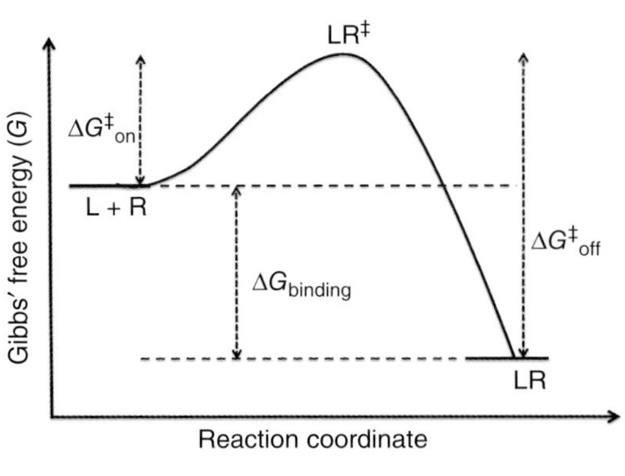

Figure 9.3 Ligand–receptor binding energy profile.

According to the transition state theory [58], the highest point on the pathway ($LR^{\ddagger}$) represents the transition state between the unbound state (L + R) and the bound configuration (LR).

As stated in Eyring's equation (Eq. (9.3)), overall kinetics constants ($K_{on/off}$) are proportional to the pre-equilibrium constant $K^{\ddagger}$ (pre-equilibrium is established between the reactants and the transition state) through a pre-exponential factor (which combines the Boltzmann's constant k_B, the Planck's constant h, and the absolute temperature T):

$$K_{on/off} = \frac{k_B T}{h} K^{\ddagger} = \frac{k_B T}{h} e^{-\frac{\Delta G^{\ddagger}{}_{on/off}}{RT}} \tag{9.3}$$

Kinetics rate constants ($K_{on/off}$) are, in turn, related to the energies of activation ($\Delta G^{\ddagger}{}_{on/off}$) of the respective transition states ($LR^{\ddagger}$). It is the result of the Gibbs' free energy difference between the unbound state (R + L) and the transition state characterizing the binding event (in the case of the K_{on}) and between the intermolecular complex and the transition state expected during the dissociation pathway from the receptor (for the K_{off}).

9.4.1 Supervised Molecular Dynamics (SuMD)

Given the high degree of biomedical relevance of GPCRs in mediating signal transduction in cells and living organisms, the increasing amount of structural information becoming available (with the Protein Data Bank (www.rcsb.org) [59] as a source for released structures) provides a solid foundation to understand the nature of key protein–ligand interactions at the final stage of the mutual recognition process. They enable SBDD, and are a starting point for computer-aided biophysical investigations, for example, using MD simulations.

High-resolution ligand–protein complex structures from multiple members of different GPCR classes/families reveal unique details of orthosteric and allosteric ligand-binding site location with respect to the TMDs. They provide structural elucidation of protein–ligand–water interactions governing SARs and pharmacological aspects which would lack explanation in the absence of any structural insights [1]. Alongside the increasing structural-biology input [1], the investigation of critical phenomena such as ligand–protein recognition have been the subject of study using computer-aided methods with the aim of drawing a parallelism between experimental observations with the atomistic-level description of the simulated system. Unbiased MD simulations of membrane-embedded pseudo-apo β adrenergic receptors (with co-crystallized ligands placed at arbitrary positions away from the binding site) demonstrated that drug molecules can spontaneously associate with GPCRs. This approach allowed describing the phenomena at an unprecedented atomistic level of detail due to the inclusion of the receptor protein in a membrane-like environment and solvated by explicit water molecules [57]. Long microsecond-timescale MD simulations describing such rare events of molecular recognition paved the way to establish a rigorous link between simulated protein–ligand conformational changes and calculated properties with observable and measurable macroscopic ones. Facts such as the convergence of the qualitative binding mode determined by MD with the experimentally determined one and agreement of ligand binding kinetic and thermodynamic calculations with macroscopically measurable parameters such as free energy of binding and ligand kinetic k_{on} and k_{off} represented a milestone in computational drug discovery [56, 60].

It has to be acknowledged that, in lead discovery or optimization of small-molecule drugs, the intrinsic potency of a compound to its target protein is a singular challenge where the success in achievement needs to be balanced with other critical factors, such as the improvement of PK properties and reducing off-target effects. Providing solid structure-based working hypotheses to guide compound optimization allows chemists to introduce targeted modifications to the lead compound while at the same time reducing the impact on, for example, intrinsic potency to the relevant target.

Therefore, for example, with many chemists working on a promising project, a high turnover of chemical inputs is expected and the need to generate solid structure-based working hypotheses to guide compound prioritization and synthesis is essential for project progression. This is especially true when chemical variation requires the *de novo* design of a synthesis route to the novel candidate molecules. At the present time, it requires all-atom long microsecond-timescale MD simulations to reproduce the experimentally determined binding conformation, which represents a solid starting point for SBDD [57], and can provide quantitative estimations on the thermodynamics and kinetics of binding. However, such simulations are extremely time consuming and may not cope with the needed *in silico* systematic exploration of multiple chemical scaffolds to inform and guide molecular design.

Supervised molecular dynamics (SuMD) addresses this shortcoming by enabling the complete ligand–protein recognition process to be elucidated. It uses unbiased all-atom MD in a reduced timescale of up to 3 orders of magnitude compared to classical MD, while ensuring rigorous sampling of receptor flexibility, with a GPCR receptor embedded in a membrane-like environment, and explicit solvation treatment [61]. A tabu-like algorithm supervises the ligand receptor distance over the course of the simulation and rejects solutions where the ligand–receptor distance is likely to increase over a period of time, thus increasing the probability of ligand–receptor binding events to occur without introducing bias to the MD simulation. In particular, if the ligand–receptor distance is likely to be shortened over the course of the simulation, the MD simulations will continue; otherwise, the simulation is restarted from the original set of coordinates and random velocities, drawn from the Boltzmann distribution, are reassigned. This has the effect of causing the restarted simulation to explore a different region of configuration space, while maintaining sampling from the correct canonical ensemble (NVT). The tabu-like supervision algorithm is repeatedly applied until ligand–receptor distance is less than a threshold defined by the user, generally when centroids of mass of ligand and the vestibule that anticipate the orthosteric binding pocket cavity match.

After placing the ligand far away from the binding pocket, SuMD simulations enable to reproduce the X-ray-determined bound conformation in hundreds of nanoseconds [61, 62]. On average, an explicitly solvated class A GPCR model, embedded in a lipid environment, constitutes a system for MD simulation of the size of approximately 60 000 atoms in a cubic simulation box, using

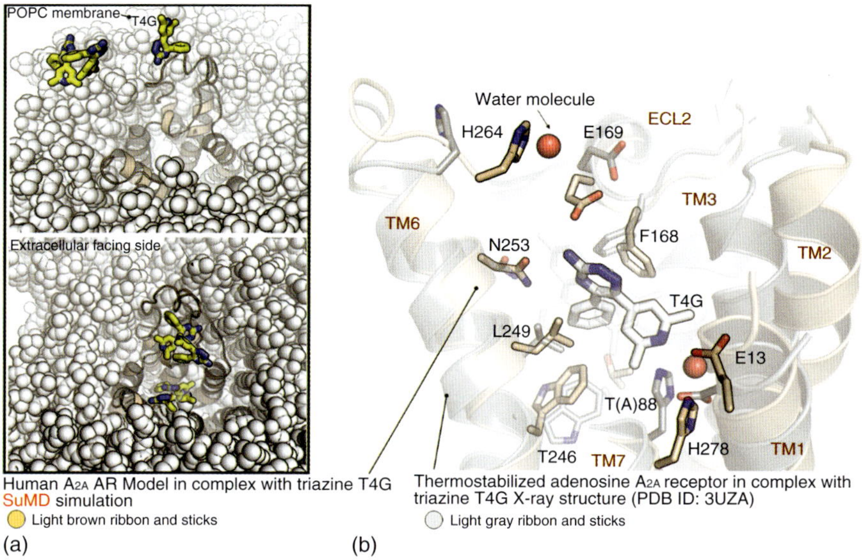

Figure 9.4 (a) Time lapse of a supervised molecular dynamics (SuMD) simulation depicting the recognition process between triazine T4G [24] and the A_{2A} adenosine receptor (AR) [6]. (b) Triazine T4G [24] in complex with the A_{2A} AR model, obtained after investigation of the ligand–receptor recognition pathway using SuMD. Molecular details of water molecules bridging ligand–receptor interactions are revealed and complement the lower resolution X-ray structure of the complex deposited in the Protein Data Bank (PDB ID: 3UZA). Stick representations are color coded and enable the comparison between the simulation output and the X-ray-determined ligand–receptor complex structure. PDB ID: 3UZA, gray sticks and SuMD simulation output, light brown sticks – T4G in white. The complexes are viewed from the membrane side facing transmembrane domains 6 (TM6) and TM7 with the view of TM7 being partially omitted. Hydrogen and protein backbone atoms are not displayed. POPC lipids are shown in white sphere and most of the solvation water molecules are hidden for clarity purposes.

periodic boundary conditions (PBCs). A recent commodity, a single NVIDIA GPU card can deliver about 100 ns/day, thus enabling probing ligand–protein recognition events with excellent throughput even on moderate-sized GPU clusters. This approach is able to retrospectively reproduce the crystallographic binding conformation of three strong A_{2A} binders such as ZM 241385 (pK_D = 9.18) [8], T4G (pK_D = 8.9, Figure 9.4) and T4E (pK_D = 9.6) [24]. It has also been applied successfully to weaker binders such as caffeine (pK_D = 5.31) [63] in complex with the A_{2A} AR. Based on these studies, it is reasonable to assume that the endpoint of a SuMD simulation can provide high-quality models of novel ligands bound to the adenosine receptor and other GPCRs for which experimental confirmation of binding mode is not yet available. Only antagonists were chosen in the validation study since the significant TMD geometrical changes associated with agonist binding to class A GPCRs are not compatible to occur in the short time span of a SuMD simulation.

As elucidated, the ability to reproduce the X-ray-determined bound conformation validates the use of SuMD as an orthogonal technique to the many existing flexible docking tools available in providing high-quality models of ligand–GPCR complexes by overcoming many of the limitations that the latter techniques possess in the treatment of receptor flexibility and explicit binding pocket solvation. In this context, it may be foreseen that ligand–GPCR models obtained through the employment of SuMD alone may provide early structural information, in support of lead optimization projects, in preparation for structural biology support through X-ray crystallography when possible. Providing high-quality homology models may be more impactful than quantitative prediction of ligand binding by enabling chemists to focus on promising regions of chemical space [64]. As for molecular docking, inspection of the output is always needed to avoid the blind acceptance of a solution that might be caused by artifacts of MM underlying principles [65] in sampling and scoring favorable ligand and protein conformational and contacts preferences. Questioning the model in the output of a SuMD simulation is essential to appreciate the usability of the technique in an industrial or academic environment. First, it is important to ensure that system preparation and MD equilibration guarantee the stereochemical quality of a protein structure and that the membrane-calculated biophysical parameters are in agreement with experimentally determined ones. Secondly, it is necessary to be aware of the caveats in the treatment of ligand conformational flexibility by judging conformational and contacts preferences of both amino acid side chains and bound ligand.

Querying small-molecule conformations and atom contact preferences from structural databases [66, 67], such as the Cambridge Structural Database (CSD) [68], and the Protein Data Bank [59] are complementary analyses that enable to question, with scientific rigor, the output of a SuMD simulation. This should regularly be carried out with the final aim of building solid hypotheses of protein–ligand interactions and drive the creative thinking process in medicinal chemistry [69].

9.4.2 Adiabatic Bias Metadynamics

Residence time is known to be important in determining the activity of GPCR ligands [70]. For example, slow dissociation increases the effect of angiotensin II AT_1 receptor antagonists [71], M_3 muscarinic (M_3) agonists residence time correlates with their intrinsic activity [72] and buprenorphine slow dissociation from the μ-opioid receptor contributes to its pharmacodynamics in humans [73]. Ligand occupancy and residence time can also help explain the difference between efficacy and toxicity. For example, the dopamine D_2 receptor antagonist haloperidol and chlorpromazine result in side effects that have been attenuated in more recent ligands with much shorter residence time [74, 75].

Despite its importance, ligand–protein residence time predictions still remain challenging for modern computational chemistry. We have developed a new computational chemistry protocol (aMetaD) based on a combination of adiabatic bias molecular dynamics (ABMD) and metadynamics to simulate ligand–protein unbinding events [55]. This automatic method is simple to use, requiring only a starting protein–ligand complex and a target unbound ligand position. This aspect, together with its speed (two ligands per day using 1 GPU), allows aMetaD to be applicable in hit-to-lead and lead-optimization phases of drug development.

aMetaD simulates a "stress test" on a ligand docked solution. It provides the energy bias required to break ligand–protein interactions and to promote the start of the unbinding event. This physics-based, fully-flexible and pose-dependent ligand scoring function evaluates the maximum energy (TS energy) required to move the ligand from the starting energy basin to the next.

Application of this method has been able to discriminate differences in residence time in two pairs of structurally related ligands, for M_3 and A_{2A}. The protocol has since been used to evaluate 12 corticotropin-releasing factor receptor 1 (CRF_1R) negative allosteric modulators (NAMs) [76]. On this test set, the resulting score (RTscore) was used successfully to categorize ligands with residence time differences above 30 minutes. CRF_1R is a class B GPCR [55, 77, 78] activated by the corticotropin-releasing factor, a 41 amino acid peptide hormone with a key role in the regulation of the stress response [79]. This receptor is implicated under pathophysiological conditions like depression, posttraumatic stress disorder, and relapse in substance abuse. The 12 NAMs considered have similar nanomolar binding affinity, but different residence time (from less than two minutes to more than seven hours) [76]. The ranking of dissociation half-life of three of the ligands considered (NBI35965 with 16 minutes, NBI34041 with 53 minutes, and NB30775 with 130 minutes) is consistent with the ranking of their sustained suppression of adrenocorticotropin *in vivo* in adrenalectomized rats (NBI35965 with less than two hours, NBI34041 with four hours, and NB30775 with six hours). The fact that these three ligands have similar fast PKs confirms that the sustained adrenocorticotropin suppression shown uniquely by NBI30775 was the result of its prolonged CRF_1 receptor occupancy.

The unbinding trajectories resulting from the aMetaD protocol are automatically analyzed and translated to atomic solvation factor (SF). These values are useful to evaluate the water dynamics during the unbinding event. SFs are stored in the PDB B factor field to be easily analyzed by any molecular modeling visualization software. The SF coloring scheme has been used to detect key differences in the degree of accessibility to water molecules during the predicted ligand unbinding events for the CRF_1R NAM test set (Figure 9.5). This simple analysis provides examples of actionable working hypothesis applicable in a discovery program to rationally optimize ligand binding kinetics.

Although more tests are certainly needed to prove the general applicability of aMetaD, RTscore and SF, the general concepts outlined here could be applied to other types of nonstandard MD methods for predicting ligand-unbinding and ligand-binding events, and could potentially provide a general framework of protocols usable in drug discovery projects to predict ligand kinetics.

When protein flexibility is important in determining ligand-binding kinetics, it is possible to apply optimized metadynamics protocols to evaluate crucial protein conformational changes. This approach has been utilized to study a series of structurally related A_{2A} antagonists with different ligand residence time [80]. The crystallographic information available highlighted a comparable ligand-binding mode for the derivatives considered in this study. Their common scaffold occupied a similar position in the binding pocket, while the different substituents

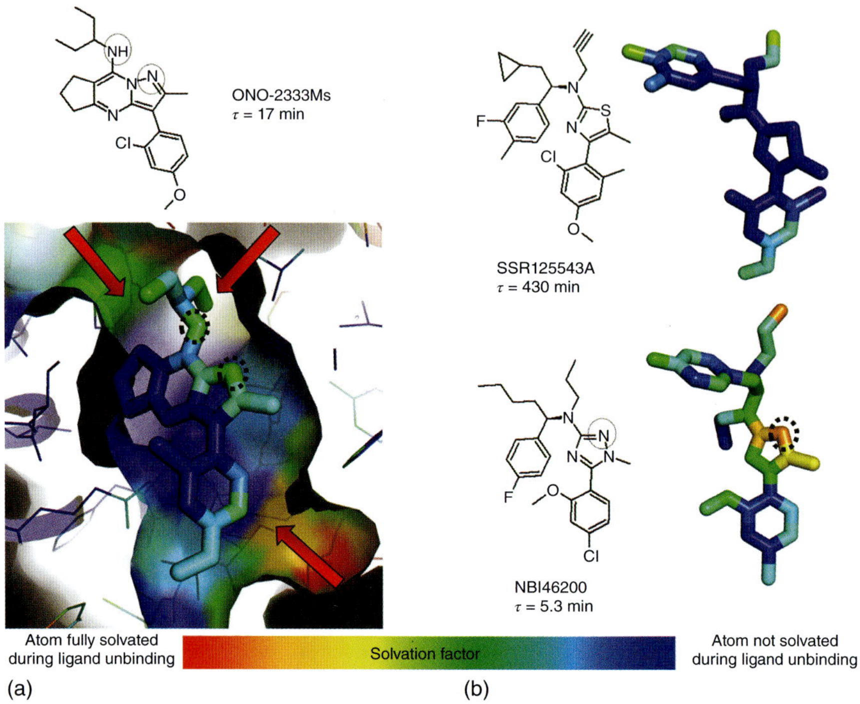

Figure 9.5 (a) At the top, the 2D structure of one of CRF_1R NAM studied is included together with its residence time τ. At the bottom, its docked conformation to CRF_1R is shown: the ligand is represented in stick, the protein in line, the binding site as surface, and they are color coded on the basis of the solvation factor in rainbow. Circles highlight the polar nitrogens involved in H-bonding interactions with water molecules during the predicted unbinding event. Arrows show the routes to the binding site accessible to the solvent. (b) Comparison of the solvation factor (ligands shown in stick representation on the right-hand side) of two structurally related CRF_1R NAMs with different residence times (2D structure included on the left-hand side).

extended to the extracellular side of the receptor, creating key interactions with the His264–Glu169 salt bridge. This salt bridge represents an important gate affecting the accessibility to the orthosteric pocket. When it is open, the site is easily reachable. On the contrary, when it is closed it can represent an important kinetic bottleneck playing a key role in the ligand-unbinding event.

We evaluated the effect of the presence of the different ligands on the stability of the His264–Glu169 salt bridge using a metadynamics protocol. During the metadynamics simulation, a history-dependent bias is added to the potential energy landscape representing the His264–Glu169 salt bridge. This free energy surface is defined using 2 degrees of freedom (collective variables): (i) the distance between the His side-chain nitrogen and the Glu acid moiety and (ii) the dihedral angle of the His side chain. When the added energy bias equals the strength of the ionic interaction, the salt bridge is broken. Long-residence-time ligands resulted in the stabilization of the His–Glu ionic interaction, while fast off-rate derivatives were generally predicted to be destabilizing the salt bridge.

9.5 Conclusion

The power of MD to impact GPCR ligand design has been illustrated in this chapter. The full potential is yet to be realized – the era of physics-based methods has only recently begun. It brings the promise of a timely and effective way to impact drug discovery projects and increase our understanding of how biomolecular systems function. The relentless demands of the gaming industry provide relatively inexpensive and extremely powerful GPUs for simulations and there are now reasonable force fields for the protein, ligand, solvent (waters), and membrane. It is sobering to reflect that a \$9 million investment in a 1985-era Cray 2 supercomputer (=\$24 million now) offered less CPU power than a single iPhone 5, using 200 kW and an external heat exchanger the size of a car! The disk space needed for MD simulations has reduced from \$50 K to ~\$0.03 per GB.

The core capability of a computational chemist to be able to predict the free energy of binding of a ligand in a way that is timely to impact project decisions and sufficiently accurate is now emerging as a reality with FEP. There are still limitations and a domain of applicability to be considered, but it is exciting days when such a calculation can be run overnight (FEP+), with the binding of a ligand to a protein binding site finally including protein flexibility and full explicit water networks in an energetically realistic way. The FEP+ method has received substantial investment (Schrodinger) and is now a reality in the arsenal for drug design, with calculations having been run on over 10 000 ligands across 100 targets.

The role of waters as the critical third dimension alongside the protein and ligand is now clear, and two complementary methods WM and WaterFLAP are discussed in this chapter. Very importantly, both use explicit waters, unlike some other approaches, that is key to evaluate network energies and perturbations thereof. An analysis of (pseudo)apo water networks in conjunction with GRID lipophilic hot spots provides a powerful druggability analysis. The use of MD in the WM approach provides a powerful physics-based method to evaluate full water networks, delineating the entropic and enthalpic components. WaterFLAP provides a fast empirical method that enables routine use for all docked structures and for seeding MD simulations with full networks including regions with trapped waters in ligand complexes.

With the ever-increasing number of multiple GPCR–ligand structures, it is found that waters play a key role in the binding of ligands, with the possibility of a polar ligand having no direct H-bonds with the receptor, just lipophilic hot spot contacts; for such ligand water molecules mediate all H-bonding. Both side-chain and backbone helix movements are observed for different ligands, making MD simulations that allow for such flexibility critical.

Speed remains an issue for longer MD simulations, such as in the useful microsecond timescale, and the SuMD approach discussed in this chapter provides much promise in addressing this shortcoming by enabling the complete ligand–protein recognition process to be elucidated, using unbiased all-atom MD, but in a reduced timescale.

Adiabatic bias metadynamics is another approach discussed, and the results for the prediction of off-rate kinetics seem very promising. For a long while, computational chemists have been asked to provide a method that gives a ligand kinetics estimation in a timely manner (e.g. overnight) that can be used by project chemists. The solvation of the ligand and protein has been found to be a key factor, and the visualization by color of this in the approach described is a simple but powerful way to help transform MD simulation results into something that a chemist can use to design a new compound.

MD simulations are thus a key capability in GPCR ligand design, enabling protein flexibility and full water networks to be included, providing estimations of potency, selectivity, and kinetics.

References

1 Cooke, R.M., Brown, A.J.H., Marshall, F.H., and Mason, J.S. (2015). Structures of G protein-coupled receptors reveal new opportunities for drug discovery. *Drug Discov. Today* 20 (11): 1355–1364.
2 Jazayeri, A., Dias, J.M., and Marshall, F.H. (2015). From G protein-coupled receptor structure resolution to rational drug design. *J. Biol. Chem.* 290 (32): 19489–19495.
3 Ciancetta, A., Sabbadin, D., Federico, S. et al. (2015). Advances in computational techniques to study GPCR-ligand recognition. *Trends Pharmacol. Sci.* 36 (12): 878–890.
4 Ball, P. (2008). Water as an active constituent in cell biology. *Chem. Rev.* 108 (1): 74–108.
5 Chaplin, M. (2006). Do we underestimate the importance of water in cell biology? *Nat. Rev. Mol. Cell Biol.* 7 (11): 861–866.
6 Ball, P. (2008). Water as a biomolecule. *ChemPhysChem* 9 (18): 2677–2685.
7 Mason, J.S., Bortolato, A., Congreve, M., and Marshall, F.H. (2012). New insights from structural biology into the druggability of G protein-coupled receptors. *Trends Pharmacol. Sci.* 33 (5): 249–260.
8 Mason, J.S., Bortolato, A., Weiss, D.R. et al. (2013). High end GPCR design: crafted ligand design and druggability analysis using protein structure, lipophilic hotspots and explicit water networks. *Silico Pharmacol.* 1 (1): 23.
9 Jastrzebska, B., Palczewski, K., and Golczak, M. (2011). Role of bulk water in hydrolysis of the rhodopsin chromophore. *J. Biol. Chem.* 286 (21): 18930–18937.
10 Pardo, L., Deupi, X., Dölker, N. et al. (2007). The role of internal water molecules in the structure and function of the rhodopsin family of G protein-coupled receptors. *ChemBioChem* 8 (1): 19–24.
11 Angel, T.E., Gupta, S., Jastrzebska, B. et al. (2009). Structural waters define a functional channel mediating activation of the GPCR, rhodopsin. *Proc. Natl. Acad. Sci. U.S.A* 106 (34): 14367–14372.
12 Tautermann, C.S. (2014). GPCR structures in drug design, emerging opportunities with new structures. *Bioorg. Med. Chem. Lett.* 24 (17): 4073–4079.
13 Sun, X., Ågren, H., and Tu, Y. (2014). Functional water molecules in rhodopsin activation. *J. Phys. Chem. B* 118 (37): 10863–10873.
14 Tehan, B.G., Bortolato, A., Blaney, F.E. et al. (2014). Unifying family a GPCR theories of activation. *Pharmacol. Ther.* 143 (1): 51–60.

15 Venkatakrishnan, A.J., Deupi, X., Lebon, G. et al. (2013). Molecular signatures of G-protein-coupled receptors. *Nature* 494 (7436): 185–194.
16 Isberg, V., De Graaf, C., Bortolato, A. et al. (2015). Generic GPCR residue numbers - aligning topology maps while minding the gaps. *Trends Pharmacol. Sci.* 36 (1): 22–31.
17 Huang, W., Manglik, A., Venkatakrishnan, A.J. et al. (2015). Structural insights into μ-opioid receptor activation. *Nature* 524 (7565): 315–321.
18 Bortolato, A., Tehan, B.G., Bodnarchuk, M.S. et al. (2013). Water network perturbation in ligand binding: adenosine A2A antagonists as a case study. *J. Chem. Inf. Model.* 53 (7): 1700–1713.
19 Goodford, P.J. (1985). A computational procedure for determining energetically favorable binding sites on biologically important macromolecules. *J. Med. Chem.* 28 (7): 849–857.
20 Abel, R., Young, T., Farid, R. et al. (2008). Role of the active-site solvent in the thermodynamics of factor Xa ligand binding. *J. Am. Chem. Soc.* 130 (9): 2817–2831.
21 Baroni, M., Cruciani, G., Sciabola, S. et al. (2007). A common reference framework for analyzing/comparing proteins and ligands. Fingerprints for ligands and proteins (FLAP): theory and application. *J. Chem. Inf. Model.* 47 (2): 279–294.
22 Tan, Q., Zhu, Y., Li, J. et al. (2013). Structure of the CCR5 chemokine receptor-HIV entry inhibitor maraviroc complex. *Science* 341 (6152): 1387–1390.
23 Wu, B., Chien, E.Y.T., Mol, C.D. et al. (2010). Structures of the CXCR4 chemokine GPCR with small-molecule and cyclic peptide antagonists. *Science* 330 (6007): 1066–1071.
24 Congreve, M., Andrews, S.P., Doré, A.S. et al. (2012). Discovery of 1,2,4-triazine derivatives as adenosine A2A antagonists using structure based drug design. *J. Med. Chem.* 55 (5): 1898–1903.
25 Sabbadin, D., Ciancetta, A., and Moro, S. (2014). Perturbation of fluid dynamics properties of water molecules during g protein-coupled receptor-ligand recognition: the human A2A adenosine receptor as a key study. *J. Chem. Inf. Model.* 54 (10): 2846–2855.
26 Michel, J., Tirado-Rives, J., and Jorgensen, W.L. (2009). Prediction of the water content in protein binding sites. *J. Phys. Chem. B* 113 (40): 13337–13346.
27 Barillari, C., Taylor, J., Viner, R., and Essex, J.W. (2007). Classification of water molecules in protein binding sites. *J. Am. Chem. Soc.* 129 (9): 2577–2587.
28 Imai, T., Hiraoka, R., Kovalenko, A., and Hirata, F. (2007). Locating missing water molecules in protein cavities by the three-dimensional reference interaction site model theory of molecular solvation. *Proteins Struct. Funct. Genet.* 66 (4): 804–813.
29 Rossato, G., Ernst, B., Vedani, A., and Smieško, M. (2011). AcquaAlta: a directional approach to the solvation of ligand–protein complexes. *J. Chem. Inf. Model.* 51 (8): 1867–1881.
30 Ross, G.A., Morris, G.M., and Biggin, P.C. (2012). Rapid and accurate prediction and scoring of water molecules in protein binding sites. *PLoS One* 7 (3): 1–11.

31 Nittinger, E., Schneider, N., Lange, G., and Rarey, M. (2015). Evidence of water molecules – a statistical evaluation of water molecules based on electron density. *J. Chem. Inf. Model.* 55 (4): 771–783.

32 Higgs, C., Beuming, T., and Sherman, W. (2010). Hydration site thermodynamics explain SARs for triazolylpurines analogues binding to the A2A receptor. *ACS Med. Chem. Lett.* 1 (4): 160–164.

33 Abel, R., Salam, N.K., Shelley, J. et al. (2011). Contribution of explicit solvent effects to the binding affinity of small-molecule inhibitors in blood coagulation factor serine proteases. *Chem. Med. Chem.* 6 (6): 1049–1066.

34 Murphy, R.B., Repasky, M.P., Greenwood, J.R. et al. (2016). WScore: a flexible and accurate treatment of explicit water molecules in ligand–receptor docking. *J. Med. Chem.* 59 (9): 4364–4384.

35 Kuhne, S., Kooistra, A.J., Bosma, R. et al. (2016). Identification of ligand binding hot spots of the histamine H1 receptor following structure-based fragment optimization. *J. Med. Chem.* 59 (19): 9047–9061.

36 Tehan, B.G. and Christopher, J.A. (2016). The use of conformationally thermostabiliseded GPCRs in drug discovery: application to fragment, structure and biophysical techniques. *Curr. Opin. Pharmacol.* 30: 8–13.

37 Warren, G.L., Andrews, C.W., Capelli, A.M. et al. (2006). A critical assessment of docking programs and scoring functions. *J. Med. Chem.* 49 (20): 5912–5931.

38 Cross, J.B., Thompson, D.C., Rai, B.K. et al. (2009). Comparison of several molecular docking programs: pose prediction and virtual screening accuracy. *J. Chem. Inf. Model.* 49 (6): 1455–1474.

39 Bortolato, A., Fanton, M., Mason, J.S., and Moro, S. (2013). Molecular docking methodologies. *Methods Mol. Biol.* 924: 339–360.

40 Li, Y., Han, L., Liu, Z., and Wang, R. (2014). Comparative assessment of scoring functions on an updated benchmark: 2. Evaluation methods and general results. *J. Chem. Inf. Model.* 54 (6): 1717–1736.

41 Michel, J. and Essex, J.W. (2010). Prediction of protein–ligand binding affinity by free energy simulations: assumptions, pitfalls and expectations. *J. Comput. Aided. Mol. Des.* 24 (8): 639–658.

42 Steinbrecher, T. and Labahn, A. (2010). Towards accurate free energy calculations in ligand protein-binding studies. *Curr. Med. Chem.* 17 (8): 767–785.

43 Chodera, J.D., Mobley, D.L., Shirts, M.R. et al. (2011). Alchemical free energy methods for drug discovery: progress and challenges. *Curr. Opin. Struct. Biol.* 21 (2): 150–160.

44 Wang, L., Berne, B.J., and Friesner, R.A. (2012). On achieving high accuracy and reliability in the calculation of relative protein–ligand binding affinities. *Proc. Natl. Acad. Sci. U.S.A* 109 (6): 1937–1942.

45 Wang, L., Wu, Y., Deng, Y. et al. (2015). Accurate and reliable prediction of relative ligand binding potency in prospective drug discovery by way of a modern free-energy calculation protocol and force field. *J. Am. Chem. Soc.* 137 (7): 2695–2703.

46 Harder, E., Damm, W., Maple, J. et al. (2016). OPLS3: a force field providing broad coverage of drug-like small molecules and proteins. *J. Chem. Theory Comput.* 12 (1): 281–296.

47 Jorgensen, W.L. (2009). Efficient drug lead discovery and optimization. *Acc. Chem. Res.* 42 (6): 724–733.

48 Lenselink, E.B., Louvel, J., Forti, A.F. et al. (2016). Predicting binding affinities for GPCR ligands using free-energy perturbation. *ACS Omega* 1 (2): 293–304.

49 Keränen, H., Gutiérrez-de-Terán, H., and Åqvist, J. (2014). Structural and energetic effects of A2A adenosine receptor mutations on agonist and antagonist binding. *PLoS One* 9 (10).

50 Copeland, R.a., Pompliano, D.L., and Meek, T.D. (2006). Drug–target residence time and its implications for lead optimization. *Nat. Rev. Drug Discov.* 5 (9): 730–739.

51 Shaw, D.E., Bowers, K.J., Chow, E. et al. (2009). Millisecond-scale molecular dynamics simulations on Anton. *Proc. Conf. High Perform. Comput. Netw. Storage Anal.* SC 09, (c), 1.

52 Barducci, A., Bonomi, M., and Parrinello, M. (2011). Metadynamics. *Wiley Interdiscip. Rev. Comput. Mol. Sci.* 1 (5): 826–843.

53 Tiwary, P., Limongelli, V., Salvalaglio, M., and Parrinello, M. (2015). Kinetics of protein–ligand unbinding: predicting pathways, rates, and rate-limiting steps. *Proc. Natl. Acad. Sci. U.S.A.* 112 (5): E386–E391.

54 Mollica, L., Decherchi, S., Zia, S.R. et al. (2015). Kinetics of protein–ligand unbinding via smoothed potential molecular dynamics simulations. *Sci. Rep.* 5: 11539.

55 Bortolato, A., Deflorian, F., Weiss, D.R., and Mason, J.S. (2015). Decoding the role of water dynamics in ligand–protein unbinding: CRF1R as a test case. *J. Chem. Inf. Model.* 55 (9): 1857–1866.

56 Buch, I., Giorgino, T., and De Fabritiis, G. (2011). Complete reconstruction of an enzyme-inhibitor binding process by molecular dynamics simulations. *Proc. Natl. Acad. Sci. U.S.A* 108 (25): 10184–10189.

57 Dror, R.O., Pan, A.C., Arlow, D.H. et al. (2011). Pathway and mechanism of drug binding to G-protein-coupled receptors. *Proc. Natl. Acad. Sci. U.S.A* 108 (32): 13118–13123.

58 Eyring, H. (1935). The activated complex in chemical reactions. *J. Chem. Phys.* 3 (2): 107–115.

59 Berman, H.M., Westbrook, J., Feng, Z. et al. (2000). The protein data bank. *Nucleic Acids Res.* 28 (1): 235–242.

60 Ferruz, N., Harvey, M.J., Mestres, J., and De Fabritiis, G. (2015). Insights from fragment hit binding assays by molecular simulations. *J. Chem. Inf. Model.* 55 (10): 2200–2205.

61 Sabbadin, D. and Moro, S. (2014). Supervised molecular dynamics (SuMD) as a helpful tool to depict GPCR-ligand recognition pathway in a nanosecond time scale. *J. Chem. Inf. Model.* 54 (2): 372–376.

62 Cuzzolin, A., Sturlese, M., Deganutti, G. et al. (2016). Deciphering the complexity of ligand–protein recognition pathways using supervised molecular dynamics (SuMD) simulations. *J. Chem. Inf. Model.* 56 (4): 687–705.

63 Doré, A.S., Robertson, N., Errey, J.C. et al. (2011). Structure of the adenosine A2A receptor in complex with ZM241385 and the xanthines XAC and caffeine. *Structure* 19 (9): 1283–1293.

64 Kuhn, B., Guba, W., Hert, J. et al. (2016). A real-world perspective on molecular design. *J. Med. Chem.* 59 (9): 4087–4102.
65 Lipkowitz, K. (1995). Abuses of molecular mechanics: pitfalls to avoid. *J. Chem. Educ.*.
66 Brameld, K.A., Kuhn, B., Reuter, D.C., and Stahl, M. (2008). Small molecule conformational preferences derived from crystal structure data. A medicinal chemistry focused analysis. *J. Chem. Inf. Model.* 48 (1): 1–24.
67 Lovell, S.C., Word, J.M., Richardson, J.S., and Richardson, D.C. (2000). The penultimate rotamer library. *Proteins Struct. Funct. Genet.* 40 (3): 389–408.
68 Allen, F.H. (2002). The cambridge structural database: a quarter of a million crystal structures and rising. *Acta Crystallogr. Sect. B Struct. Sci.* 58 (3): 380–388.
69 Stahl, M. and Baier, S. (2015). How many molecules does it take to tell a story? Case studies, language, and an epistemic view of medicinal chemistry. *ChemMedChem* 10 (6): 949–956.
70 Guo, D., Hillger, J.M., Ijzerman, A.P., and Heitman, L.H. (2014). Drug-target residence time – a case for G protein-coupled receptors. *Med. Res. Rev.* 34 (4): 856–892.
71 Verheijen, I., Vanderheyden, P.M.L., De Backer, J.P., and Vauquelin, G. (2004). AT1 receptor antagonists. *Curr. Med. Chem. Cardiovasc. Hematol. Agents* 2 (1): 69–77.
72 Dowling, M.R. and Charlton, S.J. (2006). Quantifying the association and dissociation rates of unlabelled antagonists at the muscarinic M3 receptor. *Br. J. Pharmacol.* 148: 927–937.
73 Yassen, A., Olofsen, E., Romberg, R.R. et al. (2006). Mechanism-based pharmacokinetic–pharmacodynamic modeling of the antinociceptive effect of buprenorphine in healthy volunteers. *Anesthesiology* 104 (6): 1232–1242.
74 Kapur, S. and Seeman, P. (2000). Antipsychotic agents differ in how fast they come off the dopamine D2 receptors. Implications for atypical antipsychotic action. *J. Psychiatry Neurosci.* 25 (2): 161–166.
75 Seeman, P. (2006). Targeting the dopamine D2 receptor in schizophrenia. *Expert Opin. Ther. Targets* 10 (4): 515–531.
76 Fleck, B.a., Hoare, S.R.J., Pick, R.R. et al. (2012). Binding kinetics redefine the antagonist pharmacology of the corticotropin-releasing factor type 1 receptor. *J. Pharmacol. Exp. Ther.* 341 (2): 518–531.
77 Bale, T.L. and Vale, W.W. (2004). CRF and CRF receptors: role in stress responsivity and other behaviors. *Annu. Rev. Pharmacol. Toxicol.* 44: 525–557.
78 Hollenstein, K., De Graaf, C., Bortolato, A. et al. (2014). Insights into the structure of class B GPCRs. *Trends Pharmacol. Sci.* 35 (1): 12–22.
79 Hemley, C.F., McCluskey, A., and Keller, P.A. (2007). Corticotropin releasing hormone – a GPCR drug target. *Curr. Drug Targets* 8: 105–115.
80 Segala, E., Guo, D., Cheng, R.K.Y. et al. (2016). Controlling the dissociation of ligands from the adenosine A2A receptor through modulation of salt bridge strength. *J. Med. Chem.* 59 (13): 6470–6479.

10

Ion Channel Simulations

Saurabh Pandey[1,2], *Daniel Bonhenry*[1], *and Rudiger H. Ettrich*[1,3]

[1] *Institute of Microbiology Academy of Sciences of the Czech Republic, Center for Nanobiology and Structural Biology, Zámek 136, 37333 Nové Hrady, Czech Republic*
[2] *University of South Bohemia in Ceske Budejovice, Faculty of Sciences, Zámek 136, 37333 Nové Hrady, Czech Republic*
[3] *Larkin University, College of Biomedical Sciences, 18301 North Miami Avenue, Miami, FL 33169, USA*

10.1 Introduction

Ion channels are abundant in all types of cells, from unicellular to multicellular organisms. They act as membrane-embedded pore-forming proteins that play an essential role in nerve signal transmission, maintaining the concentration of electrolytes and membrane potential in cells via connecting the inner environment of the cell to the outside, and allowing ions to pass through the channel pore; for a review, see Ref. [1]. Malfunctioning of ion channels can lead to a broad variety of neuronal and muscular diseases; hence, they have been and are a prime target for drug discovery studies [1]. Ion channels pass ions to or from the cell down their concentration gradient. In this respect, ion channels differ from ion transporters that can actively transport ions against the concentration gradient by consuming energy.

Ion channels "open" or "close" in response to various stimuli. Based on the nature of the stimuli, ion channels have been classified into voltage-gated, ligand-gated, light-gated, mechanosensitive, etc. [1]. Determination of the first crystal structure of the KcsA ion channel by Doyle et al. [2] revolutionized the molecular understanding of ion channels by giving the structural basis for ion selectivity and conduction through the pore, even though each crystal structure represents only one specific conformation of the broad conformational space those channels necessarily have to sample. Nevertheless, the availability of atomic coordinates opened the path for studies of ion channel via computational approaches; and, indeed, shortly after the first X-ray structure had been published, already the first ion channel simulations were reported [3–5]. This chapter focuses on applications and limitations of homology modeling and classical and advanced molecular dynamics (MD) simulations to study specific properties of ion channels or predict their dynamical behavior. Membrane channel systems presented in this chapter were selected to illustrate the

Biomolecular Simulations in Structure-Based Drug Discovery,
First Edition. Edited by Francesco L. Gervasio and Vojtech Spiwok.

relevance of computational simulations for medicinal chemistry research and structure-based drug design.

10.2 Overview of Computational Methods Applied to Study Ion Channels

10.2.1 Homology Modeling

Crystallization of membrane proteins is still not very straightforward and success in growing diffracting crystals is very limited; the number of available crystal structures of ion channels is small compared to the thousands of available soluble proteins [6]. The use of nuclear magnetic resonance (NMR) as an alternative method is very limited due to the comparatively large size of most ion-translocating systems. Only recently, cryo-electron microscopy (EM) promises a major breakthrough in solving a larger number of membrane protein structures [7]. For a large number of cases, there is still no suitable high-resolution crystal structure of the ion channel protein of interest available, partly due to the complexity in purifying membrane protein samples and partly due to lack of success in their consequent crystallization. However, even today there is already a good chance that at least one member of the channel family or at least the fold family is resolved, either the same channel from a different, mostly simpler organism or from a closely related ion channel [8–10]. Homology modeling can play an important role here. Homology modeling (also known as comparative modeling) is based on the fact that there is a limited number of protein folds (in the order of several hundreds, probably close to a thousand [11–14]), and thus three-dimensional (3D) structures are much more conserved than primary structures, i.e. amino acid sequences. As a rule of thumb, it can be said that proteins that share at least 25–30% primary structure do not only have the same fold but also belong to the same protein family and have a clear evolutionary relation [15]. Those proteins are called homologs and are typically divided in two groups: orthologs and paralogs. Orthologs are homologous sequences in different species that arose from a common ancestral gene during speciation; often, these are proteins having the same function. Paralogs, on the other hand, are homologous sequences within a single species that arose by gene duplication, and generally have a lower sequence identity than orthologs. If the crystal structure of either an ortholog or a paralog is available, a reasonable 3D model of the target sequence can be built on the basis of this template. A variety of software packages and web servers are available for homology modeling using either a fragment-based approach like Yasara [16], Sybyl [17], and others or a restraint-based approach such as Modeller [14, 18].

For both approaches, the first step is the search for a suitable structural template. This means an evolutionary related ion channel with more than 25–30% sequence identity and an experimentally resolved three-dimensional structure [19]. Here, it is important to note that the quality of the sequence alignment between target sequence and template as well as the quality of the available crystal structure determine the quality of the final structural model independent of the software used. A model cannot be expected to have a higher resolution than the input template and the final resolution is strongly correlated to the sequence identity with the template. The template search can be easily performed by running the query sequence with NCBI-BLAST (National Center for Biotechnology Information-Basic Local Alignment Search Tool) [20, 21] against the PDB (Protein Data Bank) database [22] that houses all available experimentally determined three-dimensional structures. In the next step, great care should be taken when aligning primary sequences of target and template as tools/servers do give an alignment with the highest score, but this is not necessarily the only alignment having the given score nor is it always the true alignment. Hence, manual observation and editing might be needed. When using a fragment-based approach [23–25], the backbone of sequence-conserved regions of the template is taken to build up the 3D scaffold of the target protein, followed by loop modeling to connect the fragments and side-chain addition and refinement to obtain a final structure containing all atoms including hydrogen [26]. In restraint-based modeling, three-dimensional structures of proteins are modeled by satisfaction of spatial restraints [14]. The restraints can operate on distances, angles, dihedral angles, pairs of dihedral angles, and other spatial features and are mostly derived from the known related structures and their alignment with the target sequence [14]. This approach allows, apart from the template structure, to include additional restraints gained from experimental structural data, as might be, for example, cross-linking, NMR, or fluorescence data. The modeling algorithm then tries to generate a three-dimensional structure that satisfies these restraints as good as possible. Whatever method is used to build the 3D structure, the final structure needs to be inspected and validated. In the optimal case, robustness and stability are tested using MD simulations. Stability in this respect means the ability of the modeled structure to maintain its fold over time in MD simulations. It is generally characterized by a low root-mean-square deviation from its initial conformation. Critical or unstable regions are identified and the preferred solution is an adjustment of the initial alignment in those regions followed by the generation of a new model. Mostly, the whole cycle is repeated several times until a stable structural model is gained and can be used in consequent simulations. Here, one has to be aware that homology modeling always leads to the same conformation that is observed in the template. A template representing a channel in a closed conformation will lead to a structural model of a channel in a closed conformation and an open conformation will lead to a model of an open channel.

10.2.2 All-atom Molecular Dynamics Simulations

Over the past few years, MD simulations have emerged as a powerful tool to study biomolecular systems [27]. Structure determination techniques provide mostly a static picture of molecules and do not give enough insight into mechanistic and dynamic details of the system. Using MD, information about small events such as interactions between amino acids or orientation of amino acid chains as well as global information about conformational changes such as movement of domains or opening and closing of ion channels is gained.

In classical MD simulations, the system is treated as hard spheres representing atoms that are connected via springs. Newton's equation of motion (force equals mass times acceleration) is numerically solved for each atom of the system in order to predict time evolution of the system. The atoms move on an empirical potential energy surface which is a sum of bonded and nonbonded terms, corresponding to different types of interatomic interaction. Newton's equation of motion is solved using numerical integration algorithms such as Verlet, velocity-Verlet, or leap-frog [28–30]. Microscopic properties (atomic positions and velocity) obtained by MD are converted to macroscopic properties (pressure, temperature, heat capacities, binding free energies, etc.) via statistical mechanics using rigorous mathematical expressions [31].

10.2.2.1 Force Fields

A force field is essentially a potential energy function, a set of equations to evaluate potential energies and their derivatives together with atom parameters which are derived empirically. Atoms in the system interact in two ways; they are either covalently bonded or/and they show nonbonded interactions. A typical potential energy function thus consists of bonded interactions, namely, covalent-bond stretching, angle bending, and proper and improper dihedral torsions, and nonbonded terms including the electrostatic interaction, represented by the Coulomb potential and the van der Waals interaction, modeled typically using the Lennard-Jones potential:

$$U_{\text{system}} = U_{\text{bonded}} + U_{\text{nonbonded}} \tag{10.1}$$

$$= \sum_{\text{bonds}} K_{\text{b}}(b - b_0)^2 + \sum_{\text{angles}} K_{\theta}(\theta - \theta_0)^2 + \sum_{\text{torsions}} K_{\chi}\{1 + \cos(n\chi - \delta)\} + \sum_{\text{impropers}} K_{\varphi}(\varphi - \varphi_0)^2 + U_{\text{nonbonded}} \tag{10.2}$$

where U is potential energy; K_{b} and K_{θ} are force constants for bond and angle, respectively, b_0 and θ_0 are equilibrium bond length and angle, respectively; K_{χ} and K_{φ} are force constants for torsional and improper angles; and n is multiplicity, δ is phase angle, and φ_0 is equilibrium improper angle.

In order to predict the time evolution of a biomolecular system, various force fields such as CHARMM [32, 33], AMBER [34], OPLS [35], GROMOS [36], etc. have been developed [37]. While some of these force fields are targeted purely to proteins, others focus on DNA or a broad range of organic molecules. A variety of force fields are available nowadays for lipids such as AMBER Lipid14, CHARMM36, GROMOS 54A7, Slipids, Martini, etc. [38–41].

Although all-atom MD simulations provide atomistic detailed information about ion channels, some properties like free energy profile along channels, ion conductance, etc. are difficult to obtain via classical MD simulations because of insufficient sampling of the conformational space. As simple prolonging of the simulations to sample a larger conformational space is often impossible in a reasonable amount of (processor) time, there are several sampling techniques available to overcome this problem. All the important sampling techniques are based on specifying collective variables (or reaction coordinates) which define the transition between different conformational states.

10.2.3 Methods for Calculation of Free Energy

Free energy of a system can be calculated as

$$A = k_{\mathrm{B}}T \ \ln \left(\iint \mathrm{d}p^N \ \mathrm{d}r^N \ \exp[-\beta H(p^N, r^N)] \right) \tag{10.3}$$

where k_{B} is the Boltzman constant, T is temperature, $\beta = 1/k_{\mathrm{B}}T$, N is the number of particles, H is the Hamiltonian, and r and p represent sets of Cartesian coordinates and conjugated momenta, respectively.

In practice, the free energy difference between different states of a system is often of higher interest than its absolute value. For two states, X and Y, a simulation can be started from state X and eventually converge to state Y. The difference in free energy of the states reads

$$\Delta A = -k_{\mathrm{B}}T \ \ln \langle \exp[-\beta(H_{\mathrm{Y}} - H_{\mathrm{X}})] \rangle_{\mathrm{X}} \tag{10.4}$$

10.2.3.1 Free Energy Perturbation

Although we can run a simulation starting from one of these states and calculate ΔA, this is only possible when there is overlap between two populations. MD is able to sample the configuration space around energy minima efficiently, but is not able to do enough sampling of higher energy states (e.g. transition states). This problem is solved by introducing intermediate states between two states. In the free energy perturbation (FEP) method, these intermediate states do not necessarily need to be physically meaningful; that is why it is also known and alchemical free energy calculation. A coupling parameter λ is introduced and the Hamiltonian of the system is modified as a linear combination of two neighboring states. For example, the constant k for bond length in an empirical force field would be modified as

$$k(\lambda) = k(\mathrm{Y}) + (1 - \lambda)k(\mathrm{X}) \tag{10.5}$$

where λ is the coupling parameter, $k(\lambda)$ is the resulting force constant for a particular value of λ, and $k(\mathrm{X})$ and $k(\mathrm{Y})$ are force constants for states X and Y.

A series of these intermediate states joining the reference and target states are generated and ΔA is calculated as

$$\Delta A = -k_{\mathrm{B}}T \sum_{i=1}^{N-1} \ln \langle \exp[-\beta \Delta H_{i,i+1}] \rangle_i \tag{10.6}$$

where i is the number of states, A is free energy, H is the Hamiltonian, T is temperature, k_{B} is the Boltzman constant, $\beta = 1/k_{\mathrm{B}}T$, and N is the number of particles.

10.2.3.2 Umbrella Sampling

One of the most popular methods to overcome this energy barrier problem for sampling the configuration space in high-energy regions is umbrella sampling [42]. In this technique, a biased potential is used to bring a system from one state to another along a reaction coordinate. This path is divided into many sampling windows and independent umbrella sampling simulations are performed over each window where the system is restrained within that window by a biased potential. Finally, the biased potential effect is removed and the potential of mean force (PMF) is calculated from each of those umbrella simulations and a combined free energy change is calculated along the reaction path using the weighted histogram analysis method (WHAM) [43].

10.2.3.3 Metadynamics

In metadynamics [44], a system is moved along collective variables (reaction coordinates) by applying a biased potential. This history-dependent biased potential is generated by adding Gaussian hills to the Hamiltonian of a system along the reaction coordinate at a definite interval. Thus, the energy basin or local minima is filled and the system is pushed away from the conformational states which have already been visited. Thus, once the potential energy surface becomes flat, the system is able to explore other regions in the free energy landscape [44].

Biasing potential is defined as

$$V_{\mathrm{G}}\left(S;N\tau_{\mathrm{G}}\right) = \omega \sum_{n=0}^{N} \exp\left(\sum_{d=1}^{D} (S_j - s_d^n)^2/\delta s^2 d\right) \tag{10.7}$$

where S is a set of collective variables, τ_{G} is a time period at which Gaussian hills are added, and ω and δs are the height and the width of the Gaussian function, respectively.

Despite being a robust and widely accepted method, metadynamics comes with certain limitations. It is difficult to decide when to end metadynamics calculations as the free energy does not converge to a definite value but fluctuates around the estimated free energy value which leads to an error proportional to the Gaussian height and hill deposition rate. One way to overcome this is to decrease the deposition rate, which means that it will take longer to fill the free energy basins. Another problem with continuing the run is that the system might visit states which are not physically relevant. *Well-tempered metadynamics* [45] offers a solution for these problems. In well-tempered metadynamics, the rate of change of potential reaches zero as the time spent in a bin gets longer. Thus, the rate of decrease is fast and allows the bias potential to converge to a finite value. The convergence issue can therefore be addressed and longer simulation time results in more accurate results. Also, overfilling of basins does not occur, which leads to computational time efficiency.

10.2.3.4 Adaptive Biased Force Method

The adaptive biased force (ABF) method [46, 47] is another way to calculate free energy of geometrical transformations. In this method, the average force acting along a reaction coordinate is estimated and integration of this average force gives the free energy. The ABF method applies an adjustment of this average force acting along a reaction coordinate. The instantaneous force acting along a reaction coordinate can be directly calculated and is the sum of the average force and random force. The average force depends only on the value of the reaction coordinate, while the random force depends on all other degrees of freedom.

The average of instantaneous force is calculated over time and the derivative of free energy working instantaneously at each point is thus estimated. Simultaneously, an external biased force is applied here to cancel out the average force estimated. Over simulation time, when this estimated force converges to average force at equilibrium, the biased force approaches zero and the system explores the flattened free energy surface. This movement is driven by random fluctuations.

Instantaneous force is defined as

$$F_\xi = -\nabla V \cdot (\partial x/\partial \xi) + \beta^{-1} (\partial \ln |J|/\partial \xi) \tag{10.8}$$

where $|J|$ is the determinant of the Jacobean matrix. x is the position along the reaction coordinate ξ.

A combination of more than one sampling techniques has also been applied in order to achieve better sampling and hence a more accurate free energy profile [48–51].

10.3 Properties of Ion Channels Studied by Computational Modeling

10.3.1 A Refined Atomic Scale Model of the *Saccharomyces cerevisiae* K^+-translocation Protein Trk1p

In yeast, [K^+] ion concentration is regulated by various membrane-embedded cation-translocation systems enabling yeast cells to grow and survive in an environment of external [K^+] ion concentrations ranging from 100 μM to 1 mM [52]. At very low concentrations of [K^+] ions, Trk1 has been identified as the most important potassium uptake system and knocking out Trk1 leads to growth deficiency of the cells [53, 54]. Using sequence comparison, it was predicted that TrK1 evolved via gene duplication of KcsA-like channels; hence, it consists of a single polypeptide chain bearing four subunits [55]. Each subunit consists of two transmembrane (TM) helices M1 and M2 connected via a pore helix and selectivity filter loop. Zayats et al. [56] proposed a dimeric model of Trk1 based on primary sequence and structural information about the interface between monomers and the multimeric state of the template crystal structures TrkH [57] and KtrB [58] that were used for generating an initial homology model of yeast Trk1. Despite the very low sequence identity of Trk1 with TrkH or KtrB (11% and 22%, respectively), the structural model has been proved to be stable in consequent MD simulations. It could provide the structural basis for addressing the question whether Trk1 is a passive or an active ion-translocation system, as the part of the ion permeation path containing the selectivity filter is shaped similar to that of ion channels. However, below the selectivity filter the pore is obstructed or regulated by a proline-containing loop [56]. In addition, MD simulations allowed a structural understanding of the role of certain conserved residues, e.g. Gly in the selectivity filter or D79/K1147 interactions on the correct protein folding or targeting to plasma membrane [56]. In summary, even a relatively low-resolution model based on weak homology can, under certain circumstances, contribute to a better structural understanding of an ion-translocation system, elucidating the role of specific amino acids even though a detailed interpretation of all atomistic detail is impossible or at least stays on the level of speculation. A smart combination of wet and computational experiments allows complementing and guiding the computations, especially in the case where several options show the same probability, allowing a structural interpretation that would not be possible otherwise.

10.3.2 Homology Modeling, Docking, and Mutagenesis Studies of Human Melatonin Receptors

Melatonin receptors are a member of the G protein–coupled receptor superfamily [57, 58] which mediate the effect of the melatonin hormone involved in various functions such as circadian rhythm, immunomodulation, scavenging free radicals, etc. [59, 60]. Human melatonin receptors consist of seven TM helices connected together by three extracellular and three intracellular linker loops [61]. A structural model of human MT2 melatonin receptor was constructed [62] on the basis of the 2.6 Å resolution bacteriorhodopsin crystal structure (PDB code 1L9H) [61]. The receptor–ligand complex was prepared by manually docking 2-iodomelatonin, 4P-PDOT, and luzindole ligands into this structural model of hMT2. The docking results were used to predict potential key residues involved in interactions that could be tested experimentally by site-directed mutagenesis and functional assays [62]. Mutations of several residues – V204A, G271T, L272A, and Y298A – either decreased or abolished the binding of 2-iodomelatonin to the hMT2 receptor. In a subsequent study, Mazna et al. [63] performed model refinement in order to investigate other possible residues involved in 2-iodomelatonin binding. The predicted residues involved in binding were experimentally mutated and the binding affinities of these residues were calculated using an adio-ligand-binding assay. Residues N268 and A275 (from TM6) and V291 and L295 (TM7) were proposed to be involved in the binding of 2-iodomelatonin with the hMT2 receptor [63]. Given the fact that bacteriorhodopsin is not in the same family as the melatonin receptor, although both share apparently the same fold family, the resulting structural model is of very low resolution and the position of side chains can be seen as only a guess. In such a case, sophisticated docking algorithms calculating interaction energies to rank poses do not make sense and cannot be reasonably applied. However, manual docking into a reasonable, analogous, or experimentally supported binding site might lead to the correct identification of additional binding residues that potentially interact with a ligand, and thus such a homology model might still be useful to support site-directed mutagenesis in combination with experimental functional studies, as demonstrated in the study mentioned.

10.3.3 Selectivity and Permeation in Voltage-Gated Sodium (Na_V) Channels

Voltage-gated sodium channels (Na_V) are important for initiation of an action potential by rapid permeation of sodium ions across the cell membrane [64, 65]. This role in action potential initiation in neuronal, cardiac, or skeletal muscle cells

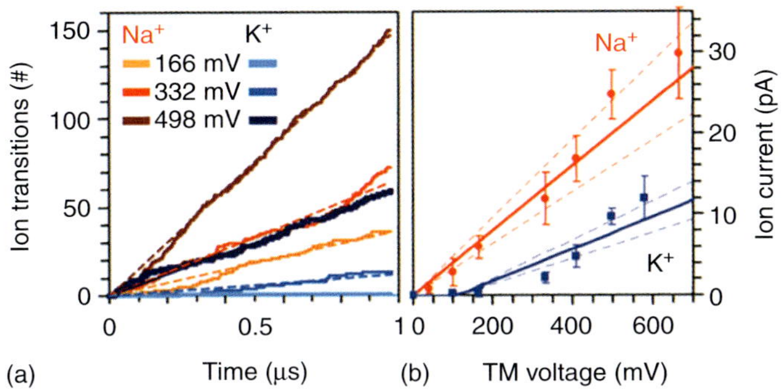

Figure 10.1 (a) Cumulative ion translocation over 1 μs at different TM voltages at 0.5 M concentrations of NaCl and KCl. (b) I–V curves for Na^+ (red) and K^+ (blue) used for calculation of single-channel conductances. The translocation of ions is voltage dependent, with a clear preference for sodium ions over potassium ions. Source: Taken from [69].

and underlying signal propagation makes mutations observed in these channels disease-relevant. Na_V mutants are found in a broad range of diseases such as epilepsy, cardiac arrhythmia, and painful congenital myotonia. [65, 66]. The bacterial Na_V channels are composed of four subunits, each consisting of six TM helices [67]. TM1–TM4 are involved in voltage sensing, whereas TM5, the first pore segment (P1), the selectivity filter, the second pore segment (P2), and TM6 form the pore. Selectivity filter is lined with TLESW residues where glutamic acid side chains recruit the Na^+ ions from the extracellular part.

The availability of a crystal structure of the open conformation of voltage-gated sodium channels from *Magnetococcus* sp. (Na_VMs) [68] provided the opportunity to investigate the full path of ion translocation via MD simulations. Ulmschneider et al. [69] performed microsecond-order MD simulations of Na_VMs under different TM potentials between 39 and 665 mV and 0.5-M concentration of NaCl or KCl. A harmonic restraint was applied to α-carbon atoms of TM helices in order to keep the pore open at the cytoplasmic gate throughout the simulations. Calculations of cumulative ion translocation for Na^+ and K^+ showed the clear preference of Na_VMs channels for Na^+ over K^+ (Figure 10.1a [69]). The selectivity ratios of the channel for Na^+ and K^+ were calculated from the current–voltage (I–V) relationship (Figure 10.1b [69]) showing that at higher TM voltage the selectivity ratio Na^+:K^+ is 3–6 : 1 and even more favoring Na^+ at lower TM voltages with a ratio of 11–18 : 1. The conductance for Na^+ calculated from the I–V relationship was ~34 pS, which is very close to the experimental value ~33 pS. Free energy profiles were calculated using the Na^+ ion binding positions throughout the simulations. Five binding sites (S_0–S_4) were identified in the selectivity filter region using the free energy profile (Figure 10.2), with S_1 being the dominant binding site. Na^+ binds to one of the four glutamic acid side-chain carboxylate group, a serine (Ser 54) hydroxyl group, and two molecules of water. The binding sites identified are in agreement with the electron density map of the crystal structure. The free energy profiles also revealed a larger barrier for K^+ at site S_1 compared to Na^+ which hinders K^+ from moving through the pore.

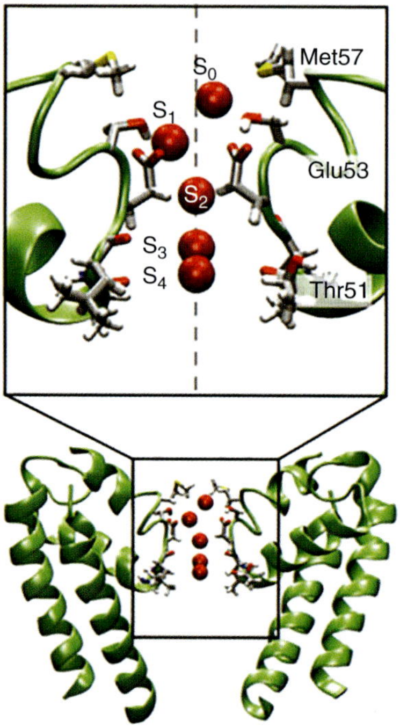

Figure 10.2 Five Na^+-binding sites, labeled S_0–S_4, identified in the equilibrium simulations. S_0 is at the vestibule of the selectivity filter, contacting Ser54 and Met57 side chains; S_1 is off the symmetry axis of the channel pore because of strong contact of the ion with one of the four Glu53 residues; S_2 is a minor site just beyond the contact distance of the Glu side chains; S_3 and S_4 are at the height of Leu52 and Thr51 backbone carbonyls, respectively. S_0, S_2, S_3, and S_4 are stabilized by full water-hydration shells. S_1 has a partial water hydration shell. Source: Taken from [69].

In order to investigate the role of the EEEE locus (glutamate side chains in selectivity filter) in selectivity and conduction, Ke et al. [70] performed MD simulations of the open Na_VMs channel (PDB: 4F4L). The double lipid bilayer scheme was used, which enabled the calculations of Na^+ influx and efflux separately. A constant number of Na^+ ions and a charge difference of 4e of Cl^- could be maintained between the two reservoirs by applying a swapping mechanism. The authors showed the inward and outward conduction mechanism with detailed interpretation of ion densities and ion dwelling time across the selectivity filter. The calculated inward conductance (27 ± 3 pS) was significantly higher than the outward conductance (15 ± 3 pS). The difference in glutamic acid side-chain flipping/rotation for inward and outward conduction was shown and the effect of Glu side-chain conformation on efflux was demonstrated. The influx was not affected by the conformation of Glu side chains, while the efflux was clearly dependent on the flipping of Glu side chains. Further PMF calculations showed that the energy barrier increased for efflux if Glu side chains were forced to remain in non-flip conformation. An inward-facing conformation (flip) of at least one Glu side chain in the selectivity filter would reduce the energy barrier for ion outward transition by providing direct coordination with the interacting Na^+ ions [70]. Summarizing, simulations of voltage-gated sodium channels using applied membrane potentials as reported for Na_VMs [69, 70] are able to reveal different conduction mechanisms for ion inward and outward transitions and hence elucidate how a specific local change can provide insights into the slow inactivation of voltage-gated sodium channels during an action potential, when the membrane potential is depolarized.

10.3.4 Study of Ion Conduction Mechanism, Favorable Translocation Path, and Ion Selectivity in KcsA Using Free Energy Perturbation and Umbrella Sampling

Potassium channels represent the most diverse group of membrane integral proteins and are key constituents of electrical signaling networks in the nervous system [71]. The prototypical K^+ channel for laboratory studies is KcsA from *Streptomyces lividans* as it can be easily expressed in *Escherichia coli* cells and handled in the laboratory [72–74]. The channel is composed of four subunits arranged around a central pore [71]. Each subunit consists of two TM helices and a pore loop (P-loop) containing a selectivity filter (Figure 10.3a). The selectivity filter contains an ion binding site formed by backbone-oxygen atoms (Figure 10.3b). Several crystal structures have been solved for KcsA. Crystal structures show four binding sites within the filter (S_1–S_4) (Figure 10.3c [71, 75]), where the potassium ion is coordinated by eight protein atoms, and two additional sites S_0 and $S_{\text{-ext}}$ (ext denotes extracellular ion binding site), where K^+ is coordinated by four backbone-oxygen atoms and four water molecules.

The existence of four binding sites within the selectivity filter region leads to the theoretical possibility of 16 binding states and hence a large number of possible translocation pathways. In order to elucidate the conduction mechanism of K^+ ions through the channel, it is essential to know the most favorable path through the selectivity filter. Aqvist et al. [76] studied the potassium channel KcsA from *S. lividans* (PDB:1bl8, [2]) using FEP calculations. The comparative energetics analysis of differently occupied binding sites (thus, different binding states) of the tetrameric channel successfully demonstrated the preference of one binding state over the others (Figure 10.4 [75]). The proposed mechanism of ion permeation involves doubly occupied binding states in the filter. The state in which ions occupy the first and third of the four potential binding sites (1010(1) state) differs

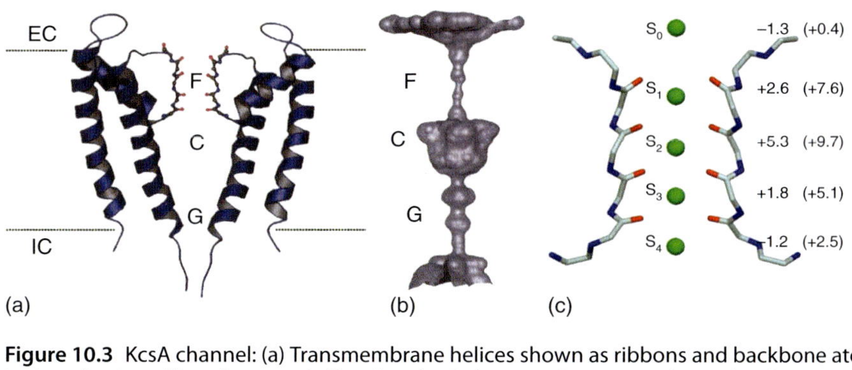

Figure 10.3 KcsA channel: (a) Transmembrane helices shown as ribbons and backbone atoms in the selectivity filter shown as ball and sticks. Only two subunits are shown for clarity. Source: Taken from [71]. (b) The pore surface of the KcsA channel calculated by the HOLE program. Source: Taken from [71]. (c) Cation binding sites in the selectivity filter of KcsA channel shown as green spheres and labeled S_0–S_4 from extracellular to intracellular sides. The numbers adjacent to the sites represent $\Delta\Delta G$(K^+ to Na^+) of binding obtained from FEP calculations. A positive value indicates a preference for potassium over sodium. Numbers in parenthesis correspond to the same quantity calculated from a restraint structure yet fitted to potassium ions. Source: Taken from [75].

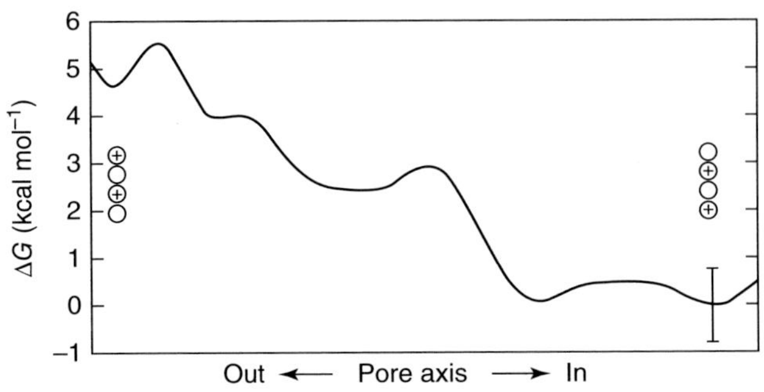

Figure 10.4 Free energy profile (potential of mean force) for the single movement between the 1010(1) and 0101(1) states. The numbers 1 and 0 denote the presence or absence of an ion at the respective site, while (1) refers to the presence of an ion occupying the cavity site. Source: Taken from [76].

in free energy by 5 kcal mol^{-1} from the state in which ions occupy the second and fourth potential binding sites (0101(1) state) [76].

In a subsequent umbrella sampling study of the potassium channel KcsA, the permeation of ions was described [77]. The ion conduction mechanism involves two main states in which two or three K^+ ions occupy the selectivity filter. The optimal path with the lowest free energy barriers corresponds to a *knock-on* mechanism proposed in earlier studies [78]. Along this pathway, the first ion reaches from the intracellular side and pushes two ions already present in the selectivity filter (Figure 10.5a). In the next step, the outermost ion in the selectivity filter is pushed out. The largest free energy barrier along this optimal ion conduction pathway was 2–3 kcal mol^{-1} [77]. The other possible pathway follows a *vacancy-diffusion* mechanism [79] where two K^+ ions in the selectivity filter move toward outer binding sites, thus leaving a vacant site behind, which in turn is occupied by the third ion coming from the inner side (Figure 10.5b). This pathway involves the largest free energy barrier of 3–4 kcal mol^{-1} [77]. With the help of the calculated free energy profile as a function of position of

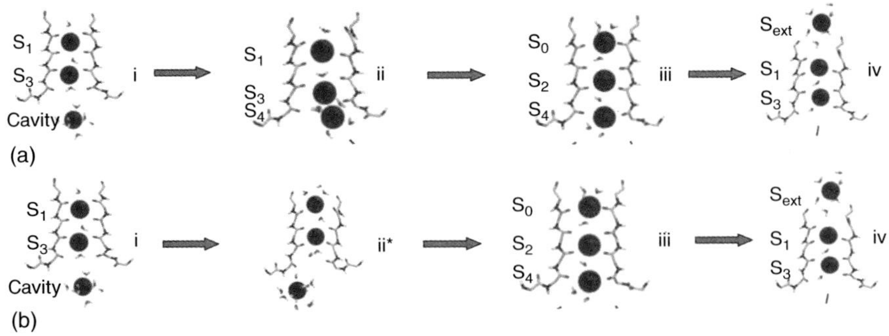

Figure 10.5 : Translocation pathway following knock-on mechanism (upper panel) and vacancy diffusion mechanism (lower panel) in KcsA [77]. Identified binding sites in the selectivity filter region are labeled S_0–S_4.

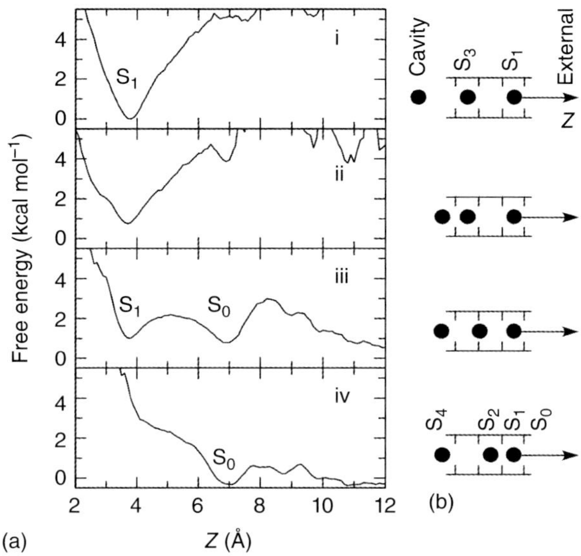

Figure 10.6 Free energy as a function of the outermost K^+ ion where two other K^+ ions are fixed in different binding sites in a selectivity filter. (a) Free energy profiles for different ion occupancies within the selectivity filter. (b) Configurations corresponding to the profiles. Source: Taken from [77].

the outermost ion in the selectivity filter (while two other K+ ion positions were fixed), the authors described favorable positions for K^+ ions in the selectivity filter that would lower the energy barrier for the exit of potassium to the extracellular cavity (Figure 10.6, [77]). The alchemical free energy calculation also showed the selectivity of the channel for K^+ over Na^+ [77].

Through KcsA, potassium permeates nearly at diffusion rate, while selectivity for K^+ ions over Na^+-ions is maintained [80]. This selectivity was earlier explained with the size of the K^+ ion, which enables it to fit between carbonyl groups lining the pore of the channel, while the smaller Na^+ does not have sufficient favorable interactions in the pore. The thermal factor in the KcsA crystal structure is 0.75 Å, which is bigger than the difference of 0.38 Å in atomic radii of K^+ and Na^+. MD simulations and free energy calculations using the FEP method were conducted in order to investigate the factors responsible for selectivity [75]. To investigate the role of local carbonyl–carbonyl electrostatic interactions in selectivity, these interactions were artificially switched off. The difference in free energy calculated here showed that the selectivity for K^+ over Na^+ was lost when these interactions were switched off. Pore flexibility seems to be essential for accommodating the K^+ ion and to maintain the selectivity. The selectivity for K^+ ions results from the flexible carbonyl groups lining the pore and the intrinsic electrostatic properties [75].

Kashahara et al. [81] simulated the Kv1.2/2.1 channel using MD and showed the effect of ion concentrations and TM potential on ion conduction. They proposed two mechanisms by which ion conduction occurs: under high-voltage condition, ion conduction occurs via a so-called III–II–III (association–dissociation) mechanism and in high-ion concentration it occurs via the III–IV–III knock on mechanism. Here, II/III/IV represents the number of ions in the pore during ion conduction. One mechanism is preferred over the other in response to change in voltage/TM potential [81].

Mouse Kv1.3 is a voltage-gated potassium channel present in T-lymphocyte cell membrane [80, 82]. MKv1.3 is a tetramer where each monomer consists of six TM helices [83, 84]. The central pore is formed from S_5, S_6 helices and P-loop segments. The blockage of the channel with various channel blockers results in reduced cytokine production and T-cell proliferation [80, 85–87], which makes channel blockers an important candidate for drug designing purpose in the treatment of autoimmune diseases. Khabiri et al. [8] studied the binding mechanism of a potential blocker belonging to the group of scorpion toxins, charybdotoxin (ChTX), to the mKv1.3 channel using umbrella sampling and free energy calculations. The simulated structural model of mKv1.3 containing S_5-pore-S_6 in addition to S_4 helix was constructed by comparative modeling using the Kv1.2 crystal structure (PDB code: 2A79) with 94% sequence identity in the aligned region as a template. Docking of the potential blocker ChTX to the model was performed with AutoDock [88] implemented in the Yasara software package [16]. Subsequently, MD simulations and umbrella sampling were performed on the equilibrated structure. Despite the fact that in these studies a homology model was used, the extremely high homology of 94% sequence identity in the aligned region allowed generating a model that came close to a crystal structure in resolution and could be used even for reliable free energy calculations. Those free energy calculations showed no effect of high versus low potassium concentrations on the binding free energy of ChTX to mKv channel but the shapes of the free energy curve were different. Although different K^+ ion concentrations do not have an effect on the binding free energy, further structural analysis of their simulations allowed the authors to conclude that the concentration of potassium is critical in the diffusional step during binding events [8].

In a subsequent study using the scorpion toxins charybdotoxin and margatoxin to block the human voltage-gated hKv1.3 channel, Nikouee and coworkers [89] built a model of hKv1.3 channel by *in silico* mutation of the abovementioned mKv1.3 model [8]. MD simulations in combination with experimental data pointed to differences in binding of these toxins to wild-type and mutant hKv1.3 channels. Here, the rigidity–flexibility in the region behind the selectivity filter played a key role in its interactions with scorpion toxins. The simulations demonstrated that a channel with a slightly more flexible selectivity filter region exhibits distinct interactions with scorpion toxins, suggesting that the toxin–channel interactions might partially restore rigidity in the selectivity filter and thereby prevent the structural rearrangements associated with C-type inactivation in the mutant [89].

10.3.5 Ion Conductance Calculations

The TM potential is the driving force for movement of ions across the membrane through the channel, which arises due to a net charge difference between intracellular and extracellular compartments. Since MD simulations use periodic boundary conditions, the solvent above and below the membrane is essentially the same, which makes it difficult to simulate different intracellular and extracellular ion concentrations. Either a vacuum or a double lipid bilayer system can be used to separate these two compartments from each other [90]. In case of a double bilayer system, the net movement of ions is governed by the TM potential generated by the charge imbalance between both compartments. The main limitation of this approach, however, is its large size and hence properties studied are limited to short computational timescales. Another approach to maintain the TM potential is to apply an external electric field normal to the TM, which in turn generates the ion flux across [91].

10.3.5.1 Voltage-Dependent Anion Channel (VDAC)

Voltage-dependent anion channels (VDACs) mediate the exchange of ions (K^+, Cl^-, and Ca^{2+}), ATP, or other small hydrophilic molecules between the cytosol and the mitochondria [92] and play an important role in the regulation of the release of apoptotic proteins. Ion conduction through the human VDAC was calculated using classical MD simulations [93]. In this study, the differences in ion conductivity of hVDAC for K^+ and Cl^- ions were measured under different TM potentials. TM potentials were maintained by applying an external constant electric field parallel to the membrane normal. The number of K^+/Cl^- ions crossing the center of the channel in the positive Z-direction in a small time interval was calculated and subsequently converted to conductance by the formula [93]

$$G = (N_K - N_{Cl})^* q / V^* \Delta t \tag{10.9}$$

where G is the conductance, N_K and N_{Cl} are the number of potassium and chloride ions crossing the channel, q is the charge of a monovalent ion, V is the TM potential, and Δt is the time interval.

Three independent sets of systems were prepared and simulations were performed at different TM potentials. The average calculated conductance was 4.8 ± 0.2 nS, which is within the range of experimentally calculated values ~3.9–4.5 nS (Figures 10.7 and 10.8).

10.3.5.2 Calculation of Ion Conduction in Low-Conductance GLIC Channel

The GLIC channel is a member of the pLGIC family that opens or closes in response to a ligand-binding stimulus. These ligand-gated ion channels mediate the neurotransmission at synapses in response to binding with neurotransmitters. Studies have shown that the members of this family such as nicotinic

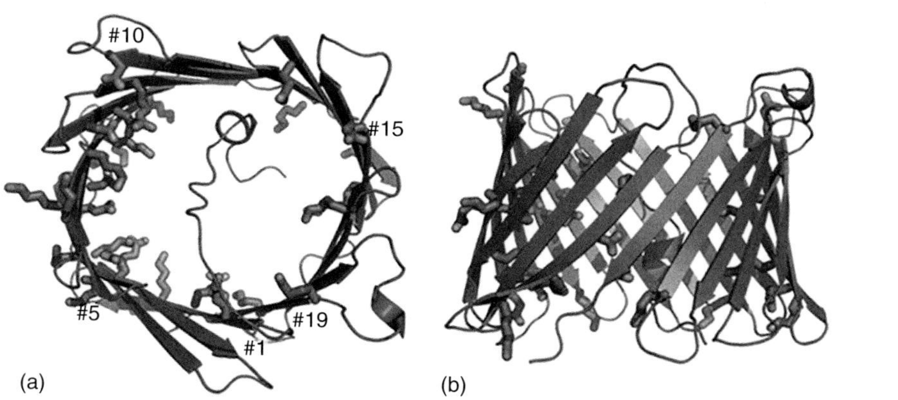

Figure 10.7 Cartoon representation of hVDAC1 from the first NMR model (PDB:2K4T). Top view (a) and side view (b). All the charged residues are highlighted with a stick representation. Source: Adapted from [93].

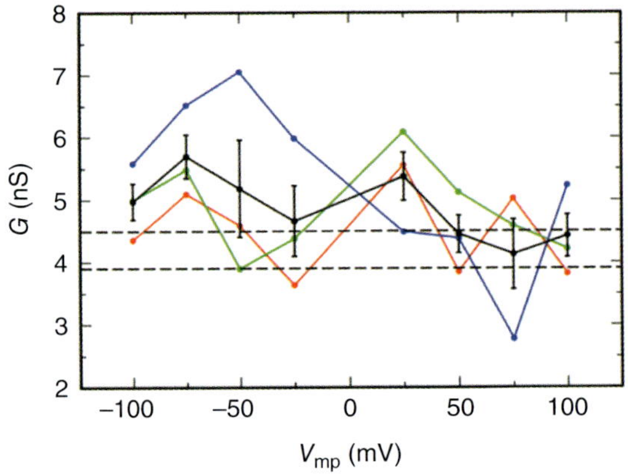

Figure 10.8 Conductance (G) of hVDAC1 calculated over last 60 ns in three sets of simulation S_1 (red), S_2 (green), and S_3 (blue) as a function of applied voltage (V_{mp}). The average conductance is shown as black line with standard error. Experimental single-channel conductance range is shown between dashed lines. Source: Taken from [93].

acetylcholine receptors (nAChRs) or gamma aminobutyric acid (GABA) receptors can be modulated by anesthetics within a range of clinical concentrations [9]. This makes pLGIC family members an important target for anesthetics. The structure of GLIC from *Gloeobacter violaceus* has been solved by X-ray crystallography [10, 94]. The overall architecture of extracellular domains and TM domains of GLIC is similar to the nAChR. The role of anesthetics in inhibiting the Na^+ currents in the GLIC channel has been shown [95, 96]. The availability of a GLIC channel crystal structure makes it a model for the pLGIC family. In order to determine the ion conductance from MD simulations, it is necessary to observe many ions passing through the channel, which is difficult in channels like GLIC with low ion conductance during a reasonable simulation time. Here, an alternative possible approach is to calculate the ion conductance using the estimated PMF of an ion along the pore channel. Zhu et al. [97] used umbrella sampling calculations with Hamiltonian replica exchange in order to determine the PMF.

The pore was held in open conformation by applying position restraints on C-α atoms. Umbrella sampling simulations were performed on 153 windows where harmonic restraints were applied on Na^+ ions along z-coordinates. Hamiltonians of the sodium ion from neighboring umbrella windows were exchanged at a definite time interval according to a Metropolis criterion. The PMF revealed a major energy barrier at the hydrophobic constriction of the channel.

Channel conductance can be estimated from simulations by combining the free energy profile and the diffusion of the ion along the reaction coordinate through the following formula

$$\gamma = q^2 \rho S / \left[k_B T \int dz \, \exp[G(z)/k_B T]/D(z) \right] \tag{10.10}$$

where γ is the conductance, q the charge of the ion, ρ the ionic density, S the cross-sectional area, and $G(z)$ and $D(z)$ the PMF and the diffusion coefficient along the reaction coordinate, respectively. The integral is taken over the whole reaction coordinate. T is temperature, and k_B is the Boltzmann constant.

The local diffusion coefficient of the Na^+ ion can be estimated from each umbrella simulation window using the formula

$$D = \mathrm{var}(z)/\tau \tag{10.11}$$

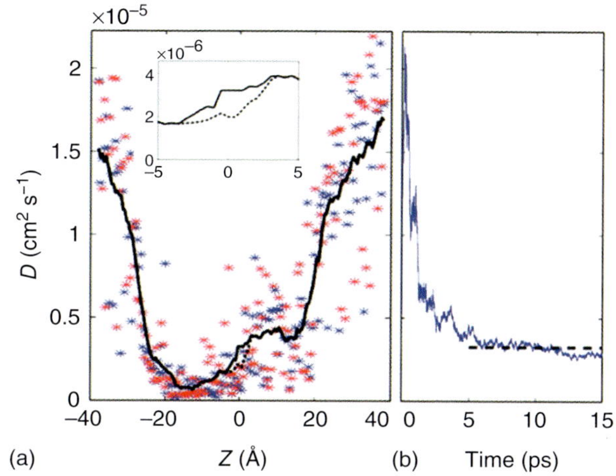

Figure 10.9 (a) Position-dependent diffusion coefficient profile for a sodium ion within the pore of GLIC calculated with Eq. (10.11). Blue and red stars correspond to the diffusion coefficient estimated from two different sets of umbrella sampling calculations. The black dashed curved corresponds to the average value of *D*(*z*). The thick black curve corresponds to the optimal diffusion coefficient obtained by using statistical inference. Inset: Zoom on the barrier region. (b) Optimal diffusion coefficient estimated at the barrier top from unrestrained simulations. An optimal diffusion coefficient is inferred by maximizing the likelihood function over 200 trajectories at each time step. The black dashed line is the average between 5 and 15 ps, used to correct the profile of *D*(*z*) at the barrier top in the inset of (a). Source: Taken from [97].

where D is the diffusion coefficient, z is the reaction coordinate, $\mathrm{var}(z)$ is the variance of z, and τ is the correlation time. τ can be calculated as

$$\tau = [(n^*\mathrm{var}(z)/\mathrm{var}(z)) - 1]\,\Delta t/2 \tag{10.12}$$

where n is the number of data points and Δt is the time interval between two data points.

The position-dependent diffusion coefficients $D(z)$ were then calculated by combining these umbrella windows (Figure 10.9). By combining the free energy profile and diffusion coefficient profile of the ion along the pore, the ion conductance was estimated around 1 ps (compared to 8 ps experimentally). The authors deduced that the crystal structure represents an open conformation.

Studies have shown that MD simulations underestimate ion currents. This underestimation could arise, on one hand, from insufficient sampling and, on the other hand, from the deficiency of force fields to accurately describe the ion and amino acid interactions or to accurately describe the membranes which are much more complex in nature than used in MD simulations [98–100].

10.3.6 Transient Receptor Potential (TRP) Channels

Transient receptor potential (TRP) channels are present in the plasma membrane of the vast majority of animal cells; for a review, see Ref. [101]. The TRP channel family includes ~30 TRP channels which are further categorized into seven subfamily groups on the basis of three-dimensional domain structure and primary amino acid sequence. A common feature of all the members of the TRP channel family is the six TM domains (S_1–S_6) with a P-loop between S_5 and S_6 helices forming a selectivity filter. TRP channels are activated by various chemical ligands, temperature, mechanical pressure, Ca^{2+} ion depletion, and many other stimuli. Being present in many sensory cells, TRP channels are related to many diseases and thus play an important role in drug discovery studies.

TRP vanilloid type 1 (TRPV1) channel is a member of the TRP channel family which opens in response to vanilloid compounds (e.g. capsaicin or resiniferatoxin), pH, and voltage. Before the TRPV1 structure was determined in 2013 (PDB id: 3J5Q) [102], Susankova et al. [103] used homology modeling and MD simulations to propose a 3D model for S_5–P–S_6 pore module of rat TRPV1 and the contribution of this pore region in gating of the channel. This model [103] of the S_5–P–S_6 pore module is very similar to that of the later determined crystal structure [102]. Since a cryo-EM structure became available, several computational studies to understand the binding mode of vanilloids to TRPV1 and energetics of ion permeation through TRPV1 channels from different organisms have been reported [104, 105]. Jorgensen et al. [105] used bias-exchange metadynamics (BE-meta) simulations to calculate the PMF of K^+ and Na^+ in rat TRPV1, successfully explaining the nonselectivity of TRPV1 channel for Na^+ and K^+ on the basis of similar binding sites as well as similar free energy profiles for these ions. This example shows that actually BE-meta simulations can be very promising for efficient sampling in order to evaluate the energetics for different possible number of permeating ions.

TRPA1 is another member of the TRP channel superfamily and has been studied as a promising drug target due to its role in pain and inflammatory diseases [106, 107]. TRPA1 can be activated by a range of stimuli such as exogenous irritants, proalgesic agents, and cold mechanical force [108–112]. The distinguishing feature of TRPA1 is its 14–18 ankyrin repeats in the N-terminal domain [106, 113, 114]. Several attempts have been made in the past to study the structure and function of ankyrin repeats by homology modeling, simulation [114, 115], and steered dynamics [116]. The ankyrin domain has specialized mechanical properties which allows it to function as a soft spring changing its end-to-end distance by almost 50% while maintaining structural integrity. Calcium binding to its EF hand increases its stiffness and constrains the end-to-end distance [115].

Recently, the human TRPA1 structure was determined to be of 4.2 Å resolution using single-particle electron cryo-microscopy [117].

10.4 Free Energy Methods Applied to Channels Bearing Hydrophobic Gates

Usually, ion channels try to mimic within their pore an environment corresponding to the one an ion would find in bulk water. However, some TM ion channels possess a singular feature. Their pore, or part of it, is lined with hydrophobic residues, thus, creating a so-called hydrophobic gate [118]. For this type of

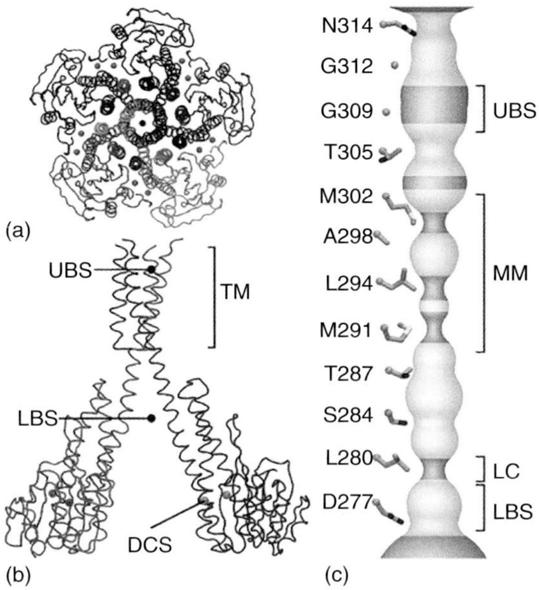

Figure 10.10 Structure of the TmCorA protein. (a) Bird's eye view from the extracellular media of the pore (code PDB databank 2HN2). Regulatory cations and within the pore are shown. (b) Side view of two protomers. Annotations show the different regions of the pore. DCS corresponds to the cytoplasmic binding region for the regulatory ions. LBS and UBS demark the lower and upper binding region, respectively. The transmembrane (TM) region is indicated by the TM bracket. (c) Surface representation of the pore diameter. The MM stretch corresponding to the hydrophobic gate is given as well as LC, the leucine constriction site. Residues lining the pore are given on the left. Source: Adapted from [120].

channels, the switching between an open and a closed state corresponds to the evacuation of the water present within this hydrophobic cavity.

Such is the case of the bacterial transporter CorA involved in the influx of magnesium ions. CorA is a homopentamer with two TM helices per monomer [119]. The upper part of the pore, giving to the extracellular media, is rather wide and lined with hydrophilic residues. Located toward the lower part, the MM stretch and further the lower leucine constriction site constitute two hydrophobic gates (see Figure 10.10). This MM stretch is suggested to be the main regulator in the gating process. Finally, a subunit called DCS (*divalent cation sensor*) present in the cytoplasm is able to bind intracellular magnesium cations.

A recent study [120] using MD simulations coupled to free energy calculations have shed light on the permeation mechanisms of ions in this transporter. The authors proposed that the wetting of the hydrophobic gate is dependent on the intracellular concentration of magnesium. Subunits of the transporter bathing in the cytoplasm are potential binding sites for magnesium ions. When the intracellular medium is depleted from magnesium ions, so is the number of bound ions; this leads to a wetting of the hydrophobic gate.

To validate their hypothesis, the authors performed massive MD simulations. The systems, with or without regulatory ions, were simulated 700 times for 35 ns, reaching a total simulation time of 54 μs sampling.

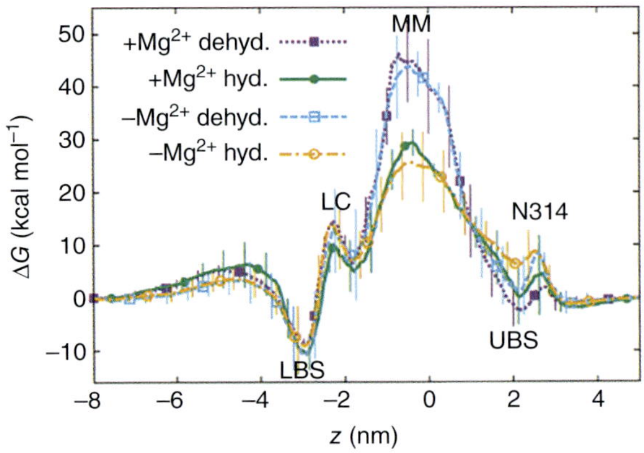

Figure 10.11 Free energy profiles for a hexahydrated magnesium ion passing through the TmCorA. The reaction coordinate is taken as the position of the ion within the pore. The four different region indicated in Figure 10.10 are reported here. Lines with circles denote a hydrated MM region, while lines with squares a dehydrated one. Dotted and solid lines correspond to a system with regulatory ions, and dash and dotted lines to one without regulatory ions. Errors were estimated from the standard deviation of repeated simulations. Source: Taken from [120].

During the simulations, it appeared that the wetting of the MM stretch, i.e. more than 20 water molecules, was rather a transient state and did not last more than 200 ps. These events are quite sudden and appeared in a stretch of the MM pore with a diameter almost double.

To quantify the energetics of the permeation, the authors used umbrella sampling calculations based on multiple starting conformations on four different states – when the DCS was loaded with magnesium or not and the hydrophobic gate was hydrated or dry (see Figure 10.11).

The calculations showed that the height of the free energy barrier for a permeating Mg^{2+} ion is correlated with the level of hydration of the hydrophobic gate. An increase in the water molecule within the pore lowers the free energy barrier. Using umbrella sampling, the free energy barrier experienced by a crossing ion through the hydrophobic gate can be lower by 20 kcal mol^{-1} when the pore is hydrated.

MD can also assist in the prediction of the open or closed state of a crystal structure. Trick et al. used MD simulations to explore the degree of hydration of a hydrophobic pore from the serotonin 5-HT$_3$R receptor as well as the glycine receptor [121]. From their calculations, the authors were able to decipher closed from open state conformations.

The 5-hydroxytryptamine type-3 (5-HT$_3$) receptor is a cation-selective ion channel. This protein belongs to the cys-loop family which encompasses the nicotinic acetylcholine receptor, the GABA$_A$ receptor, and the glycine receptor [122]. The channel is formed by five subunits, where each subunit is constituted by four TM helices (M1–M4). The pore is formed by five TM helices M2 (see Figure 10.12). The pore possesses a hydrophobic ring formed by valines (V164) and leucines (L260) and is dubbed L9′. This ring lies in the center of the

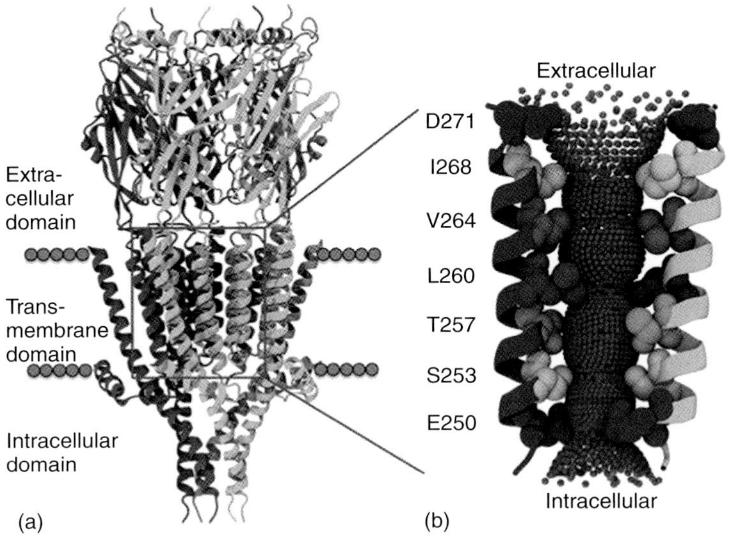

Figure 10.12 Structure of the 5-HT$_3$R protein. (a) Crystal structure (PDB databank code 4PIR) of the receptor. The five subunits are represented. The location of the membrane is given by the gray circles. Extra- and intracellular domains are given as well. (b) Pore receptor with two of the five lining M2 helices shown. Lining residues as also given. The pore surface was estimated with the HOLE software. The hydrophobic gate is formed by V164 and L260 (L9′). Source: Adapted from [121].

bilayer and its diameter of around 2.5 Å was identified as a probable hydrophobic gate. Uncertainty remains on the state in which 5-HT$_3$R was crystallized (PDB code 4PIR).

Using MD simulations and looking at water density profiles within the pore, the authors were able to suggest that the protein was crystallized in the closed state. Free energy profiles for a sodium ion, a chloride ion, and a water molecule were computed using the umbrella sampling protocols. The outcome of these calculations identified a clear barrier around the L9′ ring (see Figure 10.13). Its height was found to be 15 kJ mol^{-1} for a water molecule and 70 and 60 kJ mol^{-1} for a sodium and chloride ion, respectively. These findings further confirmed that this protein was in its closed state.

The authors applied the same line of protocols to another member of the cys-loop family, the glycine receptor. This receptor was crystallized in three different states (see Figure 10.14a). These TM helices share a common feature with the previous receptor studied, the presence of the L-9′ hydrophobic ring. Based on the computation of the water density profile, the high-energy barrier associated with L-9′ was preserved in the closed state (around 20 kJ mol^{-1}) and found to vanish in the open state (see Figure 10.14b).

Finally, steered molecular dynamics (SMD) and free energy calculations are also used to investigate the permeation of an ion through a channel. Dong et al. [123] studied the influence of hydration on the energetic crossing for a sodium ion through the store-operated Orai channel from *Drosophila* [123]. Orai is part of the stromal interaction molecule (STIM)/Orai system that

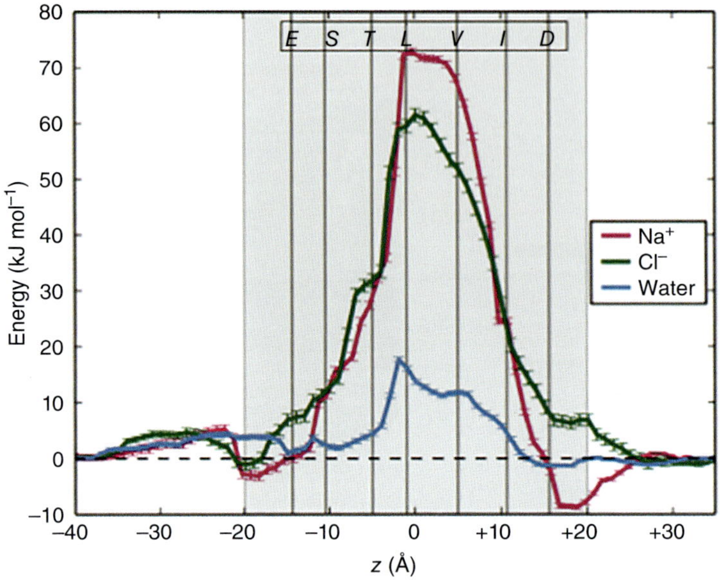

Figure 10.13 Free energy profiles for different species crossing the pore. Single ions, sodium, and chloride are shown in red and green lines, respectively. Free energy profile for a single water molecule is given in blue. The extent of the pore is represented by the gray shaded region. Approximate positions of the residues lining the pore are given by the vertical black lines. Source: Taken from [121].

provides a route of Ca^{2+} influx for physiological Ca^{2+} signaling in immune cells, skin, and muscle [124]. The system is composed of a calcium conducting channel Orai embedded in the plasma membrane, which is activated by an intracellular calcium-sensing protein called stromal interaction molecules (STIMs) embedded in the endoplasmic reticulum. Activated Orai allows the cell to be replenished with calcium ions [125].

The recent *Drosophila* Orai crystal structure revealed a hexameric arrangement around a central pore [126]. Each monomer is a four TM protein where helix M1 constitutes the pore. The selectivity filter, formed by a crown of six glutamates, provides the entry for a hydrophobic pore formed by three hydrophobic residues V174, F171, and L167. At the intracellular side, the pore ends with a basic region formed by residues K163, K159, and R155 where a high concentration of chloride ions can be found.

In their computational study, a sodium ion was dragged inside the pore using SMD and ABF calculations for two proteins – the wild type, whose crystal structure is known to represent a closed state [126], and a constitutively open mutant V174A [127].

The authors identified a much higher level of hydration of the pore in the mutant compared to the wild type. Calculations showed a correlation between the level of hydration of the pore and the free energy barriers experienced by the permeant. As expected, the free energy barrier for the ions in case of the constitutively open mutant was around $3\,\text{kcal}\,\text{mol}^{-1}$ lower than in the wild type.

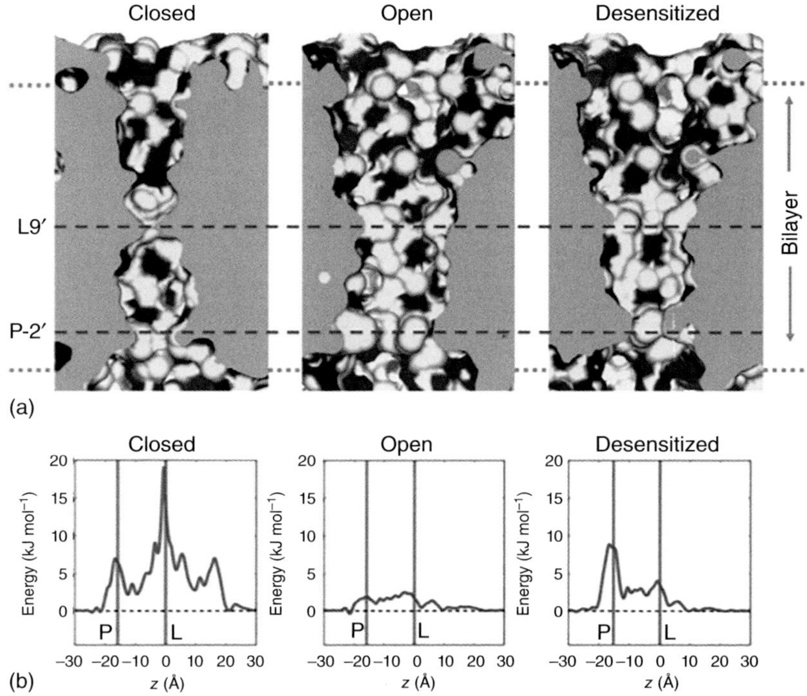

Figure 10.14 Water within the glycine receptor α1 ClyR. (a) From left to right: closed, open, and desensitized conformation of the pore (PDB databank, closed: 3JAD, open: 3JAE and desensitized: 3JAF) within the bilayer. Carbon atoms are colored gray and hydrophilic atoms are colored black. Positions of the constriction regions L9′ (L277) and P-2′ (P226) rings are indicated in black dashed lines. (b) Free energy profiles for a water molecule within the pore. The profiles correspond to the pore state given earlier. The free energy was estimated using the Boltzmann inversion of the water density within the pore. Positions of L9′ and P-2′ are given, respectively, by black lines annotated by P and L. Source: Adapted from [121].

The barrier is located at the intracellular end of the hydrophobic gate (Figure 10.15). This barrier was preceded by a sharp increase in the free energy starting from below the glutamate ring.

Furthermore, the authors investigated the potential influence of an external electric field on the calculated free energy profiles. The overall aspect of the profiles was conserved, but a decrease in the free energy barriers was noticed. At the threshold of the physiological condition, around 100 mV, the barriers were lowered by 2 kcal mol^{-1} [127].

In a recent study, Frischauf et al. [128] modeled the human Orai1 hexamer using homology modeling and the *Drosophila* crystal structure (PDB ID: 4HKR) [126] as a template. Using classical MD simulations and calcium, sodium, and chloride ion concentrations of 10 mM $CaCl_2$ and 8 mM NaCl, the authors observed binding of several Ca^{2+} ions approximately 15 A above the selectivity filter, which increased the local Ca^{2+} up to 2.5 M, and thus identified a previously unknown Ca^{2+} accumulating region (CAR) present at the extracellular entrance of the pore. The computational predictions were validated by

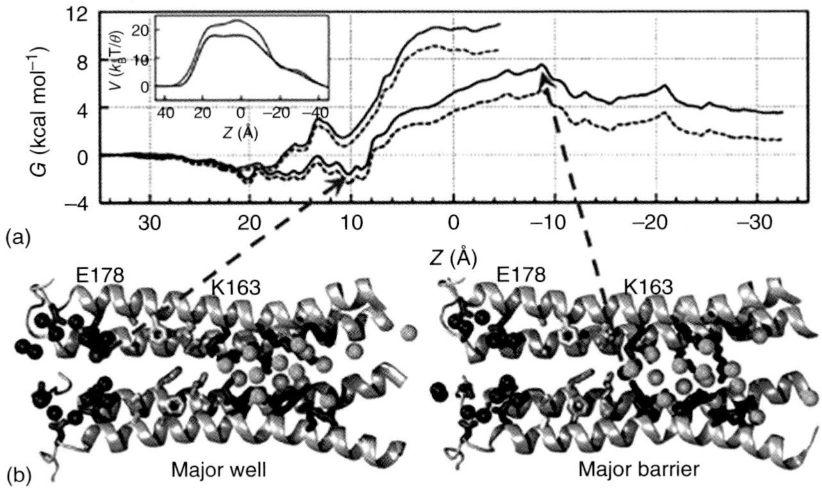

Figure 10.15 Free energy profiles for a permeating ion through the Orai pore. (a) Free energy profiles for a sodium ion crossing the wild-type (highest free energy value) channel and the V174A mutant (lowest free energy value). The dashed lines correspond to the profiles with a correction added coming from an external electric field. (b) Representation of the conformations corresponding to the major well and the major barrier given, respectively, on the left and the right. Sodium and chloride ions are represented in black and gray balls, respectively. Inset. Transmembrane potential of the two systems (wild type has the higher transmembrane potential). Source: Adapted from [127].

electrophysiological experiments and point mutations of Ca^{2+} binding residues in the CAR, which suppressed the ion currents and furthermore confirmed the computational observation that specific extracellular loop interactions compete with Ca^{2+} binding in the Orai1 CAR. Thus, CAR increases the concentration of Ca^{2+} ions in the extracellular region and thereby facilitates the permeation of Ca^{2+} ions through the pore [128]. In the living organism, this would allow to maintain Ca^{2+}-dependent gene regulation even in environments with relatively low Ca^{2+} concentrations. Indeed, STIM/Orai channels have important roles in melanocytes and melanoma cells, which reside in the basal epidermal layer where the external Ca^{2+} is much lower (0.2–0.4 mM compared to 1–2 mM to which cells in the circulation are exposed) [129, 130]. The authors therefore conclude that Orai1 CAR maintains Ca^{2+}-dependent gene regulation in various cell systems [128].

10.5 Conclusion

In the early days of ion channel simulations, researchers had to rely on the few membrane protein structures available or generate homology models, often using very distant templates with low sequence identity. In addition, simulation times were often too short to show conformational changes involved in channel activation or ion translocation. Nevertheless, as the examples of the melatonin receptor [62, 63] or the cation-translocation system Trk1 [56] have demonstrated, even a low-resolution homology modeling can give useful predictions if combined with site-directed mutagenesis and experimental functional studies. However, only the increasing number of high-resolution membrane protein structures of relevant ion-translocating systems allows to apply the full range of available modeling techniques, including precise free energy calculations that well match experimental values. The given examples so far have shown that computational simulations can give detailed atomistic information about the selectivity of a specific channel for K^+ over Na^+ or vice versa [69, 75], atomistic detail with respect to the inward and outward conduction mechanism with detailed interpretation of ion densities and ion-dwelling time across the selectivity filter [70], or the preference of one binding state over others [77], or the effect of ion concentrations and TM potential on the ion conduction [93]. It has been shown that umbrella sampling, SMD, ABF, or metadynamics can successfully be used to predict open or closed conformations [121], free energy barriers within the channel pore [127], and hydrophobic gating mechanism [120].

In summary, ion channel simulations have become more and more relevant, and seem to have gained also more and more acceptance by experimentalists given the increasing number of joint experimental and computational papers or the high citation numbers for papers reporting computational simulations. Here, it is essential to always bear in mind that even the best structural model is still a model created for a specific purpose and can be used only to support structural and dynamical features that it can describe or simulate correctly. Nevertheless, current simulations have become an important tool in understanding the structure–function relationship of ion channels, predicting disease-relevant residues or aiding structure-based drug design.

Acknowledgments

The writing of this chapter was supported by the Czech Infrastructure for Systems Biology C4SYS (project No. LM2015055), by a MOA between C4SYS and Larkin University, and by the Czech Science Foundation through a grant to RHE (13-21053S). SKP and RHE were funded in part through a European Cooperation in Science and Technology (COST) action BM1406 "IONCHAN-IMMUNRESPON" and by the program Inter-COST (project LTC17069 to RHE). DB was supported by the EFRR project Interreg Austria – Czech Republic "Czech-Austrian Center for Supracellular Medical Research (CAC-SuMeR, No# ATCZ14)."

References

1 Lipscombe, D. and Toro, C.P. (2013). Ion channels. In: *eLS*. Chichester: Wiley. doi: 10.1002/9780470015902.a0000150.pub3.

2 Doyle, D.A., Morais, C.J., Pfuetzner, R.A. et al. (1998). The structure of the potassium channel: molecular basis of K+ conduction and selectivity. *Science* 280 (5360): 69–77.

3 Roux, B. (1999). The cavity and pore helices in the KcsA K channel: electrostatic stabilization of monovalent cations. *Science* 285 (5424): 100–102. doi: 10.1126/science.285.5424.100.

4 Guidoni, L., Torre, V., and Carloni, P. (1999). Potassium and sodium binding to the outer mouth of the K channel†. *Biochemistry* 38 (27): 8599–8604. doi: 10.1021/bi990540c.

5 Allen, T.W., Kuyucak, S., and Chung, S. (1999). Molecular dynamics study of the KcsA potassium channel. *Biophys. J.* 77 (5): 2502–2516. doi: 10.1016/s0006-3495(99)77086-4.

6 Stansfeld, P.J., Goose, J.E., Caffrey, M. et al. (2015). MemProtMD: automated insertion of membrane protein structures into explicit lipid membranes. *Structure* 23 (7): 1350–1361. Web. 2 Feb 2017.

7 Rawson, S., Davies, S., Lippiat, J.D., and Muench, S.P. (2016). The changing landscape of membrane protein structural biology through developments in electron microscopy. *Mol. Membr. Biol.* 33 (1–2): 12–22. doi: 10.1080/09687688.2016.1221533.

8 Khabiri, M., Nikouee, A., Cwiklik, L. et al. (2011). Charybdotoxin unbinding from them Kv1.3 potassium channel: a combined computational and experimental study. *J. Phys. Chem. B* 115 (39): 11490–11500. doi: 10.1021/jp2061909.

9 Bocquet, N., Nury, H., Baaden, M. et al. (2008). X-ray structure of a pentameric ligand-gated ion channel in an apparently open conformation. *Nature* 457 (7225): 111–114. doi: 10.1038/nature07462.

10 Bocquet, N., Carvalho, L.P., Cartaud, J. et al. (2006). A prokaryotic proton-gated ion channel from the nicotinic acetylcholine receptor family. *Nature* 445 (7123): 116–119. doi: 10.1038/nature05371.

11 Chothia, C. (1992). One thousand families for the molecular biologist. *Nature* 357 (6379): 543–544. doi: 10.1038/357543a0.

12 Baker, D. (2001). Protein structure prediction and structural genomics. *Science* 294 (5540): 93–96. doi: 10.1126/science.1065659.

13 Chothia, C. and Lesk, A.M. (1986). The relation between the divergence of sequence and structure in proteins. *EMBO J.* 5 (4): 823–826.

14 Šali, A. and Blundell, T.L. (1993). Comparative protein modelling by satisfaction of spatial restraints. *J. Mol. Biol.* 234 (3): 779–815. doi: 10.1006/jmbi.1993.1626.

15 Vitkup, D., Melamud, E., Moult, J., and Sander, C. (2001). Completeness in structural genomics. *Nat. Struct. Biol.* 8 (6): 559–566. doi: 10.1038/88640.

16 Krieger, E., Koraimann, G., and Vriend, G. (2002). Increasing the precision of comparative models with YASARA NOVA-a self-parameterizing force field. *Proteins Struct. Funct. Bioinf.* 47 (3): 393–402. doi: 10.1002/prot.10104.

17 Sybyl-X (2012). Molecular modeling software packages, version 2.0. TRIPOS Associates, Inc., St. Louis, MO, USA.

18 Webb, B. and Sali, A. (2014). Comparative protein structure modeling using modeller. *Curr. Protoc. Bioinf.* doi: 10.1002/0471250953.bi0506s47.

19 Sternberg, M.J., Bates, P.A., Kelley, L.A., and Maccallum, R.M. (1999). Progress in protein structure prediction: assessment of CASP3. *Curr. Opin. Struct. Biol.* 9 (3): 368–373. doi: 10.1016/s0959-440x(99)80050-5.

20 Altschul, S.F., Gish, W., Miller, W. et al. (1990). Basic local alignment search tool. *J. Mol. Biol.* 215 (3): 403–410.

21 Altschul, S.F., Madden, T.L., Schäffer, A.A. et al. (1997). Gapped BLAST and PSI-BLAST: a new generation of protein database search programs. *Nucleic Acids Res.* 25 (17): 3389–3402. doi: 10.1093/nar/25.17.3389.

22 Berman, H.M., Westbrook, J., Feng, Z. et al. (2000). The protein data bank. *Nucleic Acids Res.* 28 (1): 235–242. doi: 10.1093/nar/28.1.235.

23 Sutcliffe, M., Haneef, I., Carney, D., and Blundell, T. (1987). Knowledge-based modelling of homologous proteins, part I: three-dimensional frameworks derived from the simultaneous superposition of multiple structures. *Protein Eng. Des. Sel.* 1 (5): 377–384. doi: 10.1093/protein/1.5.377.

24 Peitsch, M.C. (1996). ProMod and Swiss-model: Internet-based tools for automated comparative protein modelling. *Biochem. Soc. Trans.* 24 (1): 274–279. doi: 10.1042/bst0240274.

25 Schwede, T. (2003). SWISS-MODEL: an automated protein homology-modeling server. *Nucleic Acids Res.* 31 (13): 3381–3385. doi: 10.1093/nar/gkg520.

26 Rohl, C.A., Strauss, C.E., Chivian, D., and Baker, D. (2004). Modeling structurally variable regions in homologous proteins with rosetta. *Proteins Struct. Funct. Bioinf.* 55 (3): 656–677. doi: 10.1002/prot.10629.

27 Adcock, S.A. and McCammon, J.A. (2006). Molecular dynamics: survey of methods for simulating the activity of proteins. *Chem. Rev.* 105: 1589–1615.

28 Verlet, L. (1967). Computer "Experiments" on classical fluids. I. Thermodynamical properties of Lennard-Jones molecules. *Phys. Rev.* 159 (1): 98–103. doi: 10.1103/physrev.159.98.

29 Swope, W.C., Andersen, H.C., Berens, P.H., and Wilson, K.R. (1982). A computer simulation method for the calculation of equilibrium constants for the formation of physical clusters of molecules: application to small water clusters. *J. Chem. Phys.* 76 (1): 637–649. doi: 10.1063/1.442716.

30 Hockney, R., Goel, S., and Eastwood, J. (1974). Quiet high-resolution computer models of a plasma. *J. Comput. Phys.* 14 (2): 148–158. doi: 10.1016/0021-9991(74)90010-2.

31 Wilde, R.E. and Singh, S. (1998). *Statistical Mechanics: Fundamentals and Modern Applications*. New York: Wiley.

32 Mackerell, A.D., Feig, M., and Brooks, C.L. (2004). Extending the treatment of backbone energetics in protein force fields: limitations of gas-phase quantum mechanics in reproducing protein conformational distributions in molecular dynamics simulations. *J. Comput. Chem.* 25 (11): 1400–1415. doi: 10.1002/jcc.20065.

33 MacKerell, A.D., Bashford, D., Bellott, M. et al. (1998). All-atom empirical potential for molecular modeling and dynamics studies of proteins. *J. Phys. Chem. B.* 102 (18): 3586–3616.

34 Duan, Y., Wu, C., Chowdhury, S. et al. (2003). A point-charge force field for molecular mechanics simulations of proteins based on condensed-phase quantum mechanical calculations. *J. Comput. Chem.* 24 (16): 1999–2012. doi: 10.1002/jcc.10349.

35 Jorgensen, W.L., Maxwell, D.S., and Tirado-Rives, J. (1996). Development and testing of the OPLS all-atom force field on conformational energetics and properties of organic liquids. *J. Am. Chem. Soc.* 118 (45): 11225–11236. doi: 10.1021/ja9621760.
36 Oostenbrink, C., Villa, A., Mark, A.E., and Gunsteren, W.F. (2004). A biomolecular force field based on the free enthalpy of hydration and solvation: the GROMOS force-field parameter sets 53A5 and 53A6. *J. Comput. Chem.* 25 (13): 1656–1676. doi: 10.1002/jcc.20090.
37 MacKerell, A.D. Jr., (2004). Empirical force fields for biological macromolecules: overview and issues. *J. Comput. Chem.* 25: 1584.
38 Klauda, J.B., Venable, R.M., Freites, J.A. et al. (2010). Update of the CHARMM all-atom additive force field for lipids: validation on six lipid types. *J. Phys. Chem. B* 114 (23): 7830–7843. doi: 10.1021/jp101759q.
39 Klauda, J.B., Monje, V., Kim, T., and Im, W. (2012). Improving the CHARMM force field for polyunsaturated fatty acid chains. *J. Phys. Chem. B* 116 (31): 9424–9431. doi: 10.1021/jp304056p. Epub 2012 Jul 3.
40 Poger, D., Van Gunsteren, W.F., and Mark, A.E. (2010). A new force field for simulating phosphatidylcholine bilayers. *J. Comput. Chem.* 31: 1117–1125. doi: 10.1002/jcc.21396.
41 Berger, O., Edholm, O., and Jähnig, F. (1997). Molecular dynamics simulations of a fluid bilayer of dipalmitoylphosphatidylcholine at full hydration, constant pressure, and constant temperature. *Biophys. J.* 72 (5): 2002–2013.
42 Torrie, G.M. and Valleau, J.P. (1977). Nonphysical sampling distributions in Monte Carlo free-energy estimation: umbrella sampling. *J. Comput. Phys.* 23 (2): 187–189.
43 Kumar, S., Rosenberg, J.M., Bouzida, D. et al. (1992). The weighted histogram analysis method for free-energy calculations on biomolecules. I. The method. *J. Comput. Chem.* 13: 1011–1021. doi: 10.1002/jcc.540130812.
44 Laio, A. and Parrinello, M. (2002). Escaping free-energy minima. *Proc. Natl. Acad. Sci. U.S.A.* 99 (20): 12562–12566.
45 Barducci, A., Bussi, G., and Parrinello, M. (2008). Well-tempered metadynamics: a smoothly converging and tunable free-energy method. *Phys. Rev. Lett.* 100 (2): 020603. doi: 10.1103/PhysRevLett.100.020603.
46 Darve, E. and Pohorille, A. (2001). Calculating free energies using average force. *J. Chem. Phys.* 115: 9169–9183.
47 Comer, J., Gumbart, J.C., Henin, J. et al. (2015). The adaptive biasing force method: everything you always wanted to know but were afraid to ask. *J. Phys. Chem. B* 119 (3): 1129–1151.
48 Zhang, Y. and Voth, G.A. (2011). Combined metadynamics and umbrella sampling method for the calculation of ion permeation free energy profiles. *J. Chem. Theor. Comput.* 7 (7): 2277–2283. doi: 10.1021/ct200100e.
49 Furini, S. and Domene, C. (2016). Computational studies of transport in ion channels using metadynamics. *Biochim. Biophys. Acta (BBA) – Biomembr.* 1858 (7): 1733–1740. doi: 10.1016/j.bbamem.2016.02.015.
50 Bussi, G., Gervasio, F.L., Laio, A., and Parrinello, M. (2006). Free-energy landscape for β hairpin folding from combined parallel tempering and metadynamics. *J. Am. Chem. Soc.* 128 (41): 13435–13441. doi: 10.1021/ja062463w.

51 Piana, S. and Laio, A. (2007). A bias-exchange approach to protein folding. *J. Phys. Chem. B* 111 (17): 4553–4559. doi: 10.1021/jp067873l.

52 Ariño, J., Ramos, J., and Sychrová, H. (2010). Alkali metal cation transport and homeostasis in yeasts. *Microbiol. Mol. Biol. Rev.* 74: 95–120.

53 Gaber, R.F., Styles, C.A., and Fink, G.R. (1988). TRK1 encodes a plasma membrane protein required for high-affinity potassium transport in *Saccharomyces cerevisiae. Mol. Cell. Biol.* 8: 2848–2859.

54 Bertl, A., Ramos, J., Ludwig, J. et al. (2003). Characterization of potassium transport in wild-type and isogenic yeast strains carrying all combinations of trk1, trk2 and tok1 null mutations. *Mol. Microbiol.* 47: 767–780.

55 Durell, S.R., Hao, Y., Nakamura, T. et al. (1999). Evolutionary relationship between K(+) channels and symporters. *Biophys. J.* 77: 775–788.

56 Zayats, V., Stockner, T., Pandey, S.M. et al. (2015). A refined atomic scale model of the *Saccharomyces cerevisiae* K+-translocation protein Trk1p combined with experimental evidence confirms the role of selectivity filter glycines and other key residues. *Biochim. Biophys. Acta, Biomembr.* 1848 (5): 1183–1195.

57 Vanecek, J. (1998). Cellular mechanisms of melatonin action. *Physiol. Rev.* 78 (3): 687–721. doi: 10.1152/physrev.1998.78.3.687.

58 Dubocovich, M.L., Masana, M.I., Iacob, S., and Sauri, D.M. (1997). Melatonin receptor antagonists that differentiate between the human Mel1a and Mel1b recombinant subtypes are used to assess the pharmacological profile of the rabbit retina ML1 presynaptic heteroreceptor. *Naunyn-Schmiedebergs Arch. Pharmacol.* 355 (3): 365–375. doi: 10.1007/pl00004956.

59 Badness, T.J., Powers, J.B., Hastings, M.H. et al. (1993). The timed infusion paradigm for melatonin delivery: what has it taught us about the melatonin signal, its reception, and the photoperiodic control of seasonal responses? *J. Pineal Res.* 15 (4): 161–190. doi: 10.1111/j.1600-079x.1993.tb00903.x.

60 Maestroni, G.J. (1993). The immunoneuroendocrine role of melatonin. *J. Pineal Res.* 14 (1): 1–10. doi: 10.1111/j.1600-079x.1993.tb00478.x.

61 Okada, T., Fujiyoshi, Y., Silow, M. et al. (2002). Functional role of internal water molecules in rhodopsin revealed by X-ray crystallography. *Proc. Natl. Acad. Sci. U.S.A.* 99 (9): 5982–5987. doi: 10.1073/pnas.082666399.

62 Mazna, P., Obsilova, V., Jelinkova, I. et al. (2004). Molecular modeling of human MT2 melatonin receptor: the role of Val204, Leu272 and Tyr298 in ligand binding. *J. Neurochem.* 91 (6): 1505–1505. doi: 10.1111/j.1471-4159.2004.02965.x.

63 Mazna, P., Berka, K., Jelinkova, I. et al. (2005). Ligand binding to the human MT2 melatonin receptor: the role of residues in transmembrane domains 3, 6, and 7. *Biochem. Biophys. Res. Commun.* 332 (3): 726–734. doi: 10.1016/j.bbrc.2005.05.017.

64 Hodgkin, A.L. and Huxley, A.F. (1952). A quantitative description of membrane current and its application to conduction and excitation in nerve. *J. Physiol.* 117 (4): 500–544. doi: 10.1113/jphysiol.1952.sp004764.

65 Catterall, W.A. (2012). Voltage-gated sodium channels at 60: structure, function and pathophysiology. *J. Physiol.* 590 (11): 2577–2589. doi: 10.1113/jphysiol.2011.224204.

66 Peters, C.H. and Ruben, P.C. (2014). Introduction to sodium channels. *Handb. Exp. Pharmacol.* 1–6. doi: 10.1007/978-3-642-41588-3_1.

67 Ruiz, M.D. and Kraus, R.L. (2015). Voltage-gated sodium channels: structure, function, pharmacology, and clinical indications. *J. Med. Chem.* 58 (18): 7093–7118. doi: 10.1021/jm501981g.

68 McCusker, E.C., Bagnéris, C., Naylor, C.E. et al. (2012). Structure of a bacterial voltage-gated sodium channel pore reveals mechanisms of opening and closing. *Nat. Commun.* 3: 1102.

69 Ulmschneider, M.B., Bagnéris, C., McCusker, E.C. et al. (2013). Molecular dynamics of ion transport through the open conformation of a bacterial voltage-gated sodium channel. *PNAS* 110 (16): 6364–6369. doi: 10.1073/pnas.1214667110.

70 Ke, S., Timin, E.N., and Stary-Weinzinger, A. (2014). Different inward and outward conduction mechanisms in NaVMs suggested by molecular dynamics simulations. *PLoS Comput. Biol.* 10 (7): e1003746.

71 Sansom, M.S.P., Shrivastava, I.H., Bright, J.N. et al. (2002). Potassium channels: structures, models, simulations. *Biochim. Biophys. Acta* 1565: 294–307.

72 Dalen, A.V., Schrempf, H., Killian, J.A., and Kruijff, B.D. (2000). Efficient membrane assembly of the KcsA potassium channel in *Escherichia coli* requires the protonmotive force. *EMBO Rep.* 1 (4): 340–346. doi: 10.1093/embo-reports/kvd067.

73 Gier, J.L., Valent, Q.A., Heijne, G.V., and Luirink, J. (1997). The *E. coli* SRP: preferences of a targeting factor. *FEBS Lett.* 408 (1): 1–4. doi: 10.1016/s0014-5793(97)00402-x.

74 De Gier, J.-W.L., Scotti, P.A., Sääf, A. et al. (1998). Differential use of the signal recognition particle translocase targeting pathway for inner membrane protein assembly in *Escherichia coli*. *Proc. Natl. Acad. Sci. U.S.A.* 95 (25): 14646–14651.

75 Noskov, S.Y., Bernèche, S., and Roux, B. (2004). Control of ion selectivity in potassium channels by electrostatic and dynamic properties of carbonyl ligands. *Nature* 431: 830–834. doi: 10.1038/nature02943.

76 Aqvist, J. and Luzhkov, V. (2000). Ion permeation mechanism of the potassium channel. *Nature* 404: 881–884. doi: 10.1038/35009114.

77 Berneche, S. and Roux, B. (2001). Energetics of ion conduction through the K+ channel. *Nature* 414: 73–77. doi: 10.1038/35102067.

78 Hodgkin, A.L. and Keynes, R.D. (1955). The potassium permeability of a giant nerve fibre. *J. Physiol.* 128 (1): 61–88. doi: 10.1113/jphysiol.1955.sp005291.

79 Schumaker, M. and Mackinnon, R. (1990). A simple model for multi-ion permeation. Single-vacancy conduction in a simple pore model. *Biophys. J.* 58 (4): 975–984. doi: 10.1016/s0006-3495(90)82442-5.

80 Chandy, K.G., Decoursey, T.E., Cahalan, M.D. et al. (1984). Voltage-gated potassium channels are required for human T-cell activation. *J. Exp. Med.* 160 (2): 369–385. doi: 10.1084/jem.160.2.369.

81 Kasahara, K., Shirota, M., and Kinoshita, K. (2016). Ion concentration- and voltage-dependent push and pull mechanisms of potassium channel ion conduction. *PLoS One* 11 (3): e0150716. doi: 10.1371/journal.pone.0150716.

82 Mackinnon, R. (1991). Determination of the subunit stoichiometry of a voltage-activated potassium channel. *Nature* 350 (6315): 232–235. doi: 10.1038/350232a0.

83 Grissmer, S., Dethlefs, B., Wasmuth, J.J. et al. (1990). Expression and chromosomal localization of a lymphocyte K+ channel gene. *Proc. Natl. Acad. Sci. U.S.A.* 87 (23): 9411–9415.

84 Decoursey, T.E., Chandy, K.G., Gupta, S., and Cahalan, M.D. (1984). Voltage-gated K channels in human T lymphocytes: a role in mitogenesis? *Nature* 307 (5950): 465–468. doi: 10.1038/307465a0.

85 Decoursey, T., Chandy, K., Gupta, S., and Cahalan, M. (1985). Voltage-dependent ion channels in T-lymphocytes. *J. Neuroimmunol.* 10 (1): 71–95. doi: 10.1016/0165-5728(85)90035-9.

86 Freudenthaler, G., Axmann, M., Schindler, H. et al. (2002). Ultrasensitive pharmacological characterisation of the voltage-gated potassium channel KV1.3 studied by single-molecule fluorescence microscopy. *Histochem. Cell Biol.* 117 (3): 197–202. doi: 10.1007/s00418-001-0374-y.

87 Morris, G.M., Goodsell, D.S., Halliday, R.S. et al. (1998). Automated docking using a Lamarckian genetic algorithm and an empirical binding free energy function. *J. Comput. Chem.* 19: 1639–1662. doi: 10.1002/(SICI)1096-987X(19981115)19:14<1639::AID-JCC10>3.0.CO;2-B.

88 Nikouee, A., Khabiri, M., Grissmer, S., and Ettrich, R. (2012). Charybdotoxin and margatoxin acting on the human voltage-gated potassium channel hKv1.3 and its H399N mutant: an experimental and computational comparison. *J. Phys. Chem. B* 116 (17): 5132–5140. doi: 10.1021/jp2102463.

89 Sachs, J.N., Crozier, P.S., and Woolf, T.B. (2004). Atomistic simulations of biologically realistic transmembrane potential gradients. *J. Chem. Phys.* 121 (22): 10847. doi: 10.1063/1.1826056.

90 Delemotte, L., Dehez, F., Treptow, W., and Tarek, M. (2008). Modeling membranes under a transmembrane potential. *J. Phys. Chem. B* 112 (18): 5547–5550. doi: 10.1021/jp710846y.

91 Hodge, T. and Colombini, M. (1997). Regulation of metabolite flux through voltage-gating of VDAC channels. *J. Membr. Biol.* 157 (3): 271–279. doi: 10.1007/s002329900235.

92 Rui, H.A., Lee, K.I., Pastor, R.W., and Im, W. (2011). Molecular dynamics studies of ion permeation in VDAC. *Biophys. J.* 100: 602–610.

93 Forman, S.A. and Miller, K.W. (2011). Anesthetic sites and allosteric mechanisms of action on Cys-loop ligand-gated ion channels. *Can. J. Anesth.* 58 (2): 191–205. doi: 10.1007/s12630-010-9419-9.

94 Chandy, K.G., Gutman, G.A., and Grissmer, S. (1993). Physiological role, molecular structure and evolutionary relationships of voltage-gated potassium channels in T lymphocytes. *Semin. Neurosci.* 5 (2): 125–134. doi: 10.1016/s1044-5765(05)80007-7.

95 Nury, H., Renterghem, C.V., Weng, Y. et al. (2011). X-ray structures of general anaesthetics bound to a pentameric ligand-gated ion channel. *Nature* 469 (7330): 428–431. doi: 10.1038/nature09647.

96 Weng, Y., Yang, L., Corringer, P., and Sonner, J.M. (2010). Anesthetic sensitivity of the *Gloeobacter violaceus* proton-gated ion channel. *Anesth. Analg.* 110 (1): 59–63. doi: 10.1213/ane.0b013e3181c4bc69.

97 Zhu, F. and Hummer, G. (2012). Theory and simulation of ion conduction in the pentameric GLIC channel. *J. Chem. Theory Comput.* 8: 3759–3768. doi: 10.1021/ct2009279.

98 Roux, B. and Bernèche, S. (2002). On the potential functions used in molecular dynamics simulations of ion channels. *Biophys. J.* 82 (3): 1681–1684. doi: 10.1016/s0006-3495(02)75520-3.

99 Roux, B. (2008). The membrane potential and its representation by a constant electric field in computer simulations. *Biophys. J.* 95 (9): 4205–4216. doi: 10.1529/biophysj.108.136499.

100 Jensen, M.Ø., Jogini, V., Eastwood, M.P., and Shaw, D.E. (2013). Atomic-level simulation of current–voltage relationships in single-file ion channels. *J. Gen. Physiol.* 141 (5): 619–632. doi: 10.1085/jgp.201210820.

101 Julius, D. (2013). TRP channels and pain. *Annu. Rev. Cell Dev. Biol.* 29 (1): 355–384.

102 Cao, E., Liao, M., Cheng, Y., and Julius, D. (2013). TRPV1 structures in distinct conformations reveal mechanisms of activation. *Nature* 504 (7478): 113–118. doi: 10.1038/nature12823.

103 Susankova, K., Ettrich, R., Vyklicky, L. et al. (2007). Contribution of the putative inner-pore region to the gating of the transient receptor potential vanilloid subtype 1 channel (TRPV1). *J. Neurosci.* 27 (28): 7578–7585. doi: 10.1523/jneurosci.1956-07.2007.

104 Ohbuchi, K., Mori, Y., Ogawa, K. et al. (2016). Detailed analysis of the binding mode of vanilloids to transient receptor potential vanilloid type I (TRPV1) by a mutational and computational study. *PLoS One* 11 (9): doi: 10.1371/journal.pone.0162543.

105 Jorgensen, C., Furini, S., and Domene, C. (2016). Energetics of ion permeation in an open-activated TRPV1 channel. *Biophys. J.* 111 (6): 1214–1222. doi: 10.1016/j.bpj.2016.08.009.

106 Garrison, S.R. and Stucky, C.L. (2011). The dynamic TRPA1 channel: a suitable pharmacological pain target? *Curr. Pharm. Biotechnol.* 12 (10): 1689–1697. doi: 10.2174/138920111798357302.

107 Andrade, E., Meotti, F., and Calixto, J. (2012). TRPA1 antagonists as potential analgesic drugs. *Pharmacol. Ther.* 133 (2): 189–204. doi: 10.1016/j.pharmthera.2011.10.008.

108 Story, G.M., Peier, A.M., Reeve, A.J. et al. (2003). ANKTM1, a TRP-like channel expressed in nociceptive neurons, is activated by cold temperatures. *Cell* 112 (6): 819–829. doi: 10.1016/s0092-8674(03)00158-2.

109 Cavanaugh, E., Simkin, D., and Kim, D. (2008). Activation of transient receptor potential A1 channels by mustard oil, tetrahydrocannabinol and Ca^{2+} reveals different functional channel states. *Neuroscience* 154 (4): 1467–1476. doi: 10.1016/j.neuroscience.2008.04.048.

110 Macpherson, L.J., Dubin, A.E., Evans, M.J. et al. (2007). Noxious compounds activate TRPA1 ion channels through covalent modification of cysteines. *Nature* 445 (7127): 541–545. doi: 10.1038/nature05544.

111 Sawada, Y., Hosokawa, H., Matsumura, K., and Kobayashi, S. (2008). Activation of transient receptor potential ankyrin 1 by hydrogen peroxide. *Eur. J. Neurosci.* 27 (5): 1131–1142. doi: 10.1111/j.1460-9568.2008.06093.x.

112 Taylor-Clark, T.E., Kiros, F., Carr, M.J., and Mcalexander, M.A. (2009). Transient receptor potential ankyrin 1 mediates toluene diisocyanate–evoked respiratory irritation. *Am. J. Respir. Cell Mol. Biol.* 40 (6): 756–762. doi: 10.1165/rcmb.2008-0292oc.

113 Gaudet, R. (2008). A primer on ankyrin repeat function in TRP channels and beyond. *Mol. BioSyst.* 4 (5): 372–379. doi: 10.1039/b801481g.

114 Sotomayor, M., Corey, D.P., and Schulten, K. (2005). In search of the hair-cell gating spring elastic properties of ankyrin and cadherin repeats. *Structure* 13 (4): 669–682. doi: 10.1016/j.str.2005.03.001.

115 Zayats, V., Samad, A., Minofar, B. et al. (2012). Regulation of the transient receptor potential channel TRPA1 by its N-terminal ankyrin repeat domain. *J. Mol. Model.* 19 (11): 4689–4700. doi: 10.1007/s00894-012-1505-1.

116 Serquera, D., Lee, W., Settanni, G. et al. (2010). Mechanical unfolding of an ankyrin repeat protein. *Biophys. J.* 98 (7): 1294–1301. doi: 10.1016/j.bpj.2009.12.4287.

117 Paulsen, C.E., Armache, J., Gao, Y. et al. (2015). Structure of the TRPA1 ion channel suggests regulatory mechanisms. *Nature* 520 (7548): 511–517. doi: 10.1038/nature14367.

118 Prafulla, A., Samson, M.S.P., and Tucker, S.J. (2015). Hydrophobic gating in ion channels. *J. Mol. Biol.* 427: 121–130.

119 Lunin, V.V., Dobrovetsky, E., Khutoreskaya, G. et al. (2006). Crystal structure of the CorA Mg^{2+} transporter. *Nature* 440 (7085): 833–837.

120 Neale, C., Chakrabarti, N., Pomorski, P. et al. (2015). Hydrophobic gating of ion permeation in magnesium channel CorA. *PLoS Comput. Biol.* 440, 11 (7): e1004303.

121 Trick, J.L., Chelvaniththilan, S., Klesse, G. et al. (2016). Functional annotation of ion channel structures by molecular dynamics. *Structure* 24 (12): 2207–2216.

122 Barnes, N.M., Hales, T.G., Lummis, S.C.R., and Peters, J.A. (2009). The 5-HT_3 receptor – the relationship between structure and function. *Neuropharmacology* 56 (1): 273–284.

123 Hao, D.H., Fiorin, G., Carnevale, V. et al. (2013). Pore waters regulate ion permeation in a calcium release-activated calcium channel. *Proc. Natl. Acad. Sci. U.S.A.* 110 (43): 17332–17337.

124 Feske, S., Skolnik, E.Y., and Prakriya, M. (2012). Ion channels and transporters in lymphocyte function and immunity. *Nat. Rev. Immunol.* 12 (7): 532–547. doi: 10.1038/nri3233.

125 Shim, A.H., Tirado-Lee, L., and Prakriya, M. (2015). Structural and functional mechanisms of CRAC channel regulation. *J. Mol. Biol.* 427 (1): 77–93. doi: 10.1016/j.jmb.2014.09.021.

126 Hou, X., Pedi, L., Diver, M.M., and Long, S.B. (2012). Crystal structure of the calcium release-activated calcium channel orai. *Science* 338 (6112): 1308–1313. doi: 10.1126/science.1228757.

127 Mcnally, B.A., Somasundaram, A., Yamashita, M., and Prakriya, M. (2012). Gated regulation of CRAC channel ion selectivity by STIM1. *Nature* 482 (7384): 241–245. doi: 10.1038/nature10752.

128 Frischauf, I., Zayats, V., Deix, M. et al. (2015). A calcium-accumulating region, CAR, in the channel Orai1 enhances Ca2 permeation and SOCE-induced gene transcription. *Sci. Signaling* 8 (408): doi: 10.1126/scisignal.aab1901.

129 Stanisz, H., Stark, A., Kilch, T. et al. (2012). ORAI1 Ca(2+) channels control endothelin-1-induced mitogenesis and melanogenesis in primary human melanocytes. *J. Invest. Dermatol.* 132: 1443–1451.

130 Stanisz, H., Saul, S., Müller, C.S.L. et al. (2014). Inverse regulation of melanoma growth and migration by Orai1/STIM2-dependent calcium entry. *Pigment Cell Melanoma Res.* 27 (3): 442–453.

11

Understanding Allostery to Design New Drugs

Giulia Morra and Giorgio Colombo

CNR, Istituto di Chimica del Riconoscimento Molecolare, via Mario Bianco, 9, 20131 Milano, Italy

11.1 Introduction

The improved knowledge of disease-related genes, coupled with the technological advancements introduced by the genomic and post-genomic revolutions, boosted the discovery of new targets, creating the expectation that new drugs against them would act as "magic bullets," dismantling entire pathogenic pathways by blocking a relevant protein in the network. Such a scenario could have changed the landscape of drug development and patient care. However, it was soon realized that reality is much more difficult. Indeed, the very low numbers of new chemical entities for human therapy approved by regulatory authorities in the past decade vividly reflect the extraordinary complexity of most diseases. In the case of cancer, for instance, it is now well appreciated that a large number of mutated proteins or deregulated metabolic pathways conspire to increase the costs and to lower the success rate (typically 0.0001%) of "target-centric" drug discovery. Moreover, classical approaches of drug development aimed at targeting the active sites of specific enzymes have been approaching their limits. Due to the evolutionary and structural conservation of active sites across the proteome, the efficacy of drugs blocking them is often hampered by selectivity issues, off-target effects, and the development of drug resistance mechanisms [1, 2].

These disappointments determine an urgent necessity for new strategies to improve drug discovery, and produce, if not the "cure," at least meaningful advances in patient conditions or useful chemical tools for the investigation of biochemical pathway regulation.

In this context, allosteric modulation is attracting increasing interest from medicinal chemists and drug discovery partnerships in the search for new types of targets and strategies for drug development.

Allostery defines the feature of proteins to undergo a modulation of affinity toward a primary binding event caused by binding an "effector" in a different distant position called the allosteric site. This modulation may cause increase or decrease of protein activity and its downstream effects in the cell. As such, allostery represents one of the most common and powerful means to regulate protein function, which may offer relevant alternatives for the discovery of

Biomolecular Simulations in Structure-Based Drug Discovery,
First Edition. Edited by Francesco L. Gervasio and Vojtech Spiwok.

therapeutic molecules. In some cases, cell responses may in fact be obtained not only by inhibiting the function of an enzyme but also by activating it [3]. Moreover, in principle, allosteric drugs can be proficiently used in combination with other molecules, improving therapeutic effects.

Several factors support current interest in allosteric drug discovery. First, allosteric sites tend to be under lower sequence-conservation pressure than active sites, facilitating the design of highly specific drugs and reducing the risks of off-target effects [4]. Moreover, allosteric ligands may increase target–protein activity, enabling novel therapeutic possibilities, such as in the case of the activation of glucokinase by allosteric drugs, a potential treatment for type 2 diabetes mellitus. On the same lines, traditional drugs may be complemented by allosteric effectors, as observed in the case of aminoglycoside phosphotransferase where a previously unknown binding site could be exploited to allosterically counteract antibiotic resistance. In cancer treatment, one important marketed drug that acts through allosteric mechanisms is paclitaxel (Paxol) [5].

Finally, from the medicinal chemist and molecular designer point of view, allosteric drugs may be more manageable than active-site targeted ones: the latter often involve highly charged transition state mimics that may show poor bioavailability and absorption, distribution, metabolism, and excretion (ADME) profiles. In contrast, allosteric drugs may often not be subjected to such constraints [6].

To exploit allostery for the discovery of new molecules, one necessary preliminary step lies in the atomistic understanding of the phenomena underlying allosteric regulation. Such knowledge can then be translated into new rules for the design of novel chemical entities able to modulate the functions of their target. In this chapter, we first provide a critical overview of the current understanding of allosteric regulation, and then we provide examples of how this can be integrated with structural knowledge, binding site identification methods, and drug design approaches in the discovery of novel biologically active hits.

11.2 Protein Allostery: Basic Concepts and Theoretical Framework

The concept of allostery emerges in the context of the functional dynamics of proteins to describe the "effect at a distance," that is the influence exerted by a perturbation at a protein site (the allosteric effector) on the affinity toward a primary binding event occurring at a different distant position. The perturbation can, for instance, be another binding event, a mutation, or a posttranslational modification, which globally modulates the protein conformational dynamics or equilibrium structure, in a way that affects the thermodynamics of binding (and processing) of the primary, or orthosteric, ligand.

Examples of allosteric behavior in proteins have been observed in an increasing number of systems, ranging from the non-covalent binding of lipids, cyclic adenosine monophosphate (cAMP), drugs, other proteins, RNA, and DNA, to covalent events such as phosphorylation, point mutations, or reactions with small molecules [6–8].

The approaches to the study of allostery range from conceptual models framing this phenomenon in the context of the ligand-binding theories, such as "induced fit" or "conformational selection," to thermodynamic models aimed at a quantitative description of the affinity change. Finally, a number of experimental and theoretical methods that investigate the structural and dynamical couplings between perturbation site and primary binding site exist, which is expected to underlie allosteric communication. Some of these different approaches are discussed in the following sections.

11.2.1 The Classic View of Allostery

Two alternative conceptual models were originally developed to describe the mechanism of allostery and specifically define the relationship between the binding allosteric ligand and the modulation of affinity for the orthosteric ligand, i.e. the transition from the low-affinity to the high-affinity state. These two models, the sequential model (KNF) [9] and the concerted model (MWC) [10], were first proposed to rationalize the allosteric behavior of hemoglobin. According to the sequential model, the allosteric effector binds to the low-affinity state and then triggers a conformational change, leading to the high-affinity state. Thereby, the presence of conformational changes should only depend on the amount of allosteric ligand. In contrast, the concerted model (MWC) predicts that the low-affinity and the high-affinity states coexist at equilibrium, and the allosteric ligand stabilizes the latter, indirectly increasing the apparent affinity for the orthosteric ligand. The conformational change will therefore also depend on the equilibrium between low- (basal) and high-affinity (active) states and not only on the affinity for the allosteric ligand. While the sequential model adopts an approach resembling "induced fit" to describe allostery, as the high-affinity state is reached upon binding the effector, the MWC model is more compatible with a "conformational selection" view of ligand binding. Both conceptual models have proved effective in rationalizing the affinity changes in a number of systems [6, 11].

11.2.2 The Thermodynamic Two-State Model of Allostery

The thermodynamics of allostery can be described by means of a model without structural detail, in which the perturbed (e.g. effector-bound) and the unperturbed (basal) conformation are different equilibrium states. Both the unperturbed and the effector-bound states can bind the primary ligand L at the orthosteric site, albeit with different affinities. To elaborate, in the case of an allosteric activator, the effector induces a change that thermodynamically favors the second binding event, lowering its free energy of binding. In the case of an allosteric inhibitor, the effector-bound state has an increased free energy of binding, hence lower affinity, toward the primary ligand. These concepts can be translated into a quantitative "allosteric efficacy" by means of an allosteric two-state model (ATSM), presented in [12].

According to the model, an equilibrium between two states, the unbound or more generally inactive R and the active state R*, which is bound to its orthosteric ligand L with affinity K_L is defined by the equilibrium constant $K_L = [R^*]/[R]$. We then consider the presence of an allosteric ligand A, able to bind to both inactive and active state R and R*, with respective equilibrium constants $K_A = [RA]/[R][A]$ and $K_A^* = [R^*A]/[R^*][A]$.

The allosteric effect of A is defined as a change in the affinity toward L in the presence of A. When A is bound, the affinity for L becomes $K_{AL} = [R^*A]/[RA]$.

The allosteric efficacy α can therefore be defined as the ratio K_{AL}/K_L, so that

$$\alpha = \frac{K_{AL}}{K_L} = \frac{[R^*A][R]}{[RA][R^*]} = \frac{K_A^*}{K_A}$$

Here, the affinity change for the ligand L induced by the allosteric ligand A can be derived from the affinity for L and for A in the inactive and active states. If the allosteric efficacy α is greater than 1, which happens when A causes an increase of affinity for L, A is classified as an activator. An α value lower than 1 means that A is in fact an inhibitor.

The thermodynamic relationships can be represented in a thermodynamic cycle (Figure 11.1).

Depending on the prevalence of one direction or the other in the cycle, this scheme represents both the conformational selection and the induced-fit point of view on allostery. To elaborate, the conformation selection (MWC) view emphasizes that the microscopic path of $R \rightarrow R^* \rightarrow R^*A$ dominates the equilibrium process, whereas the induced-fit scheme (KNF) hypothesizes the prevalence of the $R \rightarrow RA \rightarrow R^*A$ path.

Moreover, we can explicitly connect the same parameters to the change in free energy of binding due to the presence of the allosteric ligand A, given that the free energy of activation (binding of the orthosteric ligand) is

$$\Delta G_A = -RT \ln \frac{[R^*] + [R^*A]}{[R] + [RA]} \sim -RT \ln \frac{[R^*A]}{[RA]}$$

when [A] tends to infinity (saturation). In the absence of A, in contrast,

$$\Delta G_0 = -RT \ln \frac{[R^*]}{[R]}$$

Therefore, the change in affinity due to the presence of the allosteric ligand can be expressed as

$$\Delta\Delta G_{0-A} = -RT \ln \alpha K_L + RT \ln K_L = -RT \ln \alpha$$

in terms of the allosteric efficacy. Hence, it is the preferred binding affinity of ligand L in the presence of A, not the overall binding affinity for A, which describes the allosteric effect.

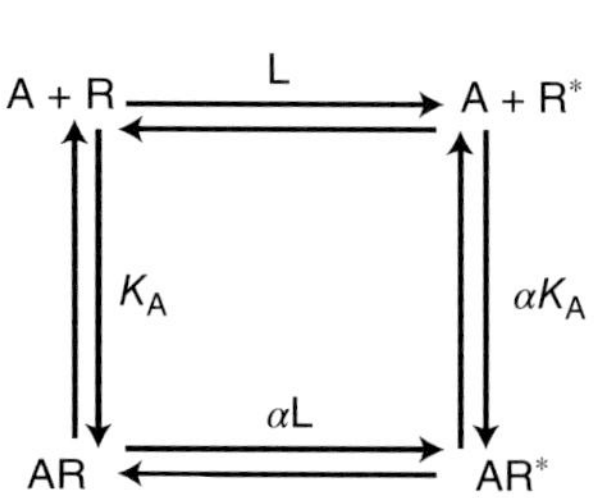

Figure 11.1 Thermodynamic cycle representing the relationship between binding an orthosteric ligand L and an allosteric ligand A. For the definition of symbols, see text.

The conformational selection model for ligand binding is not required for this thermodynamic description to hold, since all that is needed is that equilibrium is reached rapidly enough along the directions of the cycle. Yet, from that point of view, it can be assumed that both R and R* are populated at equilibrium, but the ligand A preferentially binds to one of the two states. Then the population shift, defined as the fraction of proteins in the activated state, can be re-expressed as a function of the concentration of allosteric ligand A and allosteric efficacy α and of the basal affinity:

$$f_{R^*} = \frac{[R^*] + [R^*A]}{[R^*] + [R] + [R^*A] + [RA]} = \frac{K_L + \alpha K_L K_A[A]}{1 + K_L + \alpha K_L K_A[A] + [A]}$$

Overall, this model offers a simple formal description of allostery, which can be directly connected to experimental data and is focused on the concept of allosteric efficacy as a measure for allosteric coupling. It allows to classify different allosteric ligands according to their effect as well as to connect the allosteric response to the population of the active state, thereby framing the problem of allostery in rigorous quantitative terms.

11.2.3 From Thermodynamics to Protein Structure and Dynamics

The conceptual models presented do not offer a direct connection between the allosteric states and their actual physicochemical determinants. In other words, thermodynamics does not provide insight into the molecular mechanism governing the transitions.

On the other hand, predicting any kind of molecular coupling between primary and allosteric binding site is necessary to understand allostery, even if is not sufficient to predict the activating or inhibiting character of an allosteric perturbation.

The original definition of allostery focused on purely structural features to discriminate between different allosteric states and define the coupling mechanism. According to this view, the affinity for the primary ligand L is modulated through conformational changes originating from the allosteric perturbation site and traveling through the protein till the orthosteric site. Structural rearrangements may consist of reorientation of side chains and formation or disruption of specific contacts, or hinge motions between two subdomains. The presence of a subset of residues collectively transitioning from one state to the other can be related to the cooperative character of ligand binding and allostery. The determination of high-resolution protein structures in the presence and in the absence of bound ligands has shown that, indeed, structural changes play a major role in determining allostery. For instance, the opening of the active site of the enzyme phosphoglycerate dehydrogenase (PGDH) is allosterically inhibited through rigid subdomain motions induced by binding the catalysis end product serine (apo Protein Data Bank (PDB) structure: 1psd; holo PDB structure: 1yba) [13].

A significant ligand-induced interdomain rearrangement is observed for the heat shock protein 70 (Hsp70), whose bacterial homolog DnaK has been crystallized in the ATP and ADP-bound forms. Switching the nucleotide induces docking of the substrate-binding domain (SBD) on the nucleotide-binding domain (NBD) and opening of the SBD helical lid, through a mechanism that involves a rigid rearrangement of the NBD subdomains [14] coupled to the reorientation of the flexible linker [15, 16], which modulates the affinity for the substrate.

Besides global rearrangements, subtle structural changes at the enzyme active site can also have a modulating effect [17].

However, the inspection or structural analysis of isolated extreme conformations is often not sufficient to explain the nature of the coupling between the sites, and to identify the possible preferential "propagation pathways" constituted by the residues responsible for the coordination, i.e. the allosteric mechanism.

The evolution from the purely structural model of allostery toward a more mechanistic approach includes energetics and dynamics and relies on the ensemble view of proteins, which incorporates entropy in the picture. According to the energy landscape theory, each equilibrium state can be viewed as an ensemble of conformations populating a variably broad basin on the free energy landscape. The ensemble may be characterized by spatial arrangement of minimal free energy around which atomic positions are subjected to equilibrium fluctuations. In the ensemble view of allostery, the two allosterically distinct states can, in principle, be different basins, separated by a free energy barrier, representing different global conformations. Alternatively, they can be located in the same structural minimum, but the shape of the basin can become narrower or wider due to the presence of the allosteric ligand. The latter picture would represent the case of purely dynamic allostery, where modulation of the spatial fluctuations occurs but is not accompanied by any experimentally observable structural change [18]. Hence, in those cases, dynamics might also be sufficient to account for a free energy modulation upon binding, as originally proposed by Cooper and Dryden [19].

According to this view, ligand binding induces a modulation of vibrational states, which in turn affects the entropy contribution to free energy. The modulation of vibrational entropy can, according to their hypothesis, amount to several $k_B T$/mol, where k_B is the Boltzmann constant, T is the temperature and we are using the typical energy scale for biomolecular systems. Hence, if a large number of dynamical modes of the protein are involved in ligand-induced modulation, structural fluctuations extending to fractions of 1 Å might provide free energy changes ($\Delta\Delta G$) sufficient to account for allosteric modulation.

An allosteric mechanism based on dynamic modulation cannot be revealed by structural methods such as X-ray crystallography. In contrast, a powerful technique probing protein dynamics at different timescales and contributing to these studies is nuclear magnetic resonance (NMR) spectroscopy, and in particular both relaxation dispersion NMR experiments, sensitive to protein dynamics in the microsecond to millisecond range, and spin relaxation measurements, which probe fast stochastic motions of the bond vectors on the picosecond to nanosecond timescale. Short timescales probed by spin relaxation shows that the binding of ligands can help elucidate the ligand-induced modulation of internal dynamics of proteins, and associate it with the entropy component of the binding free energy [20].

A first direct confirmation of purely dynamic allostery, i.e. modulation of binding affinity for one ligand upon binding another ligand, came from [21] where motions on both slow and fast timescales upon binding were probed for backbone motions of CAP, a homodimer binding two molecules of cAMP with different affinity. Calorimetrically measuring the entropy difference between the two sequential cAMP binding steps showed that the observed negative cooperativity in this protein is indeed due to alterations in flexibility, probed by backbone amide NMR order parameters. More generally, the global dynamic perturbation associated with ligand binding was first shown in the calcium-saturated calmodulin (CaM) system upon binding a peptide mimicking its binding partner smooth muscle myosin light chain kinase (smMLCKp) [22] and subsequently in other proteins. For instance, in the case of the allosteric model system PDZ domain (PDZ2 from the human tyrosine phosphatase 1E (hPTP1E)), Fuentes et al. [23, 24] showed changes in side-chain dynamics parameters upon binding a peptide target, indicating modulation of global protein flexibility, because of long-range dynamic effects induced by binding. For another PDZ domain, the same kind of measurements led to evaluating the allosteric effect induced by removing a structural element, a terminal helix. In the absence of the helix, the flexibility pattern of the protein was found to be modulated, even in the absence of any significant structural changes, and the affinity for the ligand was strongly impaired, demonstrating a conformational entropy mechanism inducing cooperativity between the helix, which is known to be regulated by posttranslational modification, and ligand binding at a different site.

11.2.4 Entropy in Allostery: The Ensemble Allostery Model

Despite the experimental evidence confirming the hypothesis of Cooper and Dryden, it is conceptually not straightforward to rationalize the contribution to allostery caused by modulation of dynamics and stability. To explore this problem, a statistical mechanics-based approach, called ensemble allostery model (EAM), was proposed by Hilser et al. (see [25]) with the aim of relating the allosteric response to the energetic coupling within the protein. Here, a layer of structural complexity is added relative to the thermodynamic ATSM approach proposed earlier, because the EAM introduces internal degrees of freedom and an explicit energy coupling term between different protein subdomains. This accounts for the modular structure of many allosteric proteins and the fact that multidomain regulatory proteins often segregate the binding sites for each ligand into different structural domains. In this approach, the protein is hypothesized to be able to populate the activated state even in the absence of the allosteric ligand A. The minimal model is given by a two-domain system, where each subdomain 1 or 2 has, in the simplest version, access to two states, a low (T) and a high (R) affinity state, with different energy steps ΔG_1 and ΔG_2. Moreover, the two subdomains turn on a mutual interaction term whenever any of them is in the low-affinity state. This term, Δg_{int}, represents the allosteric coupling and can be positive (rewarding agonism) or negative (antagonism), considering zero energy for the activated state (R, R). This defines a system with four microstates – (R, R), (R, T), (T, R), and (T, T) – and three energetic parameters. Given for each microstate its Boltzmann weight $S_i = e^{-\beta \Delta G_i}$ with:

$$\Delta G_{\text{RR}} = 0$$

$$\Delta G_{\text{RT}} = \Delta G_2 + \Delta g_{\text{int}}$$

$$\Delta G_{TT} = \Delta G_1 + \Delta G_2 + \Delta g_{int}$$

$$\Delta G_{TR} = \Delta G_1 + \Delta g_{int}$$

it is possible to write the intrinsic probability for, e.g. subdomain 1 to be in the high-affinity state R:

$$P_0(1 = R) = \frac{e^{-\beta\Delta G_{RR}} + e^{-\beta\Delta G_{RT}}}{Z}$$

where the partition function is $Z = e^{-\beta\Delta G_{RR}} + e^{-\beta\Delta G_{RT}} + e^{-\beta\Delta G_{TR}} + e^{-\beta\Delta G_{TT}}$

In the presence of an allosteric ligand A that binds to subdomain 2 only when 2 is in the state R, the partition function and the probability to have subdomain 1 in the state R will also depend on the stabilization energy induced by binding A on 2:

$$P_{[A]}(1 = R) = \frac{e^{-\beta\Delta G_{RR}} + e^{-\beta\Delta G_{RT}}}{Z_{[A]}}$$

where

$$Z_{[A]} = (e^{-\beta\Delta G_{RR}} + e^{-\beta\Delta G_{TR}})Z_A + e^{-\beta\Delta G_{TT}} + e^{-\beta\Delta G_{RT}}$$

Z_A being the partition function for the ligand A being complexed to 2, namely

$$Z_A = e^{-\beta G_{[A2]}} + e^{-\beta G_{[A][2]}}$$

CS can be defined as the allosteric effect of A, binding to subdomain 2, on the conformational state of subdomain 1, as the change in the probability for 1 to be in the R state relative to the same probability in the absence of A:

$$CS = P_{[A]}(1 = R) - P_0(1 = R)$$

By means of an exhaustive search in the space of the parameters $\Delta G_1, \Delta G_2, \Delta g_{int}$, Hilser demonstrated that in this model the allosteric response CS can be maximized by different combinations of values, both with negative and positive Δg_{int} (antagonism and agonism) and that there is an inverse relationship between subdomain stability and allosteric coupling that yield the highest allosteric response. In particular, a favorable interaction coupling (positive Δg_{int}) can coexist with the destabilization of the subdomains (negative ΔG_i) and promote allostery. This model therefore suggests a complex view of the relationship between structural and dynamic modulation and allosteric effect, which is compatible with many experimental observations discussed earlier.

11.3 Exploiting Allostery in Drug Discovery and Design

11.3.1 Computational Prediction of Allosteric Behavior and Application to Drug Discovery

Allostery, defined earlier as the effect of a local perturbation on the global conformational equilibrium of a protein, constitutes the basis underlying major regulatory mechanisms of function: from transcription control to activation and modulation of signaling pathways, cell death, and metabolism. To the same extent, the impact of allosteric phenomena on disease and misfunction of proteins is significant: many disease-causing mutations occur away from protein binding sites and, similarly, small molecules binding at distal regions can, in principle, affect the overall conformational and dynamical equilibrium of a protein, inducing stabilization of an active/inactive conformation [26–28].

Given a protein system of biomedical interest, understanding its allosteric properties can therefore open up the possibility of modulating its function, either in the direction of inhibition or of activation. The question is then how to exploit structural and dynamic knowledge on allostery for drug discovery purposes? First of all, one aims at predicting an allosteric site, namely, a protein region whose perturbation, by means of mutations or posttranslational modification or small molecules, has consequences on the global functionally oriented conformational population of the protein. The prediction of allosteric binding sites and of allosteric coupling mechanisms by means of computational methods attracts growing interest in the context of biomolecular simulations. This can be ideally achieved by the integrated analysis of general conformational dynamics and of possible communication pathways connecting the identified (putative) allosteric region to the orthosteric binding site, if present, and consisting of a subset of interacting residues: these are amino acids showing some degree of evolutionary, structural, or dynamic coupling – the latter typically in the form of correlated motions.

These aspects also prompt significant challenges, both in terms of optimizing the pharmacological properties of the new compounds and in terms of predicting whether the allosteric modulation will activate or inhibit the function. In this section, a number of examples are discussed, focusing on biomedically relevant protein targets. Moreover, we provide examples reporting on the identification of novel binding sites, examples of drug oriented applications, and design of novel ligands.

11.3.2 Identification of Allosteric Binding Sites Through Structural and Dynamic Approaches

Structure and dynamics-based methods have all been employed to discover novel allosteric binding sites. Daura and Panjkovich [29] integrated publicly available information on protein sequences, structures, and active sites for more than a thousand protein families to identify common pockets across representative structures of the analyzed families by measuring the pockets' structural conservation. Interestingly, within families the identified pockets showed no correlations in terms of sequence conservation, backbone flexibility, and electrostatic potential. This observation supports the general hypothesis that targeting allosteric pockets may represent a convenient venue to achieve selectivity and minimize off-target effects. Subsequent analysis of the principal changes in the dynamics of the proteins upon allosteric ligand binding showed significant modification of global flexibility compared to the unbound case. Such results are consistent with the earlier reported models, implying that allosteric mechanisms can be governed by shifts in the population of different dynamic states, without necessarily implying a change of shape.

Sali and coworkers recently developed a method to identify cryptic sites in proteins, which represents an ideal complement to investigations on allostery in the search for novel druggable spots. A cryptic site is defined as a site that forms a pocket in a holo-structure, but not in the apo-structure. Previously, cryptic sites could be detected by experimental methods depending on fragment-based approaches [30]. Sali and coworkers constructed a set of structurally defined apo–holo pairs with cryptic sites and fully characterized the cryptic sites in terms of their sequence, structure, and dynamics attributes. Based on these results, machine learning approaches were next applied to predict cryptic sites with relatively high accuracy on a benchmark dataset (true-positive and false-positive rates 73% and 29%, respectively). Cryptic site prediction was then extended to the entire structurally characterized human proteome and identified a high number of novel potentially "druggable" targets [31].

In light of the theoretical models described, besides structural characterization of putative binding sites, investigating protein plasticity can play a key role in understanding fundamental aspects of functional dynamics. Specifically, the relevance of dynamics in protein allostery supports the use of conformational sampling methods as useful tools for the identification of allosteric druggable sites. A notable example, which led to the approval of a drug by the US Food and Drug Administration (FDA), is represented by the discovery of raltegravir, an antiretroviral drug used in the treatment of HIV-AIDS infections. Schames et al. [32] carried out MD simulations of HIV integrase, a target previously considered undruggable. The simulations revealed a cryptic site, different from the enzyme active site, which was predicted as a possible allosteric pocket that could be targeted by inhibitors. The results of computational screening toward this dynamic allosteric pocket were confirmed by X-ray crystallography. Overall, this effort led to an actual drug development project by Merck & Co. and resulted in the production of an approved drug [32].

In the same conceptual framework, Carlson and coworkers exploited protein flexibility to develop ligands targeting the regulation of the flap opening/closing mechanism in HIV-1 protease. Opening and closing of the flaps controls substrate access to the active site, so that modifications in their dynamics may arguably result in a perturbation of enzyme functions via a true allosteric mechanism. MD simulations were thus used to sample multiple structures of the target and computational solvent mapping on the resulting structures was used to identify accessible pockets around the flap region. Receptor-based pharmacophore models recapitulating the complementary interactions needed for proficient ligand binding at the base of the flap were built and subsequent virtual screening identified a novel class of allosteric inhibitors, whose activity was experimentally proved [33–37].

Along these lines, over the past few years, improved simulation methods exploiting GPUs and distributed computing have consistently provided better opportunities to understand the atomistic determinants of the protein dynamics and coordinated motions that are critical to allosteric mechanisms. One notable example comes from the work of the Amaro group: using the cyclic nucleotide-binding domain of protein kinase A (PKA)'s regulatory subunit as a model system, the authors carried out long-timescale MD simulations of the cAMP-free (apo) and cAMP-bound states. The resulting simulations were then analyzed with Markov state modes, which permitted to characterize the conformational ensembles populated in the two ligand states. Both systems were found to exhibit shallow free energy landscapes that link different functional states via multiple possible transition pathways. These observations indicated that conformational selection is the general mechanism by which cAMP binding allosterically regulates this signaling protein domain. Importantly, the authors were able to define the principal structural motifs responsible for the propagation of the allosteric signal. These results can aptly be exploited to focus attention on such key motifs and integrate them with pocket identification algorithms, pharmacophore design, and virtual screening campaigns to identify new small molecules that can modify conformational populations and modulate functions as a consequence [38].

Building on the theoretical framework presented in the first part of the chapter and on the advances brought about by advances in structural and dynamic techniques, in the next paragraphs we present a critical overview of specific approaches integrating diverse lines of investigation into a consistent framework that may help bridge the understanding of protein structure and dynamics with the design of allosteric functional modulators.

11.4 Chaperones

Molecular chaperones are a class of highly conserved proteins that play a pivotal role in folding, stabilization, degradation, and general quality control of the proteome. Despite their fundamental importance, the molecular mechanisms responsible for shaping the folding energetic landscape of polypeptides remains largely unknown. The availability of chemical tools able to selectively activate or inhibit the functional dynamics of chaperones would provide a direct means to turn on/off specific cell functions by tweaking the conformational dynamics of the chaperone they target. In this framework, the new fundamental challenge for chemistry consists in identifying privileged structures capable of selectively interfering with key proteins functional substates involved in signaling pathways. In terms of molecular design, this represents a shift in paradigm that poses a number of new challenges.

Among molecular chaperones, Hsp90 is a nodal protein at the crossroad of multiple cell functions, which orchestrates different integrated cellular pathways required for cell development and maintenance. Hsp90 has been shown to play a key role in cancer, for which inhibitors are urgently needed, and in neurodegenerative diseases, for which chemical modulators (effectors) that help direct toxic aggregates toward different states may represent an important medical breakthrough. Hsp90 activities are dependent on ATP binding and hydrolysis, which directs transitions among different conformational substates endowed with different functional properties. For all these reasons, Hsp90 constitutes an ideal target to investigate the correlations between structure, dynamics, interactions, and cellular functions [39].

In this context, MD trajectories were used to investigate the mechanical coupling patterns between distinct subdomains of the *full-length* Hsp90, with the aim of characterizing signal transduction mechanisms that, upon ligand binding to a specific site, cause dynamic responses in distal regions [40]. The examination of the coordination between different domains in the apo, ATP- and ADP-bound states of *full-length* Hsp90 revealed that long-range mechanical coordination among pairs of residues extended well beyond local elements in the ATP-bound form, indicating the presence of a diffuse interaction network spanning the whole structure across the domains [40]. The functional consequence is the stabilization of the ATP-bound active closed state. In the ADP state, the positive correlation was lost and the two protomers moved in an anti-correlated way. This indicated collective motions toward opposite directions, opening up the clamp and defining the localization of molecular hinges fundamental for the regulation of functionally oriented motions.

To put these observations on a more quantitative ground, a general theoretical model was developed to identify the "hot spots" active in the allosteric communication between a certain binding site and distal regions of the protein implied in function. Allosteric coupling between two sites was defined on the basis of the communication propensity (CP) parameter. CP describes the coupling between any two residues as a function of fluctuation of their distance components: therefore, low CP values are related to efficiently communicating residues. Specific clusters of residues were identified as the key determinants (hot spots) of signal transduction from the N-terminal binding site to the C-terminal domain (CTD) in the presence of ATP. Such clusters were absent in the presence of ADP or in the apo-state. The hot spot in the C-terminal region was then subjected to structural investigation and one putative binding site in the C-terminal domain was consistently detected in all representative MD conformations. This information was then used to develop dynamic pharmacophore models for virtual screening of drug databases [41]. The resulting compounds showed affinity for the Hsp90 C-terminal domain, had specific effects on the viability of cancer cells and could disrupt Hsp90 association with clients and co-chaperones [41].

The validity of the structural model for the putative C-terminal allosteric binding site was first validated by docking a range of known C-terminal targeted drugs (the antibiotic novobiocin – the first known inhibitor to bind Hsp90-CTD – and 20 more potent related mimics), and by developing a 3D-quantitative structure–activity relationship (3D-QSAR) model that provided a quantitative correlation between the stereochemical properties of the functional groups of several novobiocin-related ligands and their antiproliferative activities. The 3D-QSAR model was next used to guide the synthesis of new novobiocin analogs, which resulted in new derivatives with antiproliferative activity at mid-nanomolar concentrations, a 1000-fold increase in potency from the initial novobiocin lead [42–44].

These findings supported the concept that specific protein residues can be identified and targeted for interfering with functional Hsp90 states.

Importantly, the reach of the dynamics-based method for allosteric drug discovery was extended to the rational design of *activators* of the ATPase enzymatic activity of the chaperone [45, 46]. The analysis of the protein's conformational response to first-generation ligands represented by 2-phenylbenzofurans was exploited to guide the design of novel derivatives with improved ability to stimulate ATP hydrolysis. Stimulation rates up to about seven times the native one could be obtained. Next, attention was focused on the cross talk between allosteric ligands and the dynamic properties of the chaperone's active states, by examining the impact of a series of designed allosteric modulators on the stability and structural and internal dynamics parameters. Such a study critically resulted in the development of a quantitative model that correlates Hsp90 activation to the presence of a certain compound, making use of information on the dynamic adaptation of protein conformations to the presence of the ligand. Finally, experimental analysis of the interaction landscape of Hsp90 in the presence of the stimulators showed increased affinity for the stimulator co-chaperone Aha1 and markedly decreased binding to the inhibitor co-chaperones Sti1. Modification of the interaction landscape of the chaperone through the acceleration of its ATPase functions negatively affected the maturation of a number of stringent clients, eventually determining cytotoxic effects in a number of cancer cell models [45, 46].

Overall, these studies on dynamic regulation of Hsp90 functions indicated an important role for the designed allosteric accelerators: they may indeed represent novel, gain-of-function chemical tools that offer the possibility of addressing the role of enzymatic and conformational dynamics in the protein endogenous environment, complementing classical active site inhibition and molecular biology experiments and opening new possibilities for the discovery of therapeutic leads with novel mechanisms of action.

11.5 Kinases

Kinase proteins constitute nearly 2% of the human coding genome (see Ref. [47] and references therein) and regulate a large amount of cellular events by transferring the gamma phosphate of ATP to the hydroxyl group of serine, threonine, and tyrosine residues. Their dysfunction is correlated to a number of diseases, from cancer to neurodegenerative conditions; therefore, they constitute valuable drug targets. Their phosphotransferase catalytic activity is not particularly efficient (ranging typically from 0.2 to 20 s^{-1}), yet they allosterically modulate their activation in a very specific manner in response to cellular signals thanks to their structural plasticity. They have a conserved core of around 250 residues, forming a smaller N-lobe and a larger C-lobe. Activation amounts to exposing the ATP binding cleft, through a conformational transition of the N-lobe. Thus, proteins switch between an inhibited, structurally rigid ground state and a more dynamic and heterogeneous active state. Pathogenic mutations have different abilities to shift this balance toward the active state. The overactivation of the enzymatic activity confers cells a growth and survival advantage, and many activating kinase mutations have been linked to cancer phenotypes.

Computational studies, such as Gaussian network models, MD simulations, as well as multiscale and enhanced sampling methods, have addressed mechanistic aspects related to the recognition and binding of ligands, and to the possible effects of mutations on kinase conformational dynamics. The current view of kinase allostery focuses on dynamic modulation and assembly of two groups of hydrophobic residues: the hydrophobic spine and the catalytic spine, whose rearrangement modulates the structure into active and inactive conformation. Network-based analysis [47] highlights a dynamic pattern with residues clustering into semirigid communities that move in microsecond to millisecond timescale, connected via the spines. Single mutations can significantly disrupt the dynamic infrastructure and thereby interfere with long-distance allosteric signaling that propagates throughout the whole molecule. Dynamics is therefore proposed to be the underlying mechanism for allosteric regulation in protein kinases. Free energy calculations and MD simulations were used to show that some oncogenic mutations in endothelial growth factor receptor (EGFR) kinases indeed lower the energy costs associated with the demanding conformational transition to the active structure, and that mutations tend to accumulate in the less stable N-lobe [48]. Allosteric ligands targeting kinases can, in principle, be able to modulate the activation mechanism, and offer the advantage of bypassing the competition with the high levels of intracellular ATP that affects ligands targeting the ATP binding site. Current efforts in allosteric drug discovery for kinases focus on three different strategies, which are reviewed in [49]. First, targeting the DFG-out pocket, which is located adjacent to the ATP binding site, represents a prototypic approach along these lines that inhibits the activation transition. Another proposed mechanism-based strategy consists of inducing the displacement of the conserved structural αC helix to modulate activation [50]. In the same area, a virtual screening approach [51] was used to identify a compound targeting a pocket on PDK1 kinase, which, in principle, could be extended to other members of the kinase family.

11.6 GPCRs

G protein–coupled receptors (GPCRs), also known as 7-transmembrane (7TM) receptors, are the largest superfamily of cell surface receptor proteins encoded by the human genome, as well as the most drug-targeted protein in the pharmacological market. GPCRs are naturally allosteric systems, as they function by propagating extracellular stimuli to the cell interior by means of a conformational transition. In the cell, the process starts when an endogenous extracellular signal binds to the primary (orthosteric) binding site of GPCR, inducing a conformational rearrangement that conveys the signal through the seven helix transmembrane region and triggers intracellular signaling cascades via heterotrimeric G proteins, kinases, and beta arrestins.

A particularly interesting phenomenon associated with allostery that has been most noted at GPCRs is the property of functional selectivity, which is the modulation of the affinity toward either G protein or beta arrestin due to specific ligands.

The attractiveness and tractability of GPCRs as drug targets, coupled with advances in drug screening at these receptors, have led to identifying different types of allosteric GPCR modulators (see for instance Table 1 in [52] for a list) targeting both the extracellular and the intracellular side of the receptors. Yet, high-throughput screening approaches face experimental challenges, due to the complexity of the system and of functional assays, whereas structure-based drug design has been prevented, until recently, due to the scarcity of high-resolution information. Nowadays in contrast , due to the remarkable growth in the number of X-ray structures solved for a range of class A, B, and C GPCRs, the information provided by crystal structures allows for more direct molecular level insights into GPCR functional mechanism and allostery, also by means of computational methods.

For example, Dror et al. used the inactive state crystal structure of the M2 mAChR as a template for performing unbiased, long-timescale molecular dynamic simulations using multiple, structurally diverse, allosteric modulators in both the absence and presence of an orthosteric antagonist. One important outcome of this study was the identification of two core "centers" that contribute to the so-called common allosteric site in the wide extracellular vestibule [53]. As the mechanism of GPCR activation is allosteric in nature, studies focusing on the propagation of the activation signal from the binding site to the intracellular side are also relevant in this context. For instance, long-term simulations of the β_2 adrenergic receptors in the presence of an agonist and starting from the active G protein–bound conformations have been performed to elucidate the deactivation mechanism and from that infer features of the activation process [54].

Other MD-based studies have addressed the problem of the modulation of allostery underlying "functional selectivity" or "biased agonism." While biased agonists have been identified and classified according to the relative bias potency (see Ref. [55]), for the most part the structural basis for this selectivity is not well understood. One way to address this gap in understanding is therefore to use MD simulations to identify the GPCR structural elements that adopt distinct ensemble conformations in response to ligands with different pharmacological properties. Analysis of microsecond MD simulations of the 5-HT2AR [56] provided a significant indication of particular agonist-mediated conformational substates of an intracellular loop, ICL2. The loop conformation is shown to adopt systematically different responses to ligands with known differences in receptor-dependent phenotypes, which can be connected to the modulation of affinity toward different downstream effectors binding to the intracellular side.

The molecular origin of allosteric activation of GPCRs has been also addressed by other computational methods such as amino acid covariation analysis in the case of dopamine receptor [57]. The method assesses the presence of coevolving residue pairs that are hypothesized to be correlated to function and again sets a structural basis to direct further rational drug design approaches.

Shedding light on the molecular mechanism underlying the bias selection, as well as in general on the allostery of GPCR activation, is the first step toward the development of more effective therapeutic drugs that target specific signaling pathways with a minimum of detrimental side effects. Along this line, McCammon and coworkers also used molecular dynamics simulations to identify druggable allosteric sites in the human β_1 and β_2 adrenergic receptors [58]. Multiple protein conformations were extracted from these simulations, and the protein surface was computationally "flooded" with small organic probes using the FTMAP algorithm [59] to identify potential binding sites. Regions of the protein surface where the organic probes could consistently and favorably bind across multiple structures were then defined as potential allosteric sites. This approach allows the identifications of five potential sites, some of which were not evident in any of the existing crystal structures.

11.7 Conclusions

A number of theoretical and technological advances in the areas of protein regulation, protein dynamics, chemical biology, and computer-aided drug design have made allosteric drug discovery an attractive and promising alternative to overcome the limits of classical active-site targeting drug-design campaigns.

Targeting allosteric sites provides new entry points into unexplored chemical spaces, since the aim is to engage pockets that may be particularly flexible or peculiarly hydrophobic. Moreover, the lower degree of conservation of allosteric sites compared to enzymatic active centers promises to address the problems of selectivity and off-target and side effects.

From the theoretical side, the progress in the atomistic understanding of the principles that underlie protein allosteric regulation combined with unprecedented advances in available computing power opens up new avenues for the investigation of more and more complex systems. The ultimate goal, in this case, would be the possibility of tuning entire biochemical pathways by modulating, and not only inhibiting, relevant hub proteins.

Finally, the ability to rationally design molecules that permit tweaking of the activities of their targets can generate new chemical tools through which the importance of specific aspects of regulation (for instance, the kinetics of conformational transitions between structures that bind to different partners in the cell) can be investigated. This would represent a powerful complement to molecular biology and genomic-based techniques.

The goal is the integration of all such knowledge in the early steps of new drug discovery for the efficient and rapid selection of new leads.

References

1 van der Greef, J. and McBurney, R.N. (2005). Innovation – rescuing drug discovery: in vivo systems pathology and systems pharmacology. *Nat. Rev. Drug. Disc.* 4: 961–967.

2 Steensma, D.P. (2009). The ordinary miracle of cancer clinical trials. *J. Clin. Oncol.* 27: 1737–1739.

3 Hardy, J.A. and Wells, J.A. (2004). Searching for new allosteric sites in enzymes. *Curr. Opin. Struct. Biol.* 14: 706–715.

4 Panjkovich, A. and Daura, X. (2010). Assessing the structural conservation of protein pockets to study functional and allosteric sites: implications for drug discovery. *BMC Struct. Biol.* 10: 9.

5 Zorn, J.A. and Wells, J.A. (2010). Turning enzymes ON with small molecules. *Nat. Chem. Biol.* 6: 179–188.

6 Cui, Q. and Karplus, M. (2008). Allostery and cooperativity revisited. *Protein Sci.* 17: 1295.

7 Csermely, P., Palotai, R., and Nussinov, R. Induced fit, conformational selection and independent dynamic segments: an extended view of binding events. *Trends Biochem. Sci.* 35: 539–546.

8 Sinha, N., Kumar, S., and Nussinov, R. (2001). Inter-domain interactions in hinge-bending transitions. *Structure* 9: 1165–1181.

9 Koshland, D.E., Nemethy, G., and Filmer, D. (1966). Comparison of experimental binding data and theoretical models in proteins containing subunits. *Biochemistry* 5: 365.

10 Monod, J., Wyman, J., and Changeux, J.P. (1965). On the nature of allosteric transitions-a plausible model. *J. Mol. Biol.* 12: 88.

11 Yu, E.W. and Koshland, D.E. Jr., (2001). Propagating conformational changes over long (and short) distances in proteins. *Proc. Natl. Acad. Sci. U.S.A.* 98: 9517–9520.

12 Tsai, C.-J. and Nussinov, R. (2014). A unified view of "how allostery works". *PLoS Comput. Biol.* 10: e1003394.

13 Thompson, J.R., Bell, J.K., Bratt, J. et al. (2005). Vmax regulation through domain and subunit changes. The active form of phosphoglycerate dehydrogenase. *Biochemistry* 44: 5763–5773.

14 Bhattacharya, A., Kurochkin, A.V., Yip, G.N.B. et al. (2009). Allostery in Hsp70 chaperones is transduced by subdomain rotations. *J. Mol. Biol.* 388: 475–490.

15 Kityk, R., Kopp, J., Sinning, I., and Mayer, M.P. (2012). Structure and dynamics of the ATP-bound open conformation of Hsp70 chaperones. *Mol. Cell* 48: 863–874.

16 Bertelsen, E.B., Chang, L., Gestwicki, J.E., and Zuiderweg, E.R. (2009). Solution conformation of wild-type *E. coli* Hsp70 (DnaK) chaperone complexed with ADP and substrate. *Proc. Natl. Acad. Sci. U.S.A.* 106: 8471–8476.

17 Laskowski, R.A., Gerick, F., and Thornton, J.M. (2009). The structural basis of allosteric regulation in proteins. *FEBS Lett.* 583: 1692–1698.

18 Swain, J.F. and Gierasch, L.M. (2006). The changing landscape of protein allostery. *Curr. Opin. Struct. Biol.* 16: 102–108.

19 Cooper, A. and Dryden, D.T. (1984). Allostery without conformational change. A plausible model. *Eur. Biophys. J.* 11: 103–109.

20 Igumenova, T.I., Frederick, K.K., and Wand, A.J. (2006). Characterization of the fast dynamics of protein amino acid side chains using NMR relaxation in solution. *Chem. Rev.* 106: 1672–1699.

21 Popovych, N., Sun, S., Ebright, R.H., and Kalodimos, C.G. (2006). Dynamically driven protein allostery. *Nat. Struct. Mol. Biol.* 13: 831–838.

22 Lee, A.L., Kinnear, S.A., and Wand, A.J. (1999). Redistribution and loss of side chain entropy upon formation of a calmodulin–peptide complex. *Nat. Struct. Mol. Biol.* 7: 72–77.
23 Fuentes, E.J., Der, C.J., and Lee, A.L. (2004). Ligand dependent dynamics of intramolecular signaling in a PDZ domain. *J. Mol. Biol.* 335: 1105–1115.
24 Fuentes, E.J., Gilmore, S.A., Mauldin, R.V., and Lee, A.L. (2006). Evaluation of energetic and dynamic coupling networks in a PDZ domain protein. *J. Mol. Biol.* 364: 337–351.
25 Hilser, V.J., Wrabl, J.O., and Motlagh, H.N. (2012). Structural and energetic basis of allostery. *Annu. Rev. Biophys.* 41: 585–609.
26 Boehr, D.D., Nussinov, R., and Wright, P.E. (2009). The role of dynamic conformational ensembles in biomolecular recognition. *Nat. Chem. Biol.* 5: 789–796.
27 Nussinov, R. and Tsai, C.-J. (2013). Allostery in disease and drug discovery. *Cell* 153: 293–305.
28 Nussinov, R., Tsai, C.-J., and Li, J. (2014). Principles of allosteric interactions in cell signaling. *J. Am. Chem. Soc.* 136: 17692–17701.
29 Panjkovich, A. and Daura, X. (2012). Exploiting protein flexibility to predict the location of allosteric sites. *BMC Bioinform.* 13: 273.
30 Hardy, J.A. and Wells, J.A. (2004). Searching for allosteric sites in enzymes. *Curr. Opin. Struct. Biol.* 14: 706–715.
31 Cimermancic, P., Weinkam, P., Rettenmaier, T.J. et al. (2016). CryptoSite: expanding the druggable proteome by characterization and prediction of cryptic binding sites. *J. Mol. Biol.* 428: 709–719.
32 Schames, J.R., Henchman, R.H., Siegel, J.S. et al. (2004). Discovery of a novel binding trench in HIV integrase. *J. Med. Chem.* 47: 1879–1881.
33 Carlson, H.A. (2002). Protein flexibility and drug design: how to hit a moving target. *Curr. Opin. Chem. Biol.* 6: 447–452.
34 Carlson, H.A., Masukawa, K.M., Rubins, K. et al. (2000). Developing a dynamic pharmacophore model for HIV-1 integrase. *J. Med. Chem.* 43: 2100–2114.
35 Damm, K.L. and Carlson, H.A. (2007). Exploring experimental sources of multiple protein conformations in structure-based drug design. *J. Am. Chem. Soc.* 129: 8225–8235.
36 Meagher, K.L. and Carlson, H.A. (2004). Incorporating protein flexibility in structure-based drug discovery: using HIV-1 protease as a test case. *J. Am. Chem. Soc.* 126: 13276–13281.
37 Meagher, K.L., Lerner, M.G., and Carlson, H.A. (2006). Refining the multiple protein structure pharmacophore method: consistency across three independent HIV-1 protease models. *J. Med. Chem.* 49: 3478–3484.
38 Malmstrom, R.D., Kornev, A.P., Taylor, S.S., and Amaro, R.E. (2015). Allostery through the computational microscope: cAMP activation of a canonical signalling domain. *Nat. Commun.* 6: 7588.
39 Krukenberg, K.A., Street, T.O., Lavery, L.A., and Agard, D.A. (2011). Conformational dynamics of the molecular chaperone Hsp90. *Q. Rev. Biophys.* 44: 229–255.

40 Morra, G., Verkhivker, G.M., and Colombo, G. (2009). Modeling signal propagation mechanisms and ligand-based conformational dynamics of the Hsp90 molecular chaperone full length dimer. *PLoS Comp. Biol.* 5: e1000323.
41 Morra, G., Neves, M.A.C., Plescia, C.J. et al. (2010). Dynamics-based discovery of allosteric inhibitors: selection of new ligands for the C-terminal domain of Hsp90. *J. Chem. Theory Comput.* 6: 2978–2989.
42 Zhao, H., Moroni, E., Colombo, G., and Blagg, B.S. (2014). Identification of a new scaffold for Hsp90 C-terminal inhibition. *ACS Med. Chem. Lett.* 5: 84–88.
43 Zhao, H., Moroni, E., Yan, B. et al. (2013). 3D-QSAR-assisted design, synthesis, and evaluation of novobiocin analogues. *ACS Med. Chem. Lett.* 4: 57–62.
44 Moroni, E., Zhao, H., Blagg, B.S., and Colombo, G. (2014). Exploiting conformational dynamics in drug discovery: design of C-terminal inhibitors of Hsp90 with improved activities. *J. Chem. Inf. Model.* 24397468.
45 Sattin, S., Tao, J., Vettoretti, G. et al. (2015). Activation of Hsp90 enzymatic activity and conformational dynamics through rationally designed allosteric ligands. *Chemistry* 21: 13598–13608.
46 Vettoretti, G., Moroni, E., Sattin, S. et al. (2016). Molecular dynamics simulations reveal the mechanisms of allosteric activation of Hsp90 by designed ligands. *Sci. Rep.* 6: 23830.
47 Kornev, A.P. and Taylor, S.S. (2015). Dynamics-driven allostery in protein kinases. *Trends Biochem. Sci.* 40: 628–647.
48 Sutto, L. and Gervasio, F.L. (2013). Effects of oncogenic mutations on the conformational free-energy landscape of EGFR kinase. *Proc. Natl. Acad. Sci. U.S.A.* 110: 10616–10621.
49 Fang, Z., Grutter, C., and Rauh, D. (2013). Strategies for the selective regulation of kinases with allosteric modulators: exploiting exclusive structural features. *ACS Chem. Biol.* 8: 58–70.
50 Palmieri, L. and Rastelli, G. (2013). alphaC helix displacement as a general approach for allosteric modulation of protein kinases. *Drug Discov. Today* 18: 407–414.
51 Rettenmaier, T.J., Fan, H., Karpiak, J. et al. (2015). Small-molecule allosteric modulators of the protein kinase PDK1 from structure-based docking. *J. Med. Chem.* 58: 8285–8291.
52 Gentry, P.R., Sexton, P.M., and Christopoulos, A. (2015). Novel allosteric modulators of G protein-coupled receptors. *J. Biol. Chem.* 290: 19478–19488.
53 Kruse, A.C., Hu, J., Pan, A.C. et al. (2012). Structure and dynamics of the M3 muscarinic acetylcholine receptor. *Nature* 482: 552–556.
54 Dror, R.O., Arlow, D.H., Maragakis, P. et al. (2011). Activation mechanism of the beta2-adrenergic receptor. *Proc. Natl. Acad. Sci. U.S.A.* 108: 18684–18689.
55 Rajagopal, S., Ahn, S., Rominger, D.H. et al. (2011). Quantifying ligand bias at seven-transmembrane receptors. *Mol. Pharmacol.* 80: 367–377.
56 Perez-Aguilar, J.M., Shan, J., LeVine, M.V. et al. (2014). A functional selectivity mechanism at the serotonin-2A GPCR involves ligand-dependent conformations of intracellular loop 2. *J. Am. Chem. Soc.* 136: 16044–16054.

57 Sung, Y.M., Wilkins, A.D., Rodriguez, G.J. et al. (2016). Intramolecular allosteric communication in dopamine D2 receptor revealed by evolutionary amino acid covariation. *Proc. Natl. Acad. Sci. U.S.A.* 113: 3539–3544.

58 Ivetac, A. and McCammon, J.A. (2010). Mapping the druggable allosteric space of G-protein coupled receptors: a fragment-based molecular dynamics approach. *Chem. Biol. Drug. Des.* 76: 201–217.

59 Brenke, R., Kozakov, D., Chuang, G.Y. et al. (2009). Fragment-based identification of druggable 'hot spots' of proteins using Fourier domain correlation techniques. *Bioinformatics* 25: 621–627.

12

Structure and Stability of Amyloid Protofibrils of Polyglutamine and Polyasparagine from Molecular Dynamics Simulations

Viet Hoang Man, Yuan Zhang, Christopher Roland, and Celeste Sagui

North Carolina State University, Department of Physics, Riddick Hall, 2401 Stinson Drive, Raleigh, NC 27695, USA

12.1 Introduction

At least nine inherited, fatal neurodegenerative diseases such as Huntington's disease (HD), dentatorubral-pallidoluysian atrophy (DRPLA), six spinocerebellar ataxias (SCA 1, 2, 3, 6, 7, and 17), and spinal-bulbar muscular atrophy (SBMA) [1] are associated with the expansion of the CAG trinucleotide repeat into an anomalously long, expanded polyglutamine (polyQ) tract in the corresponding mutant proteins. The disease age-of-onset and risk are correlated with the length of the polyQ expansion, and are associated with a pathogenic threshold length of about 35–42 Gln residues for all but two diseases. While each of the diseases has a different pathology, they are all characterized by the presence of polyQ-rich aggregated in neurons once the expansion sequence has crossed the critical threshold length. It is therefore believed that the neuronal toxicity associated with polyQ diseases is caused by the polyQ expansion itself, and that the same molecular mechanisms of neurodegeneration are shared by all of these diseases [2, 3].

PolyQ tracts have been observed as insoluble aggregates *in vitro*, in cell and in animal studies [4–6], and in diseased human brain tissue [7]. Aside from the mature fibrils, neurons also exhibit a variety of soluble aggregates, known as "protofibrils" [8–10]. A plausible scenario for the kinetics of aggregation of pure polyQ has been described as a nucleation-growth polymerization [3–5, 11–13]. This mechanism, however, can coexist with transient and metastable structures that occur before or simultaneously with fibril formation [4, 5, 8–11, 14–16], and that may be an on- or off-pathway to the formation of aggregates. While the polyQ-rich aggregates are the most visible cytological features of the polyQ disorders, at present it is unclear whether it is the formation of aggregates that causes the toxicity in neuronal cells [17], or whether the large aggregates represent a protective response of the cells against further toxicity [15–24]. Indeed, it may well be that it is the soluble monomeric or oligomeric intermediates that are toxic to the cell [2, 15–26].

These findings, common to all amyloid diseases [24], have spurred multi-decade efforts to understand the structural characteristics of soluble

Biomolecular Simulations in Structure-Based Drug Discovery,
First Edition. Edited by Francesco L. Gervasio and Vojtech Spiwok.

oligomers and the amyloidogenic precursors. In particular, many structural models have been proposed for polyQ aggregates, such as α-helical coiled coils [27], β helices [28], α-sheets [29, 30] and β sheets [31, 32]. Currently, there is evidence based on polyQ fibers and crystallites that supports the presence of cross-β structures [31–36]. In a cross-β structure, the β-strand backbones and side chains are perpendicular to the fibril axis, while the backbone hydrogen bonds are, on average, parallel to the axis. Moreover, these structures exhibit dry, "steric-zipper" interfaces [37] between the β sheets (where the residue side chains intermesh with close complementarity).

Prion diseases can also result in fatal neurodegenerative diseases such as Creutzfeldt–Jakob disease (CJD) and Straussler–Scheinkerd diseases of humans, bovine spongiform encephalopathy (BSE), and scrapie of sheep. The cellular prion protein (PrP^C) is encoded by a gene and a posttranslational modification converts it into the infectious abnormal isoform, (PrP^{Sc}) which, somewhat similar to the polyQ diseases, is rich in β-sheet content [38]. Irrespective of whether the presence of PrP^{Sc} in the cell is the product of a posttranslation mishap or whether it is an external infectious agent, PrP^{Sc} acts as a template upon which PrP^C continues to aggregate, thereby enabling the non-Mendelian inheritance characteristic of prions. Prions also cause non-Mendelian inheritance in yeast, which has become an important model system for the study of prions. Aggregation-prone, Q/N-rich domains are present in the baker's yeast *Saccharomyces cerevisiae*, and in Sup35p and Ure2p proteins, whose amyloid prions are [PSI^+] and [URE3], respectively. Although prions can cause disease in yeast, cells with certain prion forms have been shown to survive better under adverse conditions. This lends credence to the view that at least some prions developed under the impetus of positive evolutionary selection [39, 40]. For instance, fungal prions may be considered to belong to the category of functional amyloids. Other functional amyloids may be found, for instance, in biofilms of bacteria: HET-s fibrils of *Podospora anserina* which is required by heterokaryon incompatibility [41]; the chaplin peptides from the bacterium *Streptomyces coelicolor* which is associated with fimbriae formation [42]; as well as curli fibrils of *Escherichia coli* and *Salmonella enterica* which seems to serve an adhesive function [43].

In terms of prions, an important research direction involves the investigation of the relative roles of the polar, uncharged residues Q and N in Q/N-rich yeast prions [44–47]. For example, recent investigations show that Q and N play important opposing roles in the formation of the Sup35 prion-forming domain (PrD) [47]. The experiments found that if the N's in the wild-type Sup35 PrD were switched to Q's, then the formation of amyloids decreased but the toxicity to the cell increased. On the other hand, switching all Q's to N's enhanced the formation of benign, self-templating amyloids and decreased cell toxicity [47]. This correlation between the decrease in amyloid formation and increased toxicity for polyQ sequences is supported by a large body of literature in neurodegenerative polyQ diseases, where the formation of aggregates and toxicity in neuronal cells are not believed to be linked as cause–effect as was once thought. As already noted, some researchers now believe that the formation of large aggregates represents a cellular protective response against further toxicity [18, 20], and that it is the soluble monomeric or oligomeric intermediates that are the problem leading to neuronal cellular demise [2, 18, 20].

Given the experimental difficulties in obtaining atomic resolution structures of aggregates, simulations provide invaluable insights into the atomic structure and dynamics of small polyQ/polyN aggregates. Unfortunately, approximations used in simulations, although needed to make progress, can occasionally compromise the biochemical or structural identity of the system and thus inadvertently contribute to the confusion that exists in an already complex field. For example, subtle interactions (van der Waals interactions, electrostatic effects, dihedral angles, etc.) that characterize polyQ and/or polyN oligomers may not have been adequately captured in current coarse-grained models. The best way to test the validity of the various aggregate models is by contact with high-resolution experiments. In the case of polyQ aggregates, a fully atomistic description is provided in the work of Sikorski and Atkins [32] who proposed a model to explain the X-ray diffraction patterns obtained from crystalline fibers and films of a $D_2Q_{15}K_2$ polypeptide [28] (which had originally been interpreted as a hollow, water-filled nanotube [28]). The model provides for a successful explanation of the main features of the diffraction pattern and, in our estimation, became a turning point in the structural interpretation and understanding of these aggregates. Unfortunately, these aggregates do not lend themselves easily to high- (or medium-) resolution characterization. Thus, such simulations are not very common, although when they do exist, they can provide invaluable insight into the structure of protofibrils. Another case in point is provided by Ref. [48], where careful modeling of the system, involving the introduction of various mathematically well-defined crystallographic symmetries, provided very realistic models for Alzheimer's β-amyloid fibrils, in excellent agreement with NMR and X-ray fiber diffraction data.

We have carried out fully atomistic molecular dynamics (MD) simulation studies with explicit solvent and state-of-the-art force fields in order to probe the structural characteristics and conformational dynamics of various polyQ aggregates for a cumulative time of over 23 μs [49]. We first modeled the oligomeric polyQ aggregates along the lines of Ref. [32], but also characterized the possible combinations of peptide stacking and side-chain packing that can lead to different steric zipper interfaces. In addition, we have also investigated the fibril structural characteristics of Asn and Gln amino acids [50]. In this particular work, we used MD simulations to characterize all possible interfaces, structures, and the relative stability of N/Q-rich peptides – including sequences associated with the yeast Sup35 prion domain – as they form amyloid-like aggregates. The results of these investigations are summarized in the rest of this brief chapter.

12.2 Polyglutamine Protofibrils and Aggregates [49]

12.2.1 Investigations of Oligomeric Q_8 Structures

To investigate the structure and stability of polyglutamine (polyQ) protofibrils, we have carried out large-scale fully atomistic MD simulations with explicit waters and state-of-the-art force fields [49]. In particular, we map out all the possible steric interfaces that account for the interaction of a sheet with its neighbors, and the effects of different alignments of the side-chain carboxamide dipoles. The structures considered were based on assembling dimers constructed from the octapeptide Ace – $(Gln)_8$ – Nme (Q_8) with extended side chains. Specifically, the dimers were used to build $M \times N$ oligomeric "crystals," with unit cell parameters chosen in accordance with previous fiber diffraction experiments and models. Here, M refers to the number of Q_8 peptides in a sheet, and N to the number of sheets in a given aggregate. In addition, each fibril model is characterized by the relative orientation between two neighboring strands in a sheet (these can be parallel sheets (PSs) or antiparallel sheets (APSs)), and the relative orientation between two neighboring stacking sheets (which may be in a parallel packing (PP) or antiparallel packing (APP)). This construction assumes an alternating pattern between the strands in a given sheet, and between the sheets in an oligomer, so that every other component has the same orientation in the antiparallel cases or the same orientation in the parallel cases. Schematics of these structures is shown in Figure 12.1. We also considered peptides whose side-chain carboxamide groups were all oriented in the same direction, giving rise to a net total dipole (TD) in the x-direction, and structures in which the upper side-chain dipoles canceled the lower side-chain dipoles giving a net zero dipole (ZD) in the x-direction. For the packing in the y-direction, dipoles in the upper and lower side chains in neighboring planes along y are in the opposite direction. All in all, a total of 68 different oligomeric models were considered.

To probe the stability of these structures, several computational measures were used such as the population of residues in a β-sheet conformation (the larger the population, the greater the stability of the structure), various measures of hydrogen bonding, and the time it takes to break the ordered state of the structures [49]. Key results are as follows.

First, while the stability of an oligomer increases with the number of monomers, the way these monomers are assembled is important. Given a similar number of monomers, compact structures are more stable than looser structures – the latter are typically characterized by high aspect ratios. Thus, for the octamers, the more compact 4×2 (two sheets of four strands aggregates) are more stable than the 2×4 aggregate (four sheets of two strands). This makes sense insofar as compact structures maximize the number of inter- and intrapeptide hydrogen bonds and the van der Waals contacts between the side chains, all of which adds to the stability of the aggregate. One can expect that this trend is valid only for aggregates with a relatively small number of monomers with a high surface-to-volume ratio, as is the case in our investigations. Once the aggregate reaches a critical size, the fibrils or crystals should be able to grow ever larger with a high degree of stability.

Second, antiparallel β sheets are more stable than parallel β sheets. This is true not only of the single-sheet models but also for the $M \times N$ aggregates with $N > 1$. For the latter, one needs to account not only for the relative order of the strands with a hydrogen-bonded sheet (APS versus PS) but also the stacking of the sheets on top of each other. The sheets may have PP when the strands on top of each other point in the same direction, or APS when the strands on top

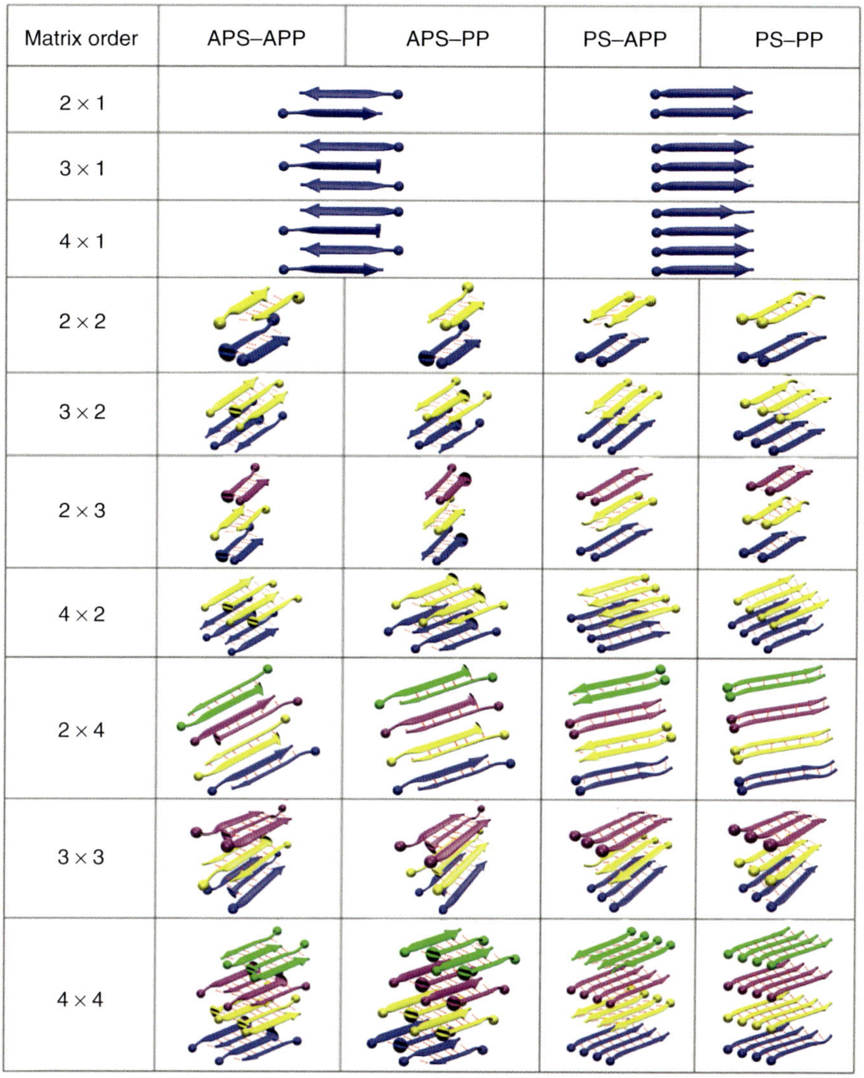

Figure 12.1 Here will show the different Q_8 protofibril models. APS and PS denote antiparallel and parallel beta sheets, while APP and PP denote antiparallel and parallel packing. The balls indicate the N-terminal of the peptides.

of each other point in the opposite direction. For the 2 × 2 systems, the order in terms of decreasing stability is APS–PP, APS–APP, PS–APP, and PS–PP. Clearly, APS systems are always more stable than PS systems. We also note that the stability of all systems increases with matrix order, so that 4 × 4 systems were found to be more stable than the 2 × 2 systems. In fact, the former was found to be the still stable after 1000 ns. In terms of side-chain packing, conformations with a nonzero dipole were found to be less stable than those with a ZD. The side-chain dipole is due to the orientation of the side-chain carboxamide groups both above and below the plane of the sheets. As may be expected, this effect is most pronounced in the single-sheet case, and diminishes as the number of sheets increases. For the 4×4 aggregates, both ZD and TD were found to be stable.

12.2.2 Time Evolution, Steric Zippers, and Crystal Structures of 4×4 Q_8 Aggregates

We investigated the stability of the 4×4 Q_8 structures by extending the simulation times of these systems to 1 μs, or longer. Specifically, systems with various initial conditions (APS/PS, APP/PP, TD/ZD) were considered. It is important to emphasize, that once the initial conditions were set, the systems were allowed to evolve naturally using MD simulations. The systems were not "over constrained." Thus, the initial side chains were fully extended and we did not implement the experimentally observed "quarter stagger" [51]. Thus, the models are not just stable or unstable insofar as they keep close to their initial conditions, but they are actually allowed to evolve in time toward more stable or unstable structures.

The time evolution of the ZD systems shows spontaneous quarter stagger (where adjacent sheets slip with respect to each other by $a/4$ in the a direction). Figure 12.2 shows initial configurations on the yz plane (a) and xy plane (b), and configuration at 1 μs on the xy plane (c) for a ZD 4×4 APS–PP system. The original fiber diffraction pattern [28, 32] seems to be sensitive only to the positions of the α-carbons, showing sharp equatorial diffraction signals for 4.75 and 8.33 Å, and a broad meridional arc of 3.60 Å. All the eight 4×4 lattices at the end of the simulation share these values, with small deviations. Thus, one can formally attribute a monoclinic lattice with $c \simeq 2 \times 3.60$ Å for all lattices, $a \simeq 4.75$ Å for the PS systems, and $a \simeq 9.50$ Å for the APS systems. The intersheet stacking is also around 8.33 Å for all systems, but whether one takes that value or twice that value for the axis b depends on the initial stacking of the side chains. In the initial conditions for the TD APS and PS systems (Figure 12.2e,f) and the ZD APS systems the side-chain rotations can be as shown, or the "+" and "−" side chains can be exchanged resulting in zippers with dipoles in the opposite direction. This results in two possible ways of stacking the systems along y. Thus, ZD PS–PP has $b \sim 8.3$ Å, and the other lattices have $b \sim 8.3$ Å or $b \sim 16.6$ Å, according to the relative directions of lower side-chain dipoles in one sheet and the upper side-chain dipoles in the adjacent sheet along y, combined with whether they are ZD or TD in the xy plane. Table 12.1 gives the crystal parameters averaged over the last 700 ns of the simulation.

Figure 12.2 shows final configurations for the 4×4 systems. The figure shows the four innermost peptides on the yz plane for an APS–PP crystal (Figure 12.2b), and four inner residues within those peptides for each of the possible packing arrangements on the xy plane (Figure 12.2a), displaying details of the side-chain packing. Notice that the N-C_α-C_β-C_γ flipping preserves the dipolar character of the side chains in the ab plane, ZD or TD. The quarter-stagger creates an additional set of hydrogen bonds.

Finally, Figure 12.2 shows the hydrogen-bond distances for the ZD 4×4 systems. The two top panels show the distance between the O_ϵ and N_ϵ atoms in

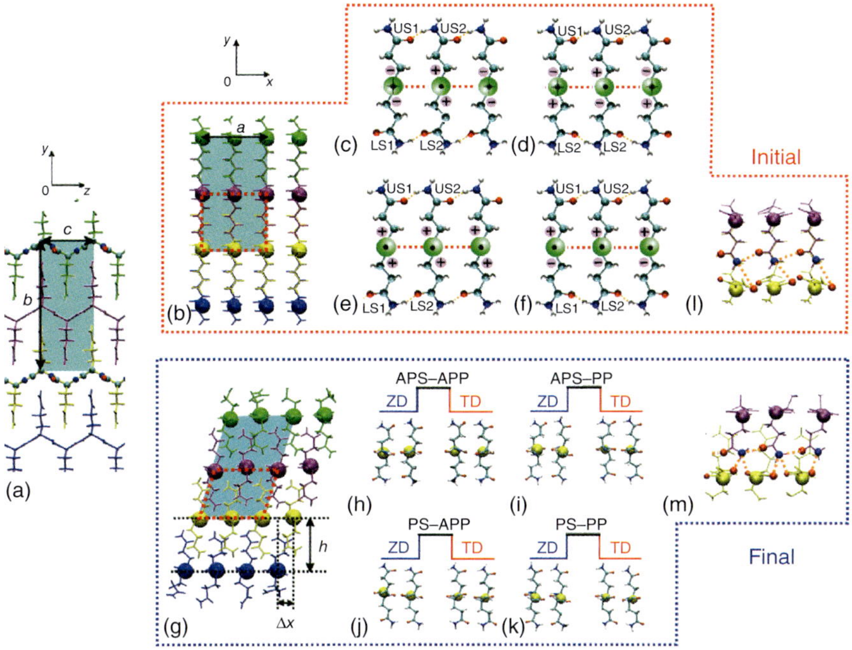

Figure 12.2 Initial polyQ crystal structure on the (a) *yz* and (b) *xy* planes with different possible side-chain packing arrangements (see c–f and m), and (g) final structure (after $t = 1\ \mu s$) on the *xy* plane with side-chain configurations (shown in h–k and n). Possible unit cells are depicted by colored boxes with lattice parameters $a = 0.950$ nm, $b = 1.660$ nm (cyan) or $b = 0.830$ nm (red dashed line) and $c = 0.657$ nm (parallel) or 0.695 nm (antiparallel). Red and blue balls represent O and N atoms. For the figure (l) and (m), the bigger (smaller) red balls indicate O atoms closer to (further from) the viewer. Hydrogen bonds are shown in orange for the selected group of atoms.

Table 12.1 *Average crystal parameters for the* 4 × 4 *ZD systems as shown in Figure* 12.2. The parameters (in Å) are obtained as an average over the last 700 ns of the 1 μs simulations. Here, *b* refers to the distance between the closest α carbons in two adjacent sheets, and Δ*x* is the displacement along *x* between adjacent β-sheet layers.

Model 4×4	*a*	*b*	*c*	Δ*x*
APS–APP	9.68 ± 0.05	8.25 ± 0.10	7.15 ± 0.02	2.42 ± 0.16
APS–PP	9.68 ± 0.05	8.25 ± 0.11	7.14 ± 0.02	2.41 ± 0.19
PS–APP	9.94 ± 0.03	8.34 ± 0.09	7.19 ± 0.05	2.41 ± 0.12
PS–PP	9.97 ± 0.03	8.14 ± 0.08	7.22 ± 0.05	1.96 ± 0.11

the US1–US2 and LS1–LS2 side chains. The two lower panels show the distance between the side chain N_ϵ of side chains US1 and LS1 and the main-chain O atom of the nearest sheet. In this figure, the distances to only one of the main-chain O atom is shown, but after the initial relaxation, the N_ϵ is equidistant to the *two* main-chain O atom in the nearest sheet, as shown schematically in Figure 12.2.

This is a consequence of the quarter-stagger conformation. This shows that the polar zipper mechanism is sustained by a stable network of hydrogen bonds, with main-chain–main-chain hydrogen bonds, side-chain–side-chain hydrogen bonds, and side-chain–main-chain hydrogen bonds.

These results are consistent with the experimental observations [28, 31, 52]. Experimentally, it has been known for a long time that Q-rich peptides exhibit hydrogen bonding not only for the backbone but also for the carboxamide groups in the side chains [52]. A detailed atomic model remained elusive for many years, until 1994 when Perutz et al. proposed the "polar zipper model" [53]. Later, Sikorski and Atkins [32] proposed a related model to explain the X-ray diffraction patterns obtained from crystalline fibers and films of a $D_2Q_{15}K_2$ polypeptide [28] (which had originally been interpreted as consisting of a hollow, water-filled nanotube [28]. This model successfully accounted for the main features of the diffraction pattern, although its stability was not tested at that time. Our simulation results are completely consistent with this model.

12.2.3 Monomeric Q_{40} Protofibrils

Having discussed the stability of different $M \times N$ aggregates, we turn to the question of the stability of a Q_{40} monomer, which is just above the pathogenic threshold length. Since the experimentally observed β-strand lengths in a fibril is 7–8 residues [14, 31, 54], a Q_{40} peptide must consist of a maximum of about four strands, including the connecting arcs and turns. We constructed a number of different models for Q_{40} based on the results obtained for the Q_8 oligomers. Specifically, we first took the 2×2 subsystems containing four central Q_8 peptides that belong to the relaxed, stable structures of the 4×4 ZD systems at 1 μs in the quarter-stagger [51] conformations. The four short peptides were connected into a long peptide by arc turns or/and hairpin turns. The topological constraint of having a single oriented peptide imposes limitations on the type of four-stranded structure that can be constructed. To label the structures, we adopted the following convention. The long peptide was divided into four strands labeled 1–4 from the amino end to the carboxyl end, and the connecting loops were denoted as arc (A) or hairpin (H). A model labeled 1H2A3H4 therefore implies that strands 1–2 and 3–4 are connected by a hairpin, while 2–3 are connected by an arc. The possible compact structures that can be built are the following: (i) an APS–APP system with 1–2 and 3–4 β sheets ("model 1," 1H2A3H4); (ii) an APS–APP system with 2–3 and 1–4 β sheets ("model 2," 1A2H3A4); (iii) an APS–PP system with 1–2 and 3–4 β sheets ("model 3," 1H2A3H4); (iv) a PS–APP system with 1–3 and 2–4 β sheets ("model 4," 1A2A3A4). As the simulations show, the APS–APP and APS–PP Q_8 aggregates used to build the monomer converge to the same structure thanks to a "quarter stagger" along the x direction. The difference between model 1 and model 3 is that in model 1 the arc connection forces strand 3 to be parallel to 2 and strand 4 parallel to 1, while in model 3 the arc forces strands 3–1 to be parallel and strands 4–2 to be parallel. We note that given the constraints, a PS–PP packing is no longer possible. Of these models, only model 4 is topologically related to a collapsed β helix. Finally, we also consider a planar arrangement ("model 5," 1H2H3H4) formed by a 4×1 APS system with three hairpins. All the Q_{40} models shown in Figure 12.3 are capped by Ace– in the

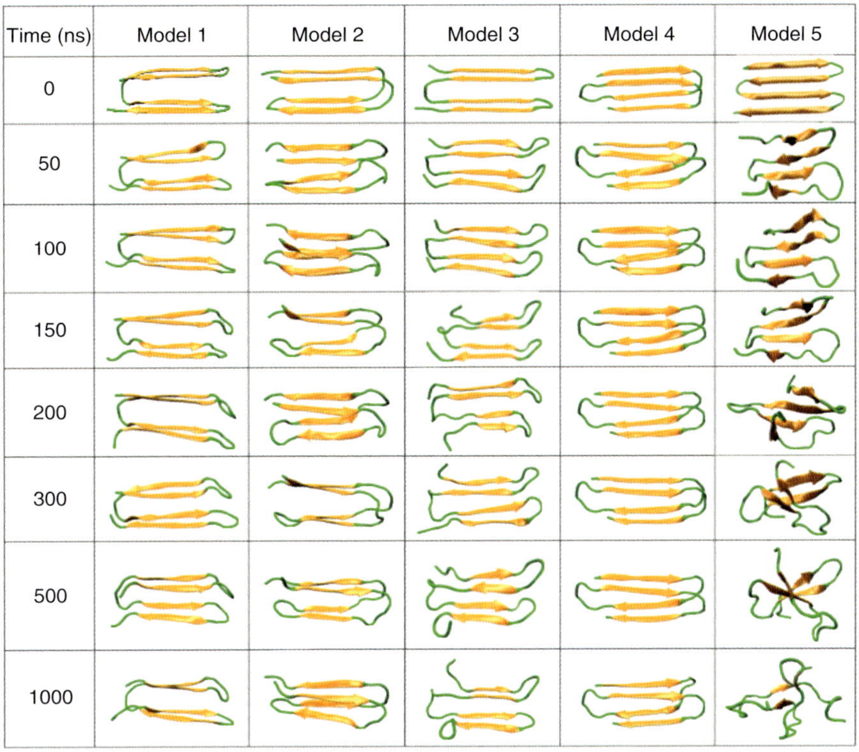

Figure 12.3 Time evolution of the Q_{40} fibril models. β strands are shown in orange color, while coil, arc and hairpin structures are shown in green.

amino end and by –Nme in the carboxyl end. In addition, in order to further probe the stability of these models, several Q_n monomers with n near the $n = 36$ threshold were created from the Q_{40} monomers via the elimination of residues from the β strands (since hairpins and arcs already have the minimum possible number of residues).

Figure 12.3 shows the initial and selected time snapshots of the evolution of the five Q_{40} aggregate models. Models 1 and 3 are "β sheet stacks." Both models contain two hairpins and one arc turn (1H2A3H4). The chief difference is that the arc in model 1 connects two strands on the same side of both sheets (thus, we call it "lateral β sheet stack"), while the arc in model 3 connects strands on the opposite side of the sheets ("diagonal β sheet stack"). Due to the quarter stagger, the "side" and the "diagonal" in this arrangement are the same length of ~ 8.3 Å, so the arcs are of the same length. Model 2, with two arcs and one hairpin (1A2H3A4), will be referred to as a "β double hook." Model 4 contains three arcs (1A2A3A4) and can be interpreted as a collapsed β helix. Finally, model 5 (1H2H3H4) is simply a single-sheet 4 × 1 β structure connected by three hairpins.

Clearly, Figure 12.3 shows that models 1–4 are stable on a 1 μs timescale, while model 5 is unstable. A more detailed investigation shows that models 1, 2, and 4 are perhaps the most stable, with model 3 being less stable. To further probe the stability of these models, we also looked at models as the number of Q residues

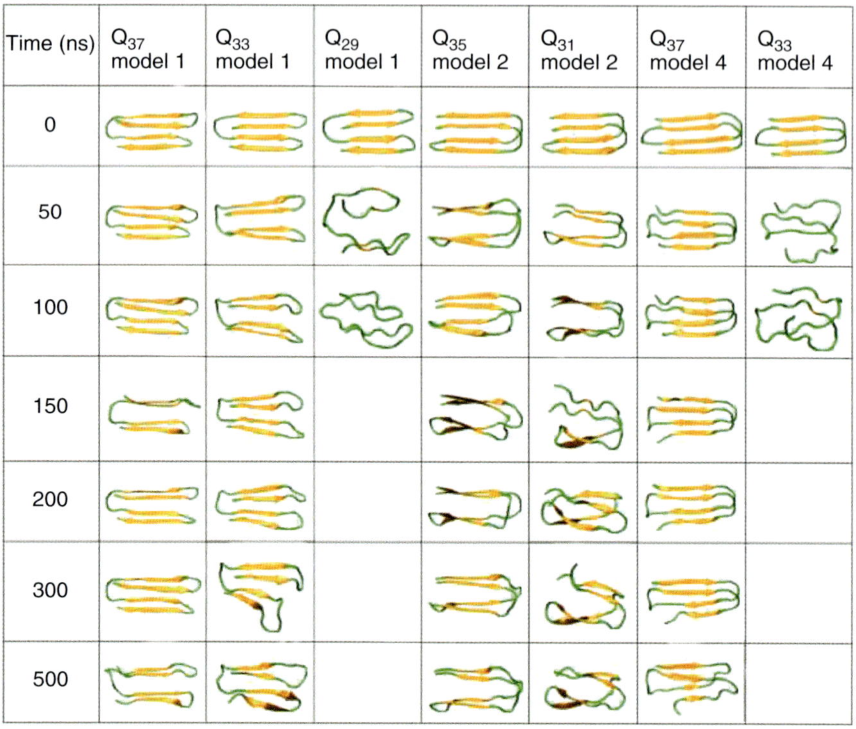

Figure 12.4 Time evolution of monomeric fibril models of various lengths. β strands are shown in orange color, while coil, arc and hairpin structures are shown in green.

oscillating around the threshold of 35–36. Figure 12.4 shows that model 1 is very stable: there is apparently no difference between 40 and 37 residues and even 33 residues appear to be stable over 500 ns. However, Q_{29} is definitely too short and the aggregate readily unfolded over 50 ns. Over a time span of 500 ns, model 2 is stable (unstable) with 35 (31) residues, while model 4 is stable (unstable) for 37 (33) residues.

The main results can be summarized as follows: the most compact models, with four β strands of approximately seven residues each and interdigitated side chains, are stable in the 1 μs timescale. This makes them excellent candidates for the monomeric misfolded structure capable of triggering aggregation. By contrast, the single-sheet 4 × 1 β structure connected by three turns is unstable, in spite of having the largest initial number of residues in a β conformation. In particular, the lateral β sheet stack (model 1), the β double hook (model 2), and the collapsed β helix (model 4) are more stable than the diagonal β sheet stack (model 3). These results confirm the previous conclusion that compact structures are more stable than the less compact ones. Out of the three stable models, the lateral β sheet stack (model 1) seems to be slightly "overly stable," as it is still stable with 33 residues, while the collapsed β helix (model 4) undergoes a sharp transition in its stability somewhere between 37 and 33 residues. Thus, it would appear that model 4 best describes the onset of a threshold around 35–36 residues.

12.3 Amyloid Models of Asparagine (N) and Glutamine(Q)

An important direction in the characterization of prions involves understanding the relative roles of the polar, uncharged residues Q and N in Q/N-rich yeast prions [44–47, 50]. For example, recent investigations showed that Q and N play opposing roles in the formation of the Sup35 PrD [47]. The experiments showed that if all the N's in the wild-type Sup35 PrD were switched to Q's, then amyloid formation decreased but cell toxicity increased. In turn, turning all Q's into N's enhanced the formation of benign, self-templating amyloids and decreased cell toxicity [47].

Currently, a detailed understanding of the mechanisms of polyQ toxicity is still elusive. Nevertheless, it is now possible to consider important structural, thermodynamic, and kinetic issues related to the amyloids themselves. For example, why do N-rich sequences form amyloid-like prions while Q-rich sequences do not? The results obtained for the Sup35 PrD [47] may be considered to be unexpected because it was previously held that Q's and N's are equivalent in terms of prion formation [55, 56]. Unfortunately, these experiments did not have an atomic resolution, and so subsequent Monte Carlo simulations were employed in the search for an atomistic explanation of these results. The results of these simulations of Q_{30} and N_{30} homopolymers suggested that the N-rich sequences increase the formation of β-hairpin turns and thus of β sheets, which in turn would lead to the formation of the N-rich amyloids.

Given this background, our investigations [50] are aimed at comparing the structural characteristic fibrils composed of Asn and Gln amino acids. To this end, we employed MD simulations that complement previous experimental and simulation studies on N/Q-rich amino acid sequences. We therefore explored all possible steric zipper interfaces for Q- and N-rich oligomers to determine which crystallographic class [57] provides the most stability to the N-rich or Q-rich fibril precursor. Before presenting our results, it is important to discuss the models considered, and the nomenclature used to describe them. These are given in Table 12.2. As in our work on polyQ protofibrils, the models are based on a set of $M \times N$ oligomeric crystals, with integer M giving the number of strands in a sheet and integer N the number of sheets in a given oligomer. The simplest structure we consider is that of a 2×1 homopolymer Ace–Asn_{23}–Nme and Ace–Gln_{23}–Nme corresponding to an antiparallel β-sheet connected by a hairpin. Most of the structures we consider here are based on 4×2 and 2×2 model aggregates which sort themselves into two main categories (4T, 4O and T2, 2O series) based on their side-chain packing. The actual aggregate structure is based on the classification proposed by Sawaya et al. [57], who proposed eight classes of rigid steric zippers according to the relative positions of the β strands and sheets. Structures are classified on the basis of whether the (i) sheets are made from parallel or antiparallel strands; whether the (ii) adjacent sheets are packed with the same ("face-to-face") or reverse ("face-to-back") surfaces; and whether the (iii) the symmetry operation used to generate the second sheet from the first preserves the order of the strands, such that the upper and lower strands in one sheet are still in that order ("up-up") in the second sheet, or inverts it ("up-down").

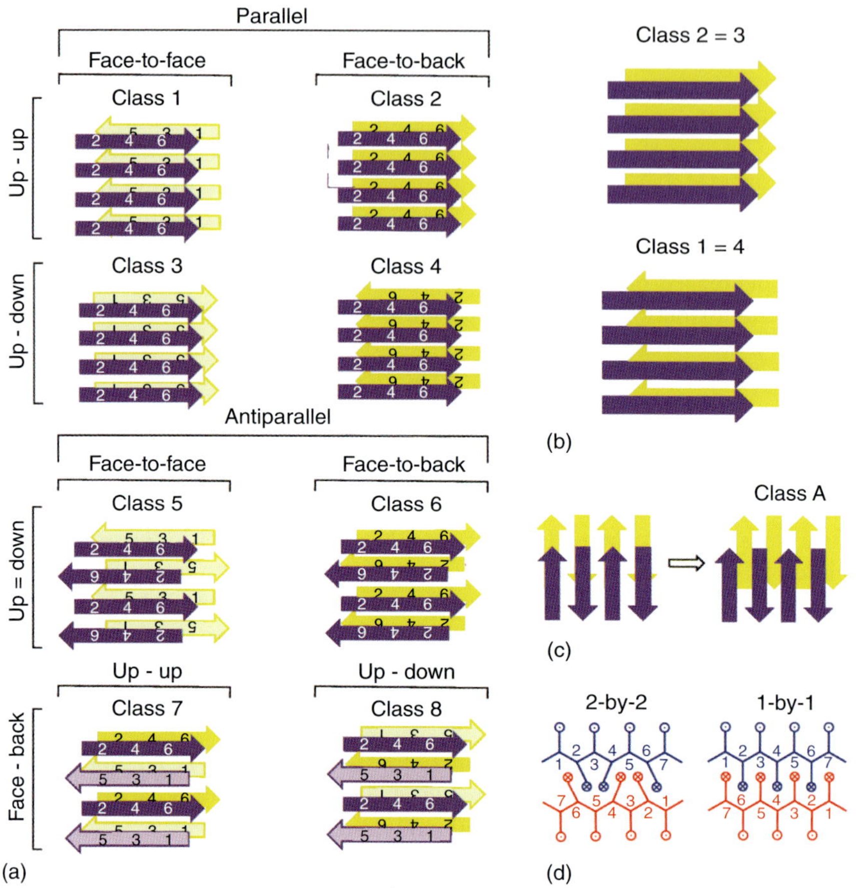

Figure 12.5 Shown here are: (a) the eight classes of steric zippers, following Sawaya et al. [57]. The sheet closer to the viewer is shown in purple, and the one behind in yellow. The solid colors show the "front side" of each sheet with the side chains identified by even numbers, and the faded colors show the "back side" of the sheets with odd-number side chains. For not too short homopeptides, some of the classes described in this figure converge; (b) for parallel sheets class 1 and 4 become the same (with sheets stacked in antiparallel fashion), and class 2 and 3 become the same (with sheets stacked in parallel fashion); (c) a schematic of the quarter-stagger displacement of one sheet with respect to the other. For antiparallel sheets undergoing a quarter-stagger, class 5–8 converge into class A (right); (d) side-chain interdigitation of steric zippers.

Schematics of the eight possible aggregate classes are shown in Figure 12.5. In addition, one also needs to take into account two more criteria based on the type of side-chain packing, as shown in the figure. The "1-by-1" packing is characterized by the side chains of two adjacent sheets intercalating with each other "one-by-one" as in a true zipper, while in the "2-by-2" packing, two side chains intercalate every second set of side chains. The "1-by-1" packing gives rise to the 4O and 2O sets of models, while the "2-by-2" packing is associated with the

4T and 2T sets. The number of steric zipper classes reduces considerably when homopeptides are considered. In this case, the "face-to-face" and "face-to-back" structures are the same for the PS, and the first four classes of Sawaya et al. [57] are reduced to two: aggregates with parallel-stranded sheets in which the sheets are either parallel (class 2 and class 3) or antiparallel (class 1 and class 4) with respect to each other. In class 1 = 4 (2 = 3), the carboxamide dipoles in polyN or polyQ are aligned at the interface between the two sheets (pointing in the direction opposite to those in the other sheet). For steric zippers with APS (class 5–8), homopeptide classes 6–8 reduce to a single structure, wherein the strands facing each other in opposite sheets point in the same direction (i.e. the two sheets are stacked in a parallel way), while class 5 (which so far has not been observed experimentally) corresponds to two sheets stacked in an antiparallel way. Furthermore, antiparallel-stranded sheets in polyQ tend to shift with respect to each other such that the strands of different sheets no longer lie on top of each other, but "in between" each other in a zipper-like manner. This is known as a "quarter stagger" in β-sheet crystals, and it has been observed both experimentally and in our polyQ simulations [32, 49]. In this case, all steric zippers involving antiparallel-stranded sheets (classes 5–8) reduce to a single structure, which we label as class A (for APS).

12.3.1 Initial Structures

Given this background, the specific models we studied are listed in Table 12.2. Models 4T1, 4T2, 4T3, and 3T4 are based on the X-ray crystal structure of the GNNQQNY peptide (PDB ID 1YJP) [58]. The model 4T1 corresponds exactly to the 1YJP crystal structure with eight GNNQQNY peptides forming two PSs with strands separated by 4.87 Å in each sheet. The sheets therefore reproduce the class 1 steric zipper with "2-by-2" side chain packing in a dry interface with an 8.5 Å separation between the sheets. To study the effects of Q versus N in the formation of Sup35 PrD, we then mutated all the Qs to Ns to obtain 4T2, which was constructed from GNNNNNY strands. We also mutated the Ns to Qs to obtain the 4T4 structure, which is based on GQQQQQY strands. Finally, in order to eliminate the effects of the terminal residues G and Y, we mutated these to N to obtain model 4T3 based on NNNNNNN strands. We also constructed one more structure 4T5, corresponding to the 2OL9 crystal structure with eight SNQNNF peptides forming two parallel sheets with strands separated by 4.79 Å in each sheet. The structure corresponds to class 2 of Sawaya et al. [57]. In order to account for side-chain packing based on "1-by-1" interdigitation, we also constructed the 4O series from pure polyQ and polyN strands made from seven residues as to be commensurate with the 4T series of models. We also constructed model 4O9 based on polyQ strands arranged in an antiparallel manner. We mention here that antiparallel models based on polyN turned out to be intrinsically unstable and therefore did not require further explicit examination. Finally, in order to gain further insight into the stability of these structures, we constructed a number of 2 × 2 models, also listed in Table 12.2.

Table 12.2 List of the initial structures and their characteristics.

System size	Model name	Sequence	Sheet	Class	Side chain
2×1 (hairpin)	Q_{23}	Ace-Gln_{23}-Nme	Antiparallel	—	—
	N_{23}	Ace-Asn_{23}-Nme	Antiparallel	—	—
4×2	4T1	GNNQQNY	Parallel	1	2-by-2
	4T2	GNNNNNY	Parallel	1	2-by-2
	4T3	NNNNNNN	Parallel	1	2-by-2
	4T4	GQQQQQY	Parallel	1	2-by-2
	4T5	SNQNNF	Parallel	2	2-by-2
4×2	4O1	NNNNNNN	Parallel	1	1-by-1
	4O2	QQQQQQQ	Parallel	1	1-by-1
	4O3	NNNNNNN	Parallel	4	1-by-1
	4O4	QQQQQQQ	Parallel	4	1-by-1
	4O5	NNNNNNN	Parallel	3	1-by-1
	4O6	QQQQQQQ	Parallel	3	1-by-1
	4O7	NNNNNNN	Parallel	2	1-by-1
	4O8	QQQQQQQ	Parallel	2	1-by-1
	4O9	NNNNNNN	Antiparallel	A	1-by-1
	4O10	QQQQQQQ	Antiparallel	A	1-by-1
2×2	2T1	GNNQQNY	Parallel	1	2-by-2
	2T2	GNNNNNY	Parallel	1	2-by-2
	2T3	NNNNNNN	Parallel	1	2-by-2
	2O2	QQQQQQQ	Parallel	1	1-by-1
	2O4	QQQQQQQ	Parallel	4	1-by-1
	2O9	NNNNNNN	Antiparallel	A	1-by-1
	2O10	QQQQQQQ	Antiparallel	A	1-by-1

12.3.2 Monomeric PolyQ β Hairpins Are More Stable than PolyN Hairpins

Having outlined the models investigated, we turn to the results. Our first simulation result shows that monomeric polyQ β-hairpins are more stable than monomeric polyN β-hairpins. Our MD simulations show that after about 100 ns Q_{23} (results not shown explicitly here) still exhibits a considerable β-sheet structure, while the comparable N_{23} hairpin has evolved into a completely disordered coil. This is an important result, and is in contradiction to the selected previous results Ref. [47], which deemed a N_{30} β-hairpin to be more well defined (leading to a much better antiparallel β sheet) than the corresponding Q-based structure. The previous simulation results [47] were attributed to the intrinsic difficulty of polyQ sequences to form tight hairpin turns. It is important to note, however, that our simulation results are in better agreement with some of the more recent data. In particular, tight hairpin turns in polyQ have been recently observed experimentally [35, 36]. Also, tight β hairpins were proposed and used [32] to successfully model the X-ray diffraction patterns obtained from crystalline fibers and films of a $D_2Q_{15}K_2$ polypeptide [28]. In our own work, we have measured a nonnegligible population of loose turns both in polyQ [13] and in polyN [59], while other simulations have also found polyQ β hairpins [60]. As discussed in the previous section, a Q_{40} monomer (which is above the aggregation threshold for polyQ) containing two tight hairpin turns and an arc turned out to be remarkably stable over a timescale longer than 1 μs [49].

12.3.3 N-rich Oligomers Are Most Stable in Class 1 Steric Zippers with 2-by-2 Interdigitation

Turning to the 4×2 structures, we find that N-rich oligomers are most stable in class 1 steric zipper configurations with 2-by-2 interdigitation. This is readily seen from the top row of Figure 12.6, which shows initial structures and final configurations after 400 ns. Model 4T1 is based on the experimental PDB 1YJP structure having a sequence of GNNQQNY, while 4T5 is based on the PDB 2OL9 crystal structure with sequence SNQNNF. These models (4T1–4T4) all correspond to class 1, with parallel stranded sheets with face-to-face and up-up packing between the sheets. Only model 4T5 corresponds to class 2: parallel stranded sheets with face-to-back and up-up packing between sheets. Figure 12.6 shows that N-rich sequences in class 1 structures are more stable than the corresponding Q-rich structures. Thus, models 4T1, 4T2, and 4T3 all retain considerable β-sheet structure, while model 4T4 is completely disordered. Model 4T5 was also found to be completely unstable. Because this model unraveled so rapidly, we believe that other class 2 structures based on polyQ and polyN will be similarly unstable.

These results are in good agreement with previous data. The stability of the heptapeptide GNNQQNY in model 4T1 (PDB ID 1YJP [58]) has been confirmed by previous simulations [61, 62]. Here, the in-register PP of GNNQQNY is consistent with the observed X-ray diffraction data; similarly, it has been noted that GNNQQNY exhibits the same amyloid properties as the full-length Sup35, as seen in the cooperative kinetics of aggregation, the binding to the dye Congo red, fibril development, and in the typical cross-β diffraction pattern [58, 63] of the structure. The surfaces of the two sheets are complementary, resulting in a dry steric zipper with tightly interdigitated side chains with the zipper stabilized by van der Waals interactions and hydrogen bonds between the polar groups. Although the π stacking of the aromatic Tyr–Tyr interactions [61] in GNNQQNY plays a role in stabilizing the parallel arrangement of the strands, the pure polyN model 4T3 also favors the formation of this kind of zipper structure. Previously, two single strands of polyN or polyQ homopeptides were shown to assemble in dimers that favor parallel (polyN) or antiparallel (polyQ) β sheets [60]. In addition, solid-state NMR experiments show that the amyloid of the full prion domain of Sup35p (residues 1–123) has an in-register parallel β-sheet structure [64].

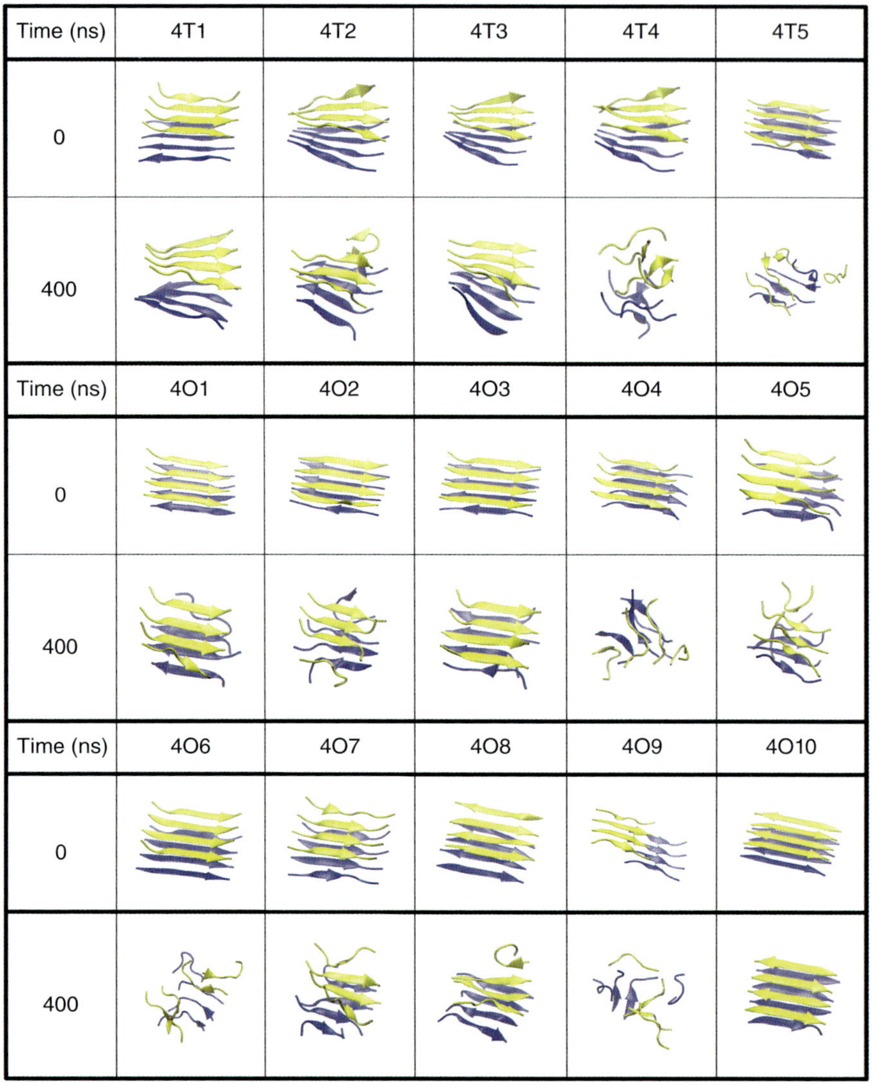

Figure 12.6 For each 4 × 2 model labeled in the figure, sample initial configurations after equilibration are shown on the top and sample final configurations after 400 ns are shown underneath.

12.3.4 PolyQ Oligomers Are Most Stable in Antiparallel Stranded β Sheets with 1-by-1 Steric Zippers

Turning to structures with one-by-one interdigitation, we find that polyQ oligomers are most stable in antiparallel stranded β sheets. Again, initial and final configurations after 400 ns MD simulations are shown in Figure 12.6. Here, the structures fall into three different classes: class 1 = 4 and class 2 = 3 based on a parallel arrangement of strands, and class A involving an antiparallel arrangement. The results show that all classes with one-by-one side chains are unstable for polyN. For polyQ, models in with steric zippers in class 1 and 4 (models 4O2 and 4O4) are more stable than class 2 and 3 (models 4O6 and 4O8), indicating that an APP of the parallel-stranded β sheets is the favored configuration. Lastly, polyQ with one-by-one side-chain packing with class A steric zippers (model 4O10) appears to be very stable. We also note polyN structures in class A (model 4O9) are also stable, but less so than model 4T3 as evidenced by the higher β sheet content of the latter. These results are in complete agreement with current evidence based on polyQ fibers and crystallites supporting the presence of cross-β structures with antiparallel β sheets [31, 32, 35, 49].

12.3.5 PolyQ Structures Show Higher Stability than Most Stable PolyN Structures

In terms of the relative stability of polyQ versus polyN structures, we find that polyQ structures show a higher stability. As previously noted, the stability of the structures increases with matrix order. Hence, to get further insight into the relative stability, we took the most stable 4×2 aggregates (for polyN the 2-by-2 class 1 structures 4T1, 4T2, 4T3, and the 1-by-1 class A model 4O9; and for polyQ the 1-by-1 structures 4O2, 4O4, and the class A 4O10) and used them to construct corresponding 2×2 models. Results are shown in Figure 12.7. Clearly, all the polyN structures unravel rather quickly and are therefore shown to be unstable. For polyQ, the 1-by-1 class 1 (2O2) and class 4 (2O4) models also become unstable. Only the 1-by-1 class A (2O10) model remains very stable.

To further elucidate this point, we used a recently developed "laser stability" method to probe the relative structural stability of the most stable aggregates [50]. The idea behind this is as follows. Free electron lasers have been used recently for melting biomolecular complexes [65, 66] Specifically, Kawasaki et al. [66] used a mid-infrared free electron laser with highly specific oscillation characteristics having a high photon density, a picosecond pulse structure, and a range of tunable frequencies for this. Such a laser pulse with frequencies tuned to the amide I bands has been used in experiments to dissociate amyloid-like fibrils of lysozyme into its native forms [66], convert insulin fibrils into soluble monomers, and dissociate a short human thyroid hormone peptide. A simulated laser pulse has also

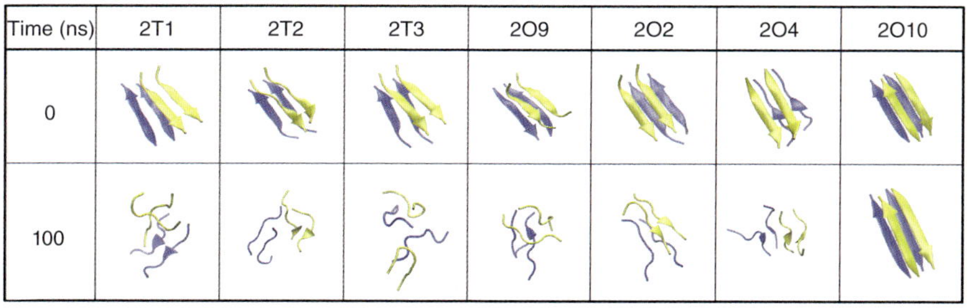

Figure 12.7 For each 2×2 model labeled in the figure, sample initial configurations after equilibration are shown on the top and sample final configurations after 100 ns are shown underneath. Top row corresponds to polyN aggregates and bottom row corresponds to polyQ aggregates.

been used in atomistic MD simulations to dissociate amyloid fibrils [67], and to study the breakup of a peptide-based nanotube [68]. In order to corroborate the results obtained for the 2×2 aggregates, we applied the laser melting techniques to the most stable 4×2 homopeptide aggregates. For polyN, this is class 1 model 4T3; and for polyQ this is class A model 4O10. Simulations were carried out using a laser frequency of 1671 cm^{-1}, which specifically targets the C=O main-chain bonds, thereby destabilizing the β sheets. Applying this laser pulse results in an almost identical energy absorption for the 4T3 and 4O10 aggregates. Thus, the initial application of the laser pulse induces a disruption of the β sheet hydrogen bonds, followed by a period of partial healing to the initial structure. Several different measures of the structural integrity show that, for different values of the electric field amplitude, there is a greater disruption of the most stable 4T3 polyN structure versus that of the most stable 4O10 polyQ structure. This provides further evidence the polyQ aggregates are more stable than the corresponding polyN ones.

12.3.6 Thermodynamic Considerations of Aggregate Formation

Good estimates of the enthalpic contribution to the free energy of the aggregates may be obtained by computing the energy difference between the aggregate and the monomers via $\Delta H = H_{\text{crystal}} - 8H_{\text{monomer}}$, where H_{crystal} is the enthalpy of the 4×2 homopeptide after equilibration and H_{monomer} the average enthalpy of an unfolded monomer. In the context of the generalized Born approximation for the solvation, we estimate $\Delta H_{\text{Q}} = -14.5$ kcal/(mol residue) in the 4O10 model and $\Delta H_{\text{N}} = -4.4$ kcal/(mol residue) for polyN's 4T3 model. Clearly, enthalpy contributions favor polyQ over polyN. The enthalpy reflects not only the main-chain hydrogen bonds but also the side-chain packing and resultant hydrogen bonds between the side chains themselves and with the main chains. The analysis shows that the shorter side-chain residues for N favor 2-by-2 side-chain packing because they form polygon structures that increase the stability of neighboring sheets, while the longer side-chain Q residues make use of side-chain/main-chain hydrogen bonds through 1-by-1 side-chain packing to form a very stable hydrogen bond network [50].

As noted, strong enthalpic terms favor the formation of polyQ aggregates. What about the entropic contributions to the free energy of zipper formation? The key difference between Q and N lies in the presence of an additional methylene group in the side chain of Q. This allows for a quick estimate of entropic free energy contribution. Asn has 2 side-chain torsion angles, while Gln has 3. The 2002 Rotamer Library [69] defines the degrees of freedom of rotameric torsions as 3 and nonrotameric torsions as 6 for N and Q, respectively. The total number of rotamers for the side chains of N and Q therefore becomes 18 and 54 for N and Q in their unfolded states. Since in the aggregate form the side chains are locked into one conformer, the total number of rotamers in the protofibril structure for both peptides is unity. Thus, the entropic loss between a side chain locked in a fibril and a free side chain can be estimated as $\Delta S = \Delta S_{\text{fibril}} - \Delta S_{\text{free}} = -R\ln(N_{\text{D}}/1)$, where N_{D} represents the number of degrees of freedom. At room temperature, this gives an entropy difference favoring the disordered state of $\Delta G_Q = -T\Delta S_Q = 2.4$ kcal/(mol residue) and $\Delta G_N = -T\Delta S_N = 1.73$ kcal/(mol residue). This estimate does not consider the degrees of freedom of the backbones, whose contribution to the free energies is probably similar for polyN and polyQ [60]. Thus, entropic effects are less favorable to the polyQ steric zippers, as disordered polyQ chains have a more extensive phase space to sample (for n_R residues, the ratio of phase spaces between polyQ and polyN grows as 3^{n_R}) before settling in the preferred conformation. There may even be considerably larger entropic barriers associated with polyQ, which would explain the very long times observed experimentally for the formation of polyQ amyloids [47, 70].

To summarize, we have carried out comprehensive MD investigations of all the possible steric interfaces of N/Q-rich peptides forming amyloids, focusing on sequences as found in the yeast Sup35 prion domain. We have probed the structural and thermodynamic stability of the aggregates in prion-like proteins. We find that different crystal classes maximize the stability of N- and Q-rich aggregates. Specifically, N-rich oligomers are most stable in parallel-stranded β sheets with 2-by-2 side-chain interdigitation. Q-rich oligomers, on the other hand, are most stable in antiparallel stranded β sheets with 1-by-1 steric zippers. Overall, polyQ oligomers are more stable than polyN oligomers. Enthalpic contributions to the free energy strongly favor the formation of polyQ protofibrils, while entropic contributions favor polyN. The considerably larger phase space that polyQ must sample during aggregation is probably the reason for the slower kinetics associated with the formation of polyQ (as opposed to polyN) aggregates.

12.4 Summary

Sequences rich in Q and N are intrinsically disordered in their monomeric form, but can aggregate into highly ordered amyloids. In particular, polyQ amyloid aggregates are characteristic of a class of neurodegenerative diseases associated with the expansion of the CAG codon. In fact, these neurodegenerative diseases are part of a larger class of diseases associated with trinucleotide repeat expansion diseases or TREDs. The sequence repeat expansion associated with these is believed to be caused by some sort of slippage during DNA replication, repair, recombination, or transcription. An important breakthrough in the understanding of TREDs has been the recognition that stable, atypical DNA secondary structure in the expanded repeats is the causative factor behind these human diseases. Hence, an important research goal aimed at understanding these diseases involves the structural characterization of the atypical DNA and/or RNA [71, 72].

In this short review chapter, we have focused not on the nucleotide structure *per se*, but rather on the structure of Q homopolymers that result from the translation of CAG repeats. Once the translated polyglutamine expansion becomes longer than about 36 residues, it triggers the formation of interneural protein aggregates that display the characteristics of β amyloid fibrils, that self-propagate through protein-conformational chain reactions. Amyloids are also characteristic of Q/N-rich prion domains as found in, for example, the yeast sup35 prion domain. We have used fully atomistic MD simulations to probe the structural characteristics, the stability and conformational dynamics of previously proposed polyQ and polyN aggregate models. We have tested the relative stability of parallel and antiparallel β sheets, their possible steric interfaces, and the effects of side-chain dipole alignments. The results are in good agreement with the experimental data. We have also probed Q_{40} monomeric models with different combinations of hairpin and arcs, and find that stable models are completely consistent with the experimental aggregation threshold. In terms of the characteristics of polyQ versus polyN aggregates, the results indicate that polyQ aggregates are the most stable. Enthalpic contributions to the free energy favor the formation of polyQ protofibrils, while entropic contributions favor polyN. The larger phase space that polyQ must sample as it aggregates is the most likely cause of the associated slower kinetics observed experimentally.

Acknowledgments

We gratefully acknowledge the financial support provided by NSF grants SSE-1534941 and XSEDE TG-MCB160064.

References

1 Zoghbi, H.Y. and Orr, H.T. (2000). Glutamine repeats and neurodegeneration. *Annu. Rev. Neurosci.* 23 (1): 217–247. doi: 10.1146/annurev.neuro.23.1.217.
2 Kayed, R., Head, E., Thompson, J.L. et al. (2003). Common structure of soluble amyloid oligomers implies common mechanism of pathogenesis. *Science* 300 (5618): 486–489.
3 Kar, K., Jayaraman, M., Sahoo, B. et al. (2011). Critical nucleus size for disease-related polyglutamine aggregation is repeat-length dependent. *Nat. Struct. Mol. Biol.* 18 (5): 328–336.
4 Scherzinger, E., Sittler, A., Schweiger, K. et al. (1999). Self-assembly of polyglutamine-containing Huntingtin fragments into amyloid-like fibrils: implications for Huntington's disease pathology. *Proc. Natl. Acad. Sci. U.S.A.* 96 (8): 4604–4609.
5 Chen, S., Berthelier, V., Yang, W., and Wetzel, R. (2001). Polyglutamine aggregation behavior in vitro supports a recruitment mechanism of cytotoxicity. *J. Mol. Biol.* 311 (1): 173–182.
6 Mangiarini, L., Sathasivam, K., Seller, M. et al. (1996). Exon 1 of the HD gene with an expanded CAG repeat is sufficient to cause a progressive neurological phenotype in transgenic mice. *Cell* 87 (3): 493–506. doi: 10.1016/s0092-8674(00)81369-0.
7 DiFiglia, M., Sapp, E., Chase, K.O. et al. (1997). Aggregation of Huntingtin in neuronal intranuclear inclusions and dystrophic neurites in brain. *Science* 277 (5334): 1990–1993. doi: 10.1126/science.277.5334.1990.

8 Poirier, M.A., Li, H., Macosko, J. et al. (2002). Huntingtin spheroids and protofibrils as precursors in polyglutamine fibrilization. *J. Biol. Chem.* 277 (43): 41032–41037.

9 Wacker, J.L., Zareie, M.H., Fong, H. et al. (2004). Hsp70 and Hsp40 attenuate formation of spherical and annular polyglutamine oligomers by partitioning monomer. *Nat. Struct. Mol. Biol.* 11: 1215–1222.

10 Lee, C.C., Walters, R.H., and Murphy, R.M. (2007). Reconsidering the mechanism of polyglutamine peptide aggregation. *Biochemistry* 46 (44): 12810–12820.

11 Scherzinger, E., Lurz, R., Turmaine, M. et al. (1997). Huntingtin-encoded polyglutamine expansions form amyloid-like protein aggregates in vitro and in vivo. *Cell* 90: 549–558.

12 Chen, S., Ferrone, F.A., and Wetzel, R. (2002). Huntington's disease age-of-onset linked to polyglutamine aggregation nucleation. *Proc. Natl. Acad. Sci. U.S.A.* 99 (18): 11884–11889.

13 Moradi, M., Babin, V., Roland, C., and Sagui, C. (2012). Are long-range structural correlations behind the aggregation phenomena of polyglutamine diseases? *PLoS Comput. Biol.* 8: e1002501.

14 Thakur, A.K. and Wetzel, R. (2002). Mutational analysis of the structural organization of polyglutamine aggregates. *Proc. Natl. Acad. Sci. U.S.A.* 99 (26): 17014–17019.

15 Muchowski, P.J. and Wacker, J.L. (2005). Modulation of neurodegeneration by molecular chaperones. *Nat. Rev. Neurosci.* 6: 11–22.

16 Nagai, Y., Inui, T., Popiel, H.A. et al. (2007). A toxic monomeric conformer of the polyglutamine protein. *Nat. Struct. Mol. Biol.* 14: 332–340.

17 Miller, J., Arrasate, M., Shaby, B.A. et al. (2010). Quantitative relationships between huntingtin levels, polyglutamine length, inclusion body formation, and neuronal death provide novel insight into Huntington's disease molecular pathogenesis. *J. Neurosci.* 30 (31): 10541–10550.

18 Arrasate, M., Mitra, S., Schweitzer, E.S. et al. (2004). Inclusion body formation reduces levels of mutant huntingtin and the risk of neuronal death. *Nature* 431: 805–810.

19 Kopito, R. (2000). Aggresomes, inclusion bodies and protein aggregation. *Trends Cell Biol.* 10 (12): 524–530.

20 Ross, C.A. and Poirier, M.A. (2005). What is the role of protein aggregation in neurodegeneration? *Nat. Rev. Mol. Cell Biol.* 6 (11): 891–898.

21 Bodner, R.A., Outeiro, T.F., Altmann, S. et al. (2006). Pharmacological promotion of inclusion formation: a therapeutic approach for Huntington's and Parkinson's diseases. *Proc. Natl. Acad. Sci. U.S.A.* 103 (11): 4246–4251.

22 Waelter, S., Boeddrich, A., Lurz, R. et al. (2001). Accumulation of mutant huntingtin fragments in aggresome-like inclusion bodies as a result of insufficient protein degradation. *Mol. Biol. Cell* 12: 1393–1407.

23 Taylor, J.P., Tanaka, F., Robitschek, J. et al. (2003). Aggresomes protect cells by enhancing the degradation of toxic polyglutamine-containing protein. *Hum. Mol. Genet.* 12 (7): 749–757.

24 Glabe, C.G. and Kayed, R. (2006). Common structure and toxic function of amyloid oligomers implies a common mechanism of pathogenesis. *Neurology* 66 (1, Suppl. 1): S74–S78.

25 Armen, R.S., Bernard, B.M., Day, R. et al. (2005). Characterization of a possible amyloidogenic precursor in glutamine-repeat neurodegenerative diseases. *Proc. Natl. Acad. Sci. U.S.A.* 102 (38): 13433–13438.

26 Tomita, K., Popiel, H.A., Nagai, Y. et al. (2011). Structure-activity relationship study on polyglutamine binding peptide QBP1. *Bioorg. Med. Chem.* 17: 1259–1263.

27 Fiumara, F., Fioriti, L., Kandel, E.R., and Hendrickson, W.A. (2010). Essential role of coiled coils for aggregation and activity of Q/N-rich prions and PolyQ proteins. *Cell* 143 (7): 1121–1135. doi: 10.1016/j.cell.2010.11.042.

28 Perutz, M.F., Finch, J.T., Berriman, J., and Lesk, A. (2002). Amyloid fibers are water-filled nanotubes. *Proc. Natl. Acad. Sci. U.S.A.* 99 (8): 5591–5595. doi: 10.1073/pnas.042681399.

29 Daggett, V. (2006). α-sheet: the toxic conformer in amyloid diseases? *Acc. Chem. Res.* 39: 594–602.

30 Babin, V., Roland, C., and Sagui, C. (2011). The α-sheet: a missing-in-action secondary structure? *Proteins Struct. Funct. Bioinf.* 79: 937–946.

31 Sharma, D., Shinchuk, L.M., Inouye, H. et al. (2005). Polyglutamine homopolymers having 8-45 residues form slablike beta-crystallite assemblies. *Proteins* 61 (2): 398–411. doi: 10.1002/prot.20602.

32 Sikorski, P. and Atkins, E. (2005). New model for crystalline polyglutamine assemblies and their connection with amyloid fibrils. *Biomacromolecules* 6 (1): 425–432. doi: 10.1021/bm0494388.

33 Makin, O.S., Atkins, E., Sikorski, P. et al. (2005). Molecular basis for amyloid fibril formation and stability. *Proc. Natl. Acad. Sci. U.S.A.* 102 (2): 315–320.

34 Schneider, R., Schumacher, M.C., Mueller, H. et al. (2011). Structural characterization of polyglutamine fibrils by solid-state NMR spectroscopy. *J. Mol. Biol.* 412 (1): 121–136. doi: 0.1016/j.jmb.2011.06.045.

35 Buchanan, L.E., Carr, J.K., Fluitt, A.M. et al. (2014). Structural motif of polyglutamine amyloid fibrils discerned with mixed-isotope infrared spectroscopy. *Proc. Natl. Acad. Sci. U.S.A.* 111 (11): 5796–5801. doi: 10.1073/pnas.1401587111.

36 Kar, K., Hoop, C.L., Drombosky, K.W. et al. (2013). β-hairpin-mediated nucleation of polyglutamine amyloid formation. *J. Mol. Biol.* 425 (7): 1183–1197. doi: 10.1016/j.jmb.2013.01.016.

37 Nelson, R., Sawaya, M.R., Balbirnie, M. et al. (2005). Structure of the cross-beta spine of amyloid-like fibrils. *Nature* 435 (7043): 773–778. doi: 10.1038/nature03680.

38 Prusiner, S.B. (1998). Prions. *Proc. Natl. Acad. Sci. U.S.A.* 95: 13363–13383.

39 True, H.L. and Lindquist, S.L. (2000). A yeast prion provides a mechanism for genetic variation and phenotypic diversity. *Nature* 407 (6803): 477–483.

40 Halfmann, R. and Lindquist, S. (2010). Epigenetics in the extreme: prions and the inheritance of environmentally acquired traits. *Science* 330 (6004): 629–632.

41 Coustou, V., Deleu, C., Saupe, S., and Bequeret, J. (1997). The protein product of the het-s heterokaryon incompatibility gene of the fungus podospora anserina behaves as a prion analog. *Proc. Natl. Acad. Sci. U.S.A.* 94 (18): 9773–9778.
42 Sawyer, E.B., Claessen, D., Haas, M. et al. (2011). The assembly of individual chaplin peptides from Streptomyces coelicolor into functional amyloid fibrils. *PLoS ONE* 6 (4): e18839.
43 Barnhart, M.M. and Chapman, M.R. (2006). Curli biogenesis and function. *Annu. Rev. Microbiol.* 60: 131–147.
44 Toombs, J.A., McCarty, B.R., and Ross, E.D. (2010). Compositional determinants of prion formation in yeast. *Mol. Cell. Biol.* 30 (1): 319–332.
45 Peters, T.W. and Huang, M. (2007). Protein aggregation and polyasparagine-mediated cellular toxicity in Saccharomyces cerevisiae. *Prion* 1 (2): 144–153.
46 Michelitsch, M.D. and Weissman, J.S. (2000). A census of glutamine/asparagine-rich regions: implications for their conserved function and the prediction of novel prions. *Proc. Natl. Acad. Sci. U.S.A.* 97 (22): 11910–11915.
47 Halfmann, R., Alberti, S., Krishnan, R. et al. (2011). Opposing effects of glutamine and asparagine govern prion formation by intrinsically disordered proteins. *Mol. Cell* 43 (1): 72–84.
48 Buchete, N.V., Tycko, R., and Hummer, G. (2005). Molecular dynamics simulations of Alzheimer's b-amyloid protofilaments. *J. Mol. Biol.* 353 (4): 804–821. doi: 10.1016/j.jmb.2005.08.066.
49 Man, V.H., Roland, C., and Sagui, C. (2015). Structural determinants of polyglutamine protofibrils and crystallites. *ACS Chem. Neurosci.* 6 (4): 632–645.
50 Zhang, Y., Man, V.H., Roland, C., and Sagui, C. (2016). Amyloid properties of asparagine and glutamine in prion-like proteins. *ACS Chem. Neurosci.* 7 (5): 576–587.
51 Fraser, R. (1973). *Conformation in Fibrous Proteins and Related Synthetic Polypeptides.* Academic Press.
52 Krull, L.H., Wall, J.S., Zobel, H., and Dimler, R.J. (1965). Synthetic polypeptides containing side-chain amide groups: water-insoluble polymers. *Biochemistry* 4 (4): 626–633. doi: 10.1021/bi00880a003.
53 Perutz, M.F., Johnson, T., Suzuki, M., and Finch, J.T. (1994). Glutamine repeats as polar zippers: their possible role in inherited neurodegenerative diseases. *Proc. Natl. Acad. Sci. U.S.A.* 91 (12): 5355–5358. doi: 10.1073/pnas.91.12.5355.
54 Darnell, G., Orgel, J.P., Pahl, R., and Meredith, S.C. (2007). Flanking polyproline sequences inhibit beta-sheet structure in polyglutamine segments by inducing PPII-like helix structure. *J. Mol. Biol.* 374 (3): 688–704. doi: 10.1016/j.jmb.2007.09.023.
55 Si, K., Lindquist, S.L., and Kandel, E.R. (2003). A neuronal isoform of the aplysia CPEB has prion-like properties. *Cell* 115 (7): 879–891.
56 Ross, E.D., Minton, A., and Wickner, R.B. (2005). Prion domains: sequences, structures and interactions. *Nat. Cell Biol.* 7: 1039–1044.

57 Sawaya, M.R., Sambashivan, S., Nelson, R. et al. (2007). Atomic structure of amyloid cross-β spines reveal varied steric zippers. *Nature* 447 (24): 453–457.
58 Nelson, R., Sawaya, M.R., Balbirnie, M. et al. (2005). Structure of the cross-β spine of amyloid-like fibrils. *Nature* 435: 773–778.
59 Zhang, Y. and Sagui, C. (2015). Secondary structure assignment for conformationally irregular peptides: comparison between DSSP, STRIDE and KAKSI. *J. Mol. Graph. Modell.* 55: 72–84.
60 Nakano, M., Ebina, K., and Tanaka, S. (2013). Study of the aggregation mechanism of polyglutamine peptides using replica exchange molecular dynamics simulations. *J. Mol. Model.* 19 (4): 1627–1639.
61 Gsponer, J., Haberthür, U., and Caflisch, A. (2003). The role of side-chain interactions in the early steps of aggregation: molecular dynamics simulations of an amyloid-forming peptide from the yeast prion sup35. *Proc. Natl. Acad. Sci. U.S.A.* 100 (9): 5154–5159.
62 Vitagliano, L., Esposito, L., Pedone, C., and Simone, A.D. (2008). Stability of single sheet GNNQQNY aggregates analyzed by replica exchange molecular dynamics: antiparallel versus parallel association. *Biochem. Biophys. Res. Commun.* 377 (4): 1036–1041.
63 Balbirnie, M., Grothe, R., and Eisenberg, D.S. (2001). An amyloid-forming peptide from the yeast prion sup35 reveals a dehydrated β-sheet structure for amyloid. *Proc. Natl. Acad. Sci. U.S.A.* 98 (5): 2375–2380.
64 Shewmaker, F., Wickner, R.B., and Tycko, R. (2006). Amyloid of the prion domain of sup35p has an in-register parallel β-sheet structure. *Proc. Natl. Acad. Sci. U.S.A.* 103 (52): 19754–19759.
65 Yagi, H., Ozawa, D., Sakuri, K. et al. (2010). Laser-induced propagation and destruction of amyloid beta fibrils. *J. Biol. Chem.* 285: 19660.
66 Kawasaki, T., Fujioka, J., Imai, T., and Tsukiyama, K. (2012). Effect of mid-infrared free-electron laser irradiation on refolding of amyloid-like fibrils of lysozyme into native form. *Protein J.* 31: 710.
67 Man, V.H., Derreumaux, P., Li, M.S. et al. (2015). Picosecond dissociation of amyloid fibrils with infrared laser: a nonequilibrium simulation study. *J. Chem. Phys.* 143 (15): 155101.
68 Man, V., Truong, P., Derreumaux, P. et al. (2015). Picosecond melting of peptide nanotube with infrared laser: a nonequilibrium study. *Phys. Chem. Chem. Phys.* 17 (41): 27275–27280.
69 Dunbrack, R.L. Jr. (2002). Rotamer libraries in the 21st century. *Curr. Opin. Struct. Biol.* 12 (4): 431–440.
70 Lu, X. and Murphy, R.M. (2015). Asparagine repeat peptides: aggregation kinetics and comparison with glutamine repeats. *Biochemistry* 54: 4784–4794.
71 Pan, F., Man, V.H., Roland, C., and Sagui, C. (2017). Structure and dynamics of DNA and RNA double helices of CAG and GAC trinucleotide repeats. *Biophys. J.* 113: 19.
72 Zhang, Y., Roland, C., and Sagui, C. (2017). Structure and dynamics of DNA and RNA double helices obtained from the GGGGCC and CCCCGG hexanucleotide repeats that are the hallmark of C9FTD/ALS diseases. *ACS Chem. Neurosci.* 8: 578.

13

Using Biomolecular Simulations to Target Cdc34 in Cancer

Miriam Di Marco, Matteo Lambrughi, and Elena Papaleo

Danish Cancer Society Research Center, Computational Biology Laboratory, Strandboulevarden, 49 DK-2100 Copenhagen, Denmark

13.1 Background

One of the hallmarks of several cancers is an uncontrollable growth and proliferation of the cells, as a result of severe defects in the cell cycle regulation. In normal cells, the cell cycle is generally regulated through a delicate balance between [1] the synthesis of proteins involved in cell cycle progression (cyclins, cyclin-dependent kinases (CDKs), and kinase inhibitors) and those that regulate their degradation [2]. The ubiquitin-mediated proteolysis pathway (ubiquitination) is the process responsible for the degradation of essential cell cycle proteins, thereby regulating cellular development and preventing cancerous growth. With the multitude of substrates targeted and the myriad of processes involved, aberrations in ubiquitination can be related to many different cancer types [3–7].

At the molecular level, the process of ubiquitination is carried out by a cascade of three classes of enzymes (Figure 13.1). In detail, it is initiated by the transfer of ubiquitin (Ub) to an E1-activating enzyme. The Ub molecule is then transferred to any one of ~40 possible E2 Ub-conjugating enzymes forming a highly unstable Ub~E2 thiolester intermediate. Labeling of a substrate protein occurs at selected lysine residues by Ub transfer from the E2~Ub intermediate via an E3 ligase enzyme and when proteasomal degradation is the final destination the cross-linking occurs on Lys48 of Ub. This process is repeated many times until a poly-Ub chain that commonly targets a substrate for degradation by the 26S proteasome is formed [8, 9].

E2 enzymes are especially central to the ubiquitination pathway [10, 11]. Indeed, they are essential for the specificity toward a particular substrate, as well as responsible for the processivity of the reaction and the topology of the Ub chain [12–15]. Some E2s are also directly catalytically active on the substrates even in the absence of E3 enzymes [16]. Virtually all the E2-mediated processes rely on protein–protein interactions with different ranges of affinity and complexity that involve all the other partners of the cascade, i.e. E1–E2, E2–E3, E2–Ub, and E2–substrate. Thus, E2 enzymes are highly sensitive to

Biomolecular Simulations in Structure-Based Drug Discovery,
First Edition. Edited by Francesco L. Gervasio and Vojtech Spiwok.

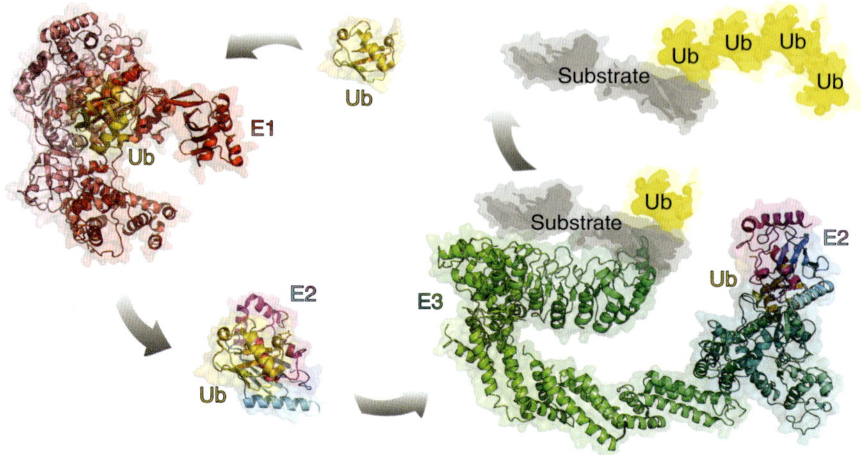

Figure 13.1 *A schematic view of the ubiquitination cascade*. The different steps for the transferring of Ub to the substrate are shown, as well as the involvement and architecture of E1, E2, and E3 classes of enzymes of the cascade.

allosteric regulation, as a result also of their modular three-dimensional (3D) structure composed of different domains [17].

Among the known 40 E2 enzymes, the human E2 enzyme Cdc34 controls the degradation of p27 [18–20], which in turn is the key inhibitor for the Cdk2, regulating its phosphorylation activity, a trigger that initiates DNA synthesis after G1 phase. In prostate, breast, and colorectal cancers, Cdk2 is overactive as a result of Cdc34 overexpression and increased ubiquitination of p27 [19, 21]. Cdc34 is thus an interesting target for therapeutic purposes [22]. The yeast orthologue Cdc34 is also used as a model system in many studies [23–28]. Nevertheless, without a clear understanding of the molecular and structural mechanisms behind Cdc34 action, it becomes difficult to design and develop suitable strategies to target it and modulate its functions in aberrant conditions. The study of Cdc34 is complicated by a complex posttranslational regulatory layer where multiple phosphorylations can target Cdc34 at different locations and with different effects, as reviewed in [29]. Moreover, Cdc34 is a multi-domain protein with large disordered regions, suggesting a large heterogeneous conformational ensemble. In this context, computational techniques that account simultaneously for protein dynamics and structure, as well as that allow modeling in fine detail the impact of mutations or modifications have been emerging as useful tools in structural biology and drug discovery [30–33]. In particular, a class of methods based on the all-atom classical molecular dynamics (MD) framework allows investigation of a highly dynamic protein such as Cdc34 and can also be used in the context of drug design. Recently, many efforts have been devoted by our colleagues and us to study with molecular simulations the Cdc34 structure and dynamics [23–26, 34], and their overview and the potential relevance for future research in biomedicine are the main subjects of this chapter. These studies could inspire more general applications of molecular simulations to the study of E2 or other similar protein–protein interactions that can be targeted for drug design.

13.2 Families of E2 Enzymes

In the first step of ubiquitin activation, the E1 enzyme binds ATP, Mg^{2+}, and ubiquitin, and catalyzes ubiquitin C-terminal acyl-adenylation. In the second step (Figure 13.1), the catalytic Cys in the E1 attacks the Ub-adenylate to form the activated Ub~thiolester-bonded complex – the tilde represents a high-energy bond between the C-terminus of Ub and AMP or with the conjugation machinery, whereas an en dash (–) represents a non-covalent complex. After this step, E1 can engage one of the many cognate E2 conjugating enzymes to initiate the downstream signaling [35].

E2s catalyze covalent attachment of Ub to the target proteins. All the E2 enzymes have in common an active site Ub-binding Cys residue and a so-called ubiquitin conjugating (UBC) domain required not only to conjugate Ub but also to recruit the cognate E3 ligases. Each E2 can interact with a multitude of different E3 ligases and thus it is ultimately involved in targeting numerous substrates [10, 11, 36, 37]. The UBC fold is a canonical α/β-fold with four α-helices and an antiparallel β-sheet formed by four β-strands and a short 3.10 helix (Figure 13.2). The catalytic cysteine, when present, is in a shallow groove formed by a loop connecting helix 2 with helix 4 and a long disordered region proximal to the active site.

E2 enzymes have been recently classified on the basis of homology of the UBC domain in a very convenient manner by Michelle et al. [39]. They identified 17 different E2 families using phylogenetic analyses of more than 200 E2 genes from different species. In this classification, both catalytically active E2 enzymes and proteins sharing the E2-like fold but lacking the catalytic cysteine were included. The analyses of the different family members allowed to the identification of common features of E2 enzymes, as well as some family-specific properties. For example, the His–Pro–Asn (HPN) signature, followed by an invariant Trp at residue 16 (up to 29) downstream of the HPN motif, was highly conserved. In particular, the histidine and asparagine of the HPN motif can have an important role for the E2 catalysis [40].

E2s generally recognize E3 ligases through the L1 and L2 loops and the N-terminal helix 1 on the E2 surface. Sequence variations in these regions contribute to the specificity of E3 binding. When a specific E2 interacts with multiple E3 ligases, the E2 residues involved in E3 recognition are not necessarily the same. For example, UBE2N uses Arg6 and Lys10 to recognize residues

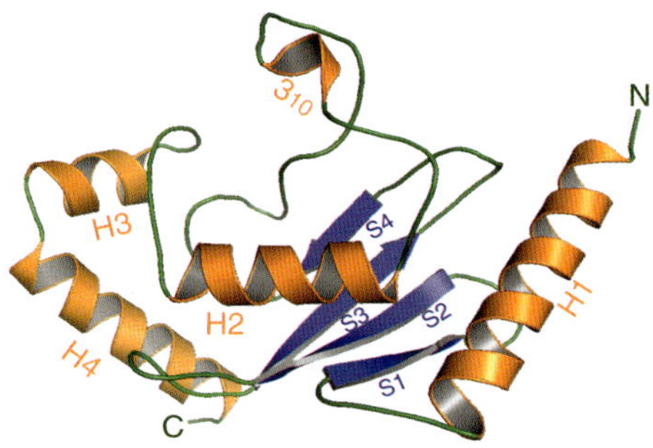

Figure 13.2 *3D architecture of the family 3 of E2 enzymes*. The structure of human Ube2g2 (PDB entry 2CYX, chain A) is shown as an example of members of the family 3 [38]. It is possible to observe the mixed UBC α/β fold, as well as the long disordered insertion between the strand S4 and the helix H2, where some of the E2 enzymes, a short 3.10 helix is included.

upstream of the RING domain of the E3 enzyme TrAF6, whereas it employs Arg7 and Lys10 to contact the Ubox of the C-terminal region of HSC70 [36].

Among the different families of E3 ligases, the Really Interesting New Gene (RING) containing E3s are often associated with Cdc34. The RING domain consists of a short motif rich in cysteine and histidine residues to coordinate two zinc metal ions. The main role of the RING domains of E3 enzymes is suggested to be a scaffolding role to increase the probability of reaction by bringing together the substrate lysine of the target substrate and the E2~Ub intermediate [41–45].

13.3 Cdc34 Protein Sequence and Structure

As stated, Cdc34 stimulates cellular proliferation by promoting the degradation of p53 and p27, which both act as inhibitors of cell cycle progression. As also mentioned in the Background Session, Cdc34 is highly modulated by phosphorylations.

The *Saccharomyces cerevisiae* Cdc34 (ScCdc34) enzyme is encoded by the gene CDC34 (UBC3) and is a 295-residue protein; it is responsible for the ubiquitination and subsequent degradation of several targets, including the Cdk1/Clb inhibitor Sic1 [46], which is an intrinsically disordered protein also modulated by posttranslational phosphorylations [47, 48].

In both human and yeast, Cdc34 is a multidomain protein characterized by a folded domain with the E2-like fold (Figure 13.3) and a large C-terminal intrinsically disordered domain [34]. It is characterized also by a long 12-residue disordered insertion (aa 103–114 in ScCdc34) called the "acidic loop," due to its enrichment in acidic residues. The acidic insertion and its sequence identity with other E2s classify Cdc34 as a member of the family 3 of E2 enzymes.

ScCdc34 is often chosen as a model system due to its high sequence similarity to the human variant (*Homo Sapiens* Cdc34, HsCdc34), as well as the possibility to perform simple assays in yeast cells. We thus started our research using this variant and we just recently turned our attention to the human counterpart for future development and consolidation of our findings and hypothesis. When we

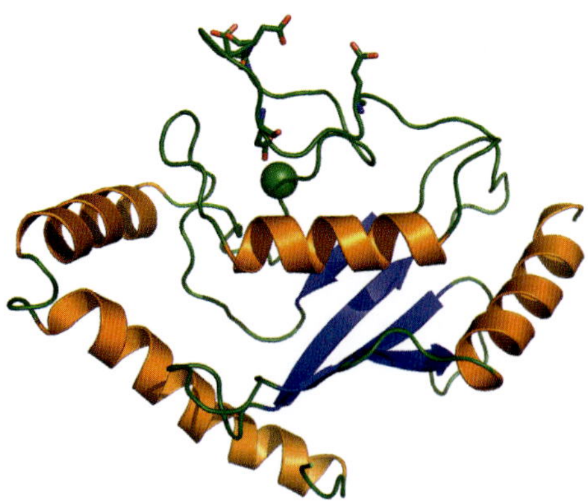

Figure 13.3 *Cdc34 3D architecture.* The structure of the human Cdc34 (PDB entry 2OB4 [49]) generated in our group is shown and the catalytic cysteine and the acidic residues of the insertion are shown as Cα sphere and sticks, respectively.

report the results of our studies subsequently, we thus refer mainly to the yeast variant and use the Cdc34 name for sake of clarity.

13.4 Cdc34 Heterogeneous Conformational Ensemble in Solution

NMR studies [50] and our all-atom MD investigations [23, 24] point in the direction that Cdc34 UBC is a very flexible structure and characterized by a heterogeneous ensemble of conformations in solution where most of the variability is due to different conformational states of the long disordered regions where the acidic insertion is included. Part of the structure could, in principle, fold in a 3.10 helix as the one observed for other E2 enzymes belonging to family 3 (Figure 13.2), such as Ube2g2, even if we could not observe it stably formed in MD simulations [23].

Under normal conditions and in the absence of phosphorylation from CK2 or other kinases, Cdc34 is expected to be in a major state where the acidic loop is in a closed conformation and the catalytic cysteine buried and not accessible for ubiquitination (Figure 13.4). Although, thanks to its high flexibility, partially open states of the loop are likely to be populated even in the absence of phosphorylation and can thus explain the low basal Ub-charging activity observed *in vitro* [23, 51]. Important structural elements that guide the conformational changes are (i) the alternation of hydrophobic and negatively charged residues in the acidic loop, which allow either the stabilization of the open or closed states, respectively, as well as (ii) two conserved hinge residues which include the invariant tryptophan of the UBC fold and an aliphatic residue that in Cdc34 is a leucine [23, 24].

The pattern of hydrophobic/acidic residues in the L7 insertion is indeed an invariant feature of the members of the family 3 of E2 enzymes, as we showed by phylogenetic analyses [24]; it is described by the short linear motif ILHxPG-DPxuhEhxxERW, where x indicates non-conserved amino acids, h a hydrophobic residue, and u a tiny residue (A, G, or S). The negatively charged

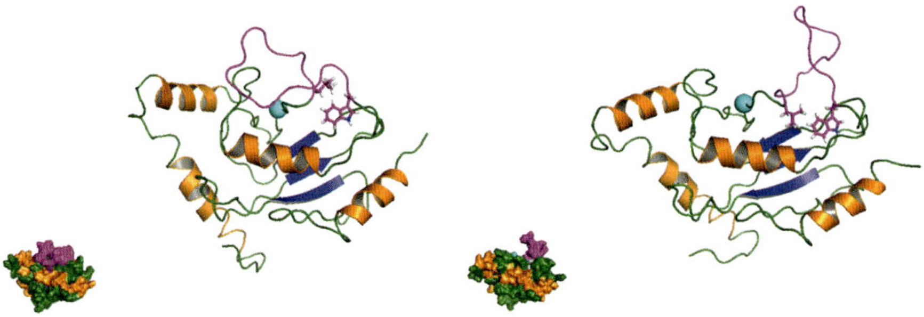

Figure 13.4 *Conformational changes of Cdc34 acidic loop.* The conformational changes observed in the simulations [23] from closed (catalytic cysteine not accessible to the solvent) and open (catalytic cysteine accessible to the solvent) of the disordered acidic loop are illustrated in the figure. The catalytic cysteine and the hinge residues [24] are shown as Cα sphere and sticks, respectively. We illustrate the structures using both the cartoon and the surface (bottom left) to better appreciate the conformational changes.

residues are sensitive to the phosphorylation by the CK2 kinase, being at the heart of an electrostatic repulsion mechanism which disfavors closed states of the loop and stabilizes the active open states [23], described in detail below.

We also showed using both coarse-grained and atomistic simulations that L7 dynamics is tightly correlated to the dynamics of another loop in the catalytic cleft of Cdc34 and Cdc34-like proteins, i.e. L8. The closed inactive states of Cdc34 are characterized by interactions between the two loops. The hydrophobic L8 loop is also very conserved across the family and can be described by a short linear motif PSxxS(P,G)AN [24].

Loops in proteins are not merely disordered and generally are deputed to key functions or can be an important modulator of dynamics [52]. In the case of the acidic loop, we showed that it can be classified as a regulatory omega loop [24].

We have been able to appreciate the different conformations that populate the very rugged free energy landscape of Cdc34, as well as to study the effects of the phosphorylations, using a multi-replicate approach to all-atom MD simulations. This means that different starting models have been used for independent MD simulations so that we increase the likelihood of exploring different portions of the free energy landscape [53–55]. This method often turns out to be more efficient than running just one single very long simulation to speed up the first exploration of the conformational landscape of a protein. It would not be possible to highlight the influence of the phosphorylation on the conformational states of the loops using only a comparison between two simulations of each variant and we thus discourage similar approaches also in other applications. Moreover, we have now reached a stage in which computational resources and software for MD simulations have been capable of supporting the sampling of microsecond timescale routinely and this means that we could thus benefit from more extensive studies with multiple replicates over longer timescales than what was published just 5 or 10 years ago. Efficient methods for enhanced sampling are also accessible at present and turn out to be useful, after a first exploration with unbiased classical MD, to more accurate and quantitative calculations [52, 56–60].

To add to the complexity, it has also to be considered that the full-length variant of Cdc34 is a multi-domain protein, characterized by a C-terminal disordered domain which includes additional ubiquitin-binding sites [34, 50]; and it is likely to wrap around the UBC folded domain to stabilize the binding to the Ub moiety within the E2 catalytic cleft [34]. The different conformational states of the Cdc34 C-terminal domain, along with the structural changes in the acidic loop thus cooperate to generate a large conformational heterogeneity, which the protein exploits for the recognition and interactions with the biological partners, including the cognate E3 and the substrates.

13.5 Long-Range Communication in Family 3 Enzymes: A Structural Path from the Ub-Binding Site to the E3 Recognition Site

Using methods inspired by graph theory and applied to the analyses of conformational ensembles of Ube2g2 as a representative member, within the framework of the so-called protein structure network [33, 61–66], we have been able to shed new light on structure–function relationship of the family 3 E2 members [67]. In this approach, we analyze a protein conformational ensemble using graph theory, i.e. treating the protein as a network where the centers of mass of the side chain of the residues are the nodes and the contacts between them are the edges that can be weighted according the correlation of the fluctuations of side chains during dynamics [68, 69] or for the persistence of the contacts in the ensemble [66].

In the PSN study, we selected Ube2g2 due to the availability of both X-ray and NMR structures. It would have been difficult to carry out similar analyses on simulations from a model, such as the ones that we had so far for Cdc34. Indeed, Protein Structure Network (PSN) approaches require a very stable sampling around the experimental starting structure [33, 62, 65] and a high-resolution structure. Due to the high similarity between E2 family 3 members, the major findings highlighted by these studies are likely to be transferable to Cdc34 and other members. Indeed, we identified preexisting pathways of long-range communication from the catalytic site and the acidic loop in particular and the regions for interaction with E3 enzymes, such as the α-helix 1. These paths of communication, occurring through a cascade of atomic contacts between protein residues are an intrinsic property of the E2 protein structure and its ensemble of conformations in the unbound state, but they are strengthened in the active forms of the enzymes when the acidic loop is displaced in an open conformation.

These analyses can thus suggest sensitive and druggable regions if the final goal is to inhibit steps of the ubiquitination cascade that are downstream the Ub-charging into the catalytic site of the E2, such as the recruitment of the E3 enzymes, and deserve further investigation.

13.6 Cdc34 Modulation by Phosphorylation: From Phenotype to Structure

Cdc34 is heavily regulated by phosphorylation both in the human and yeast variants and it is the target of different kinases. In particular, ScCdc34 phosphorylated by the casein kinase 2 (CK2) on residues Ser130 and Ser167 have been discovered with a key role on its activity *in vitro* and *in vivo* [51].

Thanks to the integration of homology modeling, multi-replicate molecular dynamics simulations and biochemical assays of unphosphorylated and phosphorylated variants of the Cdc34 UBC domain, we revealed the details behind the effects that phosphorylation of Ser130 and Ser167 have on the enzyme activity [23]. A single phosphorylation at Ser130 is sufficient to stabilize the acidic loop in an open conformation, allowing the catalytic cysteine to be accessible for conjugation with Ub. The loop displacement is promoted by the electrostatic repulsion between the negatively charged phosphate on pSer130 and the acidic residues of the loop. A double phosphorylation, interestingly also the Ser167 site, has the effect of enhancing further the stabilization of open conformations; but the structural details of this long-range effects are still unknown. A Cdc34 variant lacking the 12-residue acidic insertion is constitutively active independently of phosphorylation, as shown *in vitro* and *in vivo*, further probing the mechanism of regulation of the Cdc34 catalytic site through the acidic loop conformational changes.

The acidic insertion and the phospho sites of the Cdc34 UBC domain are two co-evolving signatures that are likely to act together as a posttranslationally controlled switch to modulate the opening and closing of the catalytic cleft of Cdc34.

We have only scratched the surface of the complex regulation by phosphorylation of Cdc34 and other events could be important such as, for example, phosphorylation by other kinases, such as the CDK. For example, we recently identified an invariant signature in another loop close to the catalytic site in catalytically active members of the E2 superfamily [70] using phylogenetic analyses, accompanied by molecular modeling and simulations to unveil the structural details and the impact on activity. The signature consists of a conserved acidic residue in the proximity of the catalytic site, which is, in some cases, for example in Cdc34, replaced by a phosphorytable serine. The role of this phosphorylation was widely investigated in other E2 enzymes, but it could exert similar effects in Cdc34 and will deserve closer attention.

13.7 The Dual Role of the Acidic Loop of Cdc34: Regulator of Activity and Interface for E3 Binding

The acidic loop, which is the signature of Cdc34-like enzymes (family 3 of E2 enzymes), is central to Cdc34 functions. Indeed, we showed that it has a dual role. First, as stated earlier, it is a regulatory element for Ub-charging activity. In this context, the acidic loop is involved in the first E2-mediated step, i.e. the conjugation of the activated Ub molecules into the E2 catalytic site. Until the acidic loop is properly displaced from the catalytic cysteine, the conjugation cannot occur and phosphorylations are deputed to tightly modulate the conformation of the loop to ensure a fully active enzyme only in certain cellular conditions or cell cycle phases.

Once the complex between E2 and Ub is formed, the acidic loop can readjust its conformation around the bound Ub molecules, interacting and stabilizing the Ub positively charged face through electrostatic interactions [24].

The invariant Trp residue is located after the acidic insertion of the family 3 of E2 enzymes. Crystallographic structures of the complexes between E2s and E3s showed that the Trp is located in close proximity to a proline at the tip of L7, as well as to the two prolines at the base of the Pro-rich L4 loop [71]. According to our studies, the pro-rich motif in L4 of family 3 can be summarized by the short linear motif FPxn(Y,F)Ph(S,R)PP, where x indicates any amino acid, h hydrophobic residues, and n negatively charged residues [24]. Interactions between the Trp and the prolines were speculated to stabilize the L7 loop and contribute to the correct reciprocal orientation of L4 and L7 loop for E3 recognition mediated by L4 specific residues [39].

The importance of the acidic loop does not end then with the role in the Ub charging step; indeed, it can also interact with the E3 enzyme providing an interaction interface for the Rbx1 (RING-box protein) subunit, which is again stabilized by electrostatic interactions among the acidic residues of the Cdc34 loop and key arginine residues on the E3 subunit, as we recently suggested with modeling and simulations [24].

The data collected from our modeling and simulation approaches are in fairly good agreement with experimental data that have been published by Kleiger's group in 2013 [72]. Indeed, they showed that the acidic loop can promote the interaction between Cdc34 and the ubiquitin ligase partner, SCF. NMR studies also point in the direction that Rbx1 is very sensitive to the binding of Cdc34 in the area where the arginine residues mentioned are located [73]. The experimental works that followed our modeling studies thus contribute to consolidate the mechanism that we proposed.

Moreover, Kleiger and colleagues also showed that the two glutamic acid residues located on the distal site of the loop can cooperate with the invariant histidine of the HPN motif on the proximal side of the loop to suppress the pK_a of ionizing species in the proximity of the catalytic site, thus contributing to Cdc34 catalysis [72].

13.8 Different Strategies to Target Cdc34 with Small Molecules

In light of the role exerted by Cdc34, the discovery of potential inhibitory compounds would provide new means to modulate the activity of the ubiquitination pathway in a more specific manner than just targeting the proteasome or E3s, directly.

CDK inhibitor p27 plays a crucial role in the regulation of mammalian cell cycle progression. The p27 protein induces cell cycle exit at the G1 phase. Ubiquitination and degradation of p27 during S-phase progression is mediated by the ubiquitin ligase SCF-Skp2, and Cdc34 is the main E2 conjugating enzyme to label p27 for degradation. Interestingly, it has been demonstrated that low p27 protein levels are common in tumors such as colorectal carcinomas and breast cancer and they are associated with a poor prognosis for the patient, as mentioned in the Background session of this chapter. E3 enzymes have been selected as therapeutic targets but the lack of specificity of action and the complexity of their architecture have represented serious obstacles to the development of effective molecules [1]. Moreover, SCF ligase does not contain an evident central enzymatic active site to which small molecules could bind since it has a scaffolding function.

Since E2 enzymes are catalytically active and often play a role in the determination of the Ub linkage, their inhibition through small molecules could allow to control over different Ub signals.

Despite their potential as targets, E2 enzymes have been poorly exploited to develop small molecules. This might be partially due to the challenges of targeting protein–protein interaction interfaces.

We and other colleagues have identified different druggable areas of the enzyme and small molecules have been proposed by *in silico* approaches for experimental validations or already investigated experimentally, as described in detail subsequently.

Two similar molecules (BAY 11-708268 and NSC697923 [74, 75]) have been shown to inhibit a large group of E2 enzymes by covalently reacting with them and, as a consequence, blocking the catalytic cysteine residue. These molecules have the limitation of suffering of lack of selectivity, even if they did not report any side effects *in vivo*.

A first inhibitor of Cdc34 catalytic activity, CC0651 was identified using a p27 ubiquitination assay. CC0651 inhibits Cdc34 enzyme in a dose-dependent manner with an IC50 equal to 1.72 μM. Structural studies demonstrated that the compound acts allosterically on the E2 enzyme. Indeed, in the crystal structure, its binding site is located 19 Å far from the catalytic site in a hydrophobic pocket exposed on the protein surface, which is mostly buried in the apo form of the enzyme. CC0651 thus mainly functions by stabilizing the interaction between Cdc34 and the charged Ub molecule as attested by NMR [76, 77]. It does not have an effect on SCF complexes and it specifically targets the E2 enzyme.

Stabilization of E2–Ub interaction can thus be an effective strategy for drug design, as also showed, in a broader context, for other small molecules acting on protein–protein interaction interfaces as stabilizers [78].

Due to the crucial role of the L7 disordered insertion in two important steps of E2 activity (i.e. in Ub binding and then in the recognition of the conjugate E3 enzyme), the acidic loop of Cdc34 also represents an interesting potential drug target. We, thus, recently applied virtual screening, MD, and docking simulations to select small molecules that could inhibit ScCdc34's activity by binding its acidic loop and keeping it in the proximity of the catalytic cysteine (i.e. in a closed conformation), thus preventing Ub conjugation and interaction with E3 at the same time [25]. The identification of a group of suitable molecules was carried out by modeling Cdc34 structure and performing all-atom MD simulations to shed light on the conformational ensemble of the protein in solution, as well as to isolate the most representative conformations suitable for docking calculations. Virtual screening of libraries with thousands of molecules and docking simulations allowed us then to select the compounds and to define a common scaffold to inhibit Cdc34 catalytic site acting on the acidic loop. In particular, the molecules that we identified with our study have been designed with the goal of developing a molecular zipper that can bind the acidic loop in its close conformation. The pharmacophore model for the molecular zipper [25] is characterized by two amino groups in the central part of the model and two aromatic chains, which establish interactions with the acidic loop (Asp108 and Glu109) residues and hydrogen bonds with the putative phosphorylation site Ser139 [70], which is also an important residue for Cdc34 activity [79].

13.9 Conclusions and Perspectives

In conclusion, the computational studies on Cdc34 carried out in the past five years by our group and other colleagues clearly demonstrate a very complex

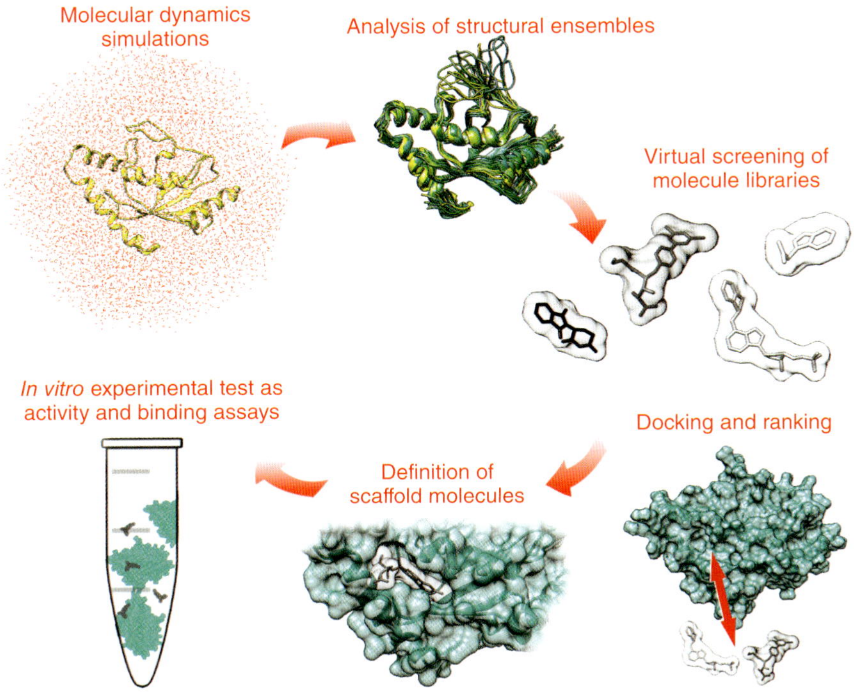

Figure 13.5 A schematic view of our computational workflow.

but intriguing mechanism where conformational heterogeneity of the enzyme in the proximity of the catalytic site plays a key role for its different functions in virtually all the steps of the ubiquitination cascade. Cdc34 turned out also as a very promising target for design of small molecules with antitumoral activity. Computational studies such as the ones here illustrated (Figure 13.5) are thus a useful complementary tool to experimental drug design and discovery and we expect to see the field growing more and more in the future thanks to the development of very accurate physical models and computational approaches to simulate such a plastic and dynamic target.

The first structural studies on full-length Cdc34 also emphasize the importance of developing comprehensive models in which the influence of the disordered C-terminal domain can be considered. Indeed, this additional disordered region is likely to play a role on the conformational states of the acidic loop and the surrounding of the catalytic site of the UBC domain. This could, for example, be achieved using new methods that integrate molecular dynamics simulations and NMR-derived experimental data [33, 52, 80–83]. The understanding of the interplay between the UBC and the C-terminal domains is important to be able to more accurately predict the effects of small molecules that could interact with the acidic loop or the active site itself, as well as to account for allosteric inhibitors considering the high number of long-range structural communication paths between the Ub-binding sites and the E2 interface for interaction with the E3 [67].

Our pioneering computational studies on Cdc34 also emphasize the notion that the mechanisms involving the acidic loop are amenable to be targeted to design molecules that act as stabilizers of the Ub–E2 interaction or, vice-versa, can inhibit it, depending on the context where Cdc34 is in play or its down-/overexpression in cancer. Moreover, we envision that new efforts should be devoted to the understanding of the effects induced over long distances by phosphorylations that are not in close proximity to the active site, inspired, for example, by our recent achievements on phosphorylation of p53 and its distal effects on DNA binding [58]. This knowledge will be fundamental not only at the basic level but also to be effectively capable of interfering with Cdc34 cellular functions by the use of small molecules.

Acknowledgments

We thank all the former team members (and especially Valeria Ranzani, Alberto Arrigoni, Ilaria Valimberti, Barbara Grillo, Alessandro Vitriolo and Matteo Sensi) and colleagues who collaborated with us during the years (Paola Coccetti, Marco Vanoni, Farida Tripodi, Luca De Gioia, Luca Bertini, Gary Shaw, and Ruth Nussinov) on the studies on Cdc34 and contributed to unveil its complex mechanisms. This work is also supported by a KBVU pre-graduate fellowship (R157-A10330) and Erasmus Traineeship fellowship from the University of Bari in 2015–2016 awarded to MDM to carry out research in EP group.

References

1 Nakayama, K.I. and Nakayama, K. (2006). Ubiquitin ligases: cell-cycle control and cancer. *Nat. Rev. Cancer* 6: 369–381.
2 Yamasaki, L. and Pagano, M. (2004). Cell cycle, proteolysis and cancer. *Curr. Opin. Cell Biol.* 16: 623–628.
3 Ohta, T. and Fukuda, M. (2004). Ubiquitin and breast cancer. *Oncogene* 23: 2079–2088.
4 Mani, A. and Gelmann, E.P. (2005). The ubiquitin-proteasome pathway and its role in cancer. *J. Clin. Oncol.* 23: 4776–4789.
5 Gallo, L.H., Ko, J., and Donoghue, D.J. (2017). The importance of regulatory ubiquitination in cancer and metastasis. *Cell Cycle* 16: 634–648.
6 Ma, J., Guo, W., and Li, C. (2017). Ubiquitination in melanoma pathogenesis and treatment. *Cancer Med.* 6: 1362–1377.
7 Kleiger, G. and Mayor, T. (2014). Perilous journey: a tour of the ubiquitin–proteasome system. *Trends Cell Biol.* 24: 352–359.
8 Pickart, C.M. (2001). Mechanisms underlying ubiquitination. *Annu. Rev. Biochem.* 70: 503–533.
9 Passmore, L.A. and Barford, D. (2004). Getting into position: the catalytic mechanisms of protein ubiquitylation. *Biochem. J* 379: 513–525.
10 Stewart, M.D., Ritterhoff, T., Klevit, R.E. et al. (2016). E2 enzymes: more than just middle men. *Cell Res.* 26: 423–440.

11 van Wijk, S.J.L. and Timmers, H.T.M. (2010). The family of ubiquitin-conjugating enzymes (E2s): deciding between life and death of proteins. *FASEB J.* 24: 981–993.
12 Petroski, M.D. and Deshaies, R.J. (2005). Mechanism of lysine 48-linked ubiquitin-chain synthesis by the cullin-RING ubiquitin-ligase complex SCF-Cdc34. *Cell* 123: 1107–1120.
13 Christensen, D.E., Brzovic, P.S., and Klevit, R.E. (2007). E2-BRCA1 RING interactions dictate synthesis of mono- or specific polyubiquitin chain linkages. *Nat. Struct. Mol. Biol.* 14: 941–948.
14 Hyoung, T.K., Kwang, P.K., Lledias, F. et al. (2007). Certain pairs of ubiquitin-conjugating enzymes (E2s) and ubiquitin-protein ligases (E3s) synthesize nondegradable forked ubiquitin chains containing all possible isopeptide linkages. *J. Biol. Chem.* 282: 17375–17386.
15 Sowa, M.E. and Harper, J.W. (2006). From loops to chains: unraveling the mysteries of polyubiquitin chain specificity and processivity. *ACS Chem. Biol.* 1: 20–24.
16 David, Y., Ziv, T., Admon, A. et al. (2010). The E2 ubiquitin-conjugating enzymes direct polyubiquitination to preferred lysines. *J. Biol. Chem.* 285: 8595–8604.
17 Paiva, S., Silva, S.R., De Araujo, E.D. et al. (2017). Regulating the master regulator: controlling ubiquitination by thinking outside the active site. *J Med Chem.* 61: 405–421.
18 Pagano, M., Tam, S.W., Theodoras, A.M. et al. (1995). Role of the ubiquitin-proteasome pathway in regulating abundance of the cyclin-dependent kinase inhibitor p27. *Science* 269: 682–685.
19 Block, K., Appikonda, S., Lin, H.-R. et al. (2005). The acidic tail domain of human Cdc34 is required for p27Kip1 ubiquitination and complementation of a cdc34 temperature sensitive yeast strain. *Cell Cycle* 4: 1421–1427.
20 Shirane, M., Harumiya, Y., Ishida, N. et al. (1999). Down-regulation of p27(Kip1) by two mechanisms, ubiquitin-mediated degradation and proteolytic processing. *J. Biol. Chem.* 274: 13886–13893.
21 Chauhan, D., Li, G., Hideshima, T. et al. (2004). Blockade of ubiquitin-conjugating enzyme CDC34 enhances anti-myeloma activity of Bortezomib/Proteasome inhibitor PS-341. *Oncogene* 23: 3597–3602.
22 Burslem, G.M. and Crews, C.M. (2017). Small-molecule modulation of protein homeostasis. *Chem. Rev.* 117: 11269–11301.
23 Papaleo, E., Ranzani, V., Tripodi, F. et al. (2011). An acidic loop and cognate phosphorylation sites define a molecular switch that modulates ubiquitin charging activity in Cdc34-like enzymes. *PLoS Comput. Biol.* 7: e1002056.
24 Papaleo, E., Casiraghi, N., Arrigoni, A. et al. (2012). Loop 7 of E2 enzymes: an ancestral conserved functional motif involved in the E2-mediated steps of the ubiquitination cascade. *PLoS One* 7: e40786.
25 Arrigoni, A., Bertini, L., De Gioia, L. et al. (2014). Inhibitors of the Cdc34 acidic loop: a computational investigation integrating molecular dynamics, virtual screening and docking approaches. *FEBS Open Bio* 4: 473–484.

26 Suryadinata, R., Holien, J.K., Yang, G. et al. (2013). Molecular and structural insight into lysine selection on substrate and ubiquitin lysine 48 by the ubiquitin-conjugating enzyme Cdc34. *Cell Cycle* 12: 1732–1744.
27 Lass, A., Cocklin, R., Scaglione, K.M. et al. (2011). The loop-less tmCdc34 E2 mutant defective in polyubiquitination in vitro and in vivo supports yeast growth in a manner dependent on Ubp14 and Cka2. *Cell Div.* 6: 7.
28 Cocklin, R., Heyen, J., Larry, T. et al. (2011). New insight into the role of the Cdc34 ubiquitin-conjugating enzyme in cell cycle regulation via Ace2 and Sic1. *Genetics* 187: 701–715.
29 Papaleo, E., Casiraghi, N., Arrigoni, A. et al. (2012). Atomistic insights into the regulatory mechanisms mediated by post-translational modifications: molecular dynamics investigations. *Curr. Phys. Chem.* 2: 344–362.
30 Dror, R.O., Dirks, R.M., Grossman, J.P. et al. (2012). Biomolecular simulation: a computational microscope for molecular biology. *Annu. Rev. Biophys.* 41: 429–452.
31 Piana, S., Klepeis, J.L., and Shaw, D.E. (2014). Assessing the accuracy of physical models used in protein-folding simulations: quantitative evidence from long molecular dynamics simulations. *Curr. Opin. Struct. Biol.* 24: 98–105.
32 Vendruscolo, M. (2007). Determination of conformationally heterogeneous states of proteins. *Curr. Opin. Struct. Biol.* 17: 15–20.
33 Papaleo, E. (2015). Integrating atomistic molecular dynamics simulations, experiments, and network analysis to study protein dynamics: strength in unity. *Front. Mol. Biosci.* 2: 28.
34 Arrigoni, A., Grillo, B., Vitriolo, A. et al. (2012). C-terminal acidic domain of ubiquitin-conjugating enzymes: a multi-functional conserved intrinsically disordered domain in family 3 of E2 enzymes. *J. Struct. Biol.* 178: 245–259.
35 Schulman, B.A. and Wade Harper, J. (2009). Ubiquitin-like protein activation by E1 enzymes: the apex for downstream signalling pathways. *Nat. Rev. Mol. Cell Biol.* 10: 319–331.
36 Ye, Y. and Rape, M. (2009). Building ubiquitin chains: E2 enzymes at work. *Nat. Rev. Mol. Cell Biol.* 10: 755–764.
37 Wenzel, D.M., Stoll, K.E., and Klevit, R.E. (2011). E2s: structurally economical and functionally replete. *Biochem. J.* 433: 31–42.
38 Arai, R., Yoshikawa, S., Murayama, K. et al. (2006). Structure of human ubiquitin-conjugating enzyme E2 G2 (UBE2G2/UBC7). *Acta Crystallogr., Sect. F: Struct. Biol. Cryst. Commun.* 62: 330–334.
39 Michelle, C., Vourc'h, P., Mignon, L. et al. (2009). What was the set of ubiquitin and ubiquitin-like conjugating enzymes in the eukaryote common ancestor? *J. Mol. Evol.* 68: 616–628.
40 Cook, B.W. and Shaw, G.S. (2012). Architecture of the catalytic HPN motif is conserved in all E2 conjugating enzymes. *Biochem. J.* 445: 167–174.
41 Liu, J. and Nussinov, R. (2009). The mechanism of ubiquitination in the cullin-RING E3 ligase machinery: conformational control of substrate orientation. *PLoS Comput. Biol.* 5: e1000527.
42 Spratt, D.E., Walden, H., and Shaw, G.S. (2014). RBR E3 ubiquitin ligases: new structures, new insights, new questions. *Biochem. J.* 458: 421–437.

43 Duda, D.M., Scott, D.C., Calabrese, M.F. et al. (2011). Structural regulation of cullin-RING ubiquitin ligase complexes. *Curr. Opin. Struct. Biol.* 21: 257–264.

44 Metzger, M.B., Pruneda, J.N., Klevit, R.E. et al. (2014). RING-type E3 ligases: master manipulators of E2 ubiquitin-conjugating enzymes and ubiquitination. *Biochim. Biophys. Acta, Mol. Cell. Res.* 1843: 47–60.

45 Zheng, N. and Shabek, N. (2017). Ubiquitin ligases: structure, function, and regulation. *Annu. Rev. Biochem.* 86: 129–157.

46 Verma, R., Feldman, R.M., and Deshaies, R.J. (1997). SIC1 is ubiquitinated in vitro by a pathway that requires CDC4, CDC34, and cyclin/CDK activities. *Mol. Biol. Cell* 8: 1427–1437.

47 Lambrughi, M., Papaleo, E., Testa, L. et al. (2012). Intramolecular interactions stabilizing compact conformations of the intrinsically disordered kinase-inhibitor domain of Sic1: a molecular dynamics investigation. *Front. Physiol.* 3: 1–17.

48 Brocca, S., Samalíková, M., Uversky, V.N. et al. (2009). Order propensity of an intrinsically disordered protein, the cyclin-dependent-kinase inhibitor Sic1. *Proteins* 76: 731–746.

49 Sheng, Y., Hong, J.H., Doherty, R. et al. (2012). A human ubiquitin conjugating enzyme (E2)-HECT E3 ligase structure-function screen. *Mol. Cell. Proteomics* 11: 329–341.

50 Spratt, D.E. and Shaw, G.S. (2011). Association of the disordered C-terminus of CDC34 with a catalytically bound ubiquitin. *J. Mol. Biol.* 407: 425–438.

51 Coccetti, P., Tripodi, F., Tedeschi, G. et al. (2008). The CK2 phosphorylation of catalytic domain of Cdc34 modulates its activity at the G_1 to S transition in *Saccharomyces cerevisiae*. *Cell Cycle* 7: 1391–1401.

52 Papaleo, E., Saladino, G., Lambrughi, M. et al. (2016). The role of protein loops and linkers in conformational dynamics and allostery. *Chem. Rev.* 116: 6391–6423.

53 Papaleo, E., Mereghetti, P., Fantucci, P. et al. (2009). Free-energy landscape, principal component analysis, and structural clustering to identify representative conformations from molecular dynamics simulations: the myoglobin case. *J. Mol. Graphics Modell.* 27: 889–899.

54 Mereghetti, P., Riccardi, L., Brandsdal, B.O. et al. (2010). Near native-state conformational landscape of psychrophilic and mesophilic enzymes: probing the folding funnel model. *J. Phys. Chem. B* 114: 7609–7619.

55 Caves, L.S.D., Evanseck, J.D., and Karplus, M. (1998). Locally accessible conformations of proteins: multiple molecular dynamics simulations of crambin. *Protein Sci.* 7: 649–666.

56 Laio, A. and Gervasio, F.L. (2008). Metadynamics: a method to simulate rare events and reconstruct the free energy in biophysics, chemistry and material science. *Rep. Prog. Phys.* 71: 126601.

57 Papaleo, E., Sutto, L., Gervasio, F.L. et al. (2014). Conformational changes and free energies in a proline isomerase. *J. Chem. Theory Comput.* 10: 4169–4174.

58 Lambrughi, M., De Gioia, L., Gervasio, F.L. et al. (2016). DNA-binding protects p53 from interactions with cofactors involved in transcription-independent functions. *Nucleic Acids Res.* 44: 9096–9109.

59 Abrams, C. and Bussi, G. (2013). Enhanced sampling in molecular dynamics using metadynamics, replica-exchange, and temperature-acceleration. *Entropy* 16: 163–199.

60 Spiwok, V., Sucur, Z., and Hosek, P. (2015). Enhanced sampling techniques in biomolecular simulations. *Biotechnol. Adv.* 33: 1130–1140.

61 Papaleo, E., Renzetti, G., and Tiberti, M. (2012). Mechanisms of intramolecular communication in a hyperthermophilic acylaminoacyl peptidase: a molecular dynamics investigation. *PLoS One* 7: e35686.

62 Salamanca Viloria, J., Allega, M.F., Lambrughi, M. et al. (2016). An optimal distance cutoff for contact-based Protein Structure Networks using side chain center of masses. *Sci. Rep.* 7: 2838.

63 Di Paola, L. and Giuliani, A. (2015). Protein contact network topology: a natural language for allostery. *Curr. Opin. Struct. Biol.* 31: 43–48.

64 Di Paola, L., De Ruvo, M., Paci, P. et al. (2013). Protein Contact Networks: an emerging paradigm in chemistry. *Chem. Rev.* 113: 1598–1613.

65 Invernizzi, G., Tiberti, M., Lambrughi, M. et al. (2014). Communication routes in ARID domains between distal residues in helix 5 and the DNA-binding loops. *PLoS Comput. Biol.* 10: e1003744.

66 Tiberti, M., Invernizzi, G., Lambrughi, M. et al. (2014). PyInteraph: a framework for the analysis of interaction networks in structural ensembles of proteins. *J. Chem. Inf. Model.* 54: 1537–1551.

67 Papaleo, E., Lindorff-Larsen, K., and De Gioia, L. (2012). Paths of long-range communication in the E2 enzymes of family 3: a molecular dynamics investigation. *Phys. Chem. Chem. Phys.* 14: 12515.

68 Angelova, K., Felline, A., Lee, M. et al. (2011). Conserved amino acids participate in the structure networks deputed to intramolecular communication in the lutropin receptor. *Cell. Mol. Life Sci.* 68: 1227–1239.

69 Vishveshwara, S., Ghosh, A., and Hansia, P. (2009). Intra and inter-molecular communications through protein structure network. *Curr. Protein Pept. Sci.* 10: 146–160.

70 Valimberti, I., Tiberti, M., Lambrughi, M. et al. (2015). E2 superfamily of ubiquitin-conjugating enzymes: constitutively active or activated through phosphorylation in the catalytic cleft. *Sci. Rep.* 5: 14849.

71 Martinez-Noel, G., Müller, U., and Harbers, K. (2001). Identification of molecular determinants required for interaction of ubiquitin-conjugating enzymes and RING finger proteins. *Eur. J. Biochem.* 268: 5912–5919.

72 Ziemba, A., Hill, S., Sandoval, D. et al. (2013). Multimodal mechanism of action for the Cdc34 acidic loop: a case study for why ubiquitin-conjugating enzymes have loops and tails. *J. Biol. Chem.* 288: 34882–34896.

73 Spratt, D.E., Wu, K., Kovacev, J. et al. (2012). Selective recruitment of an E2~ubiquitin complex by an E3 ubiquitin ligase. *J. Biol. Chem.* 287: 17374–17385.

74 Strickson, S., Campbell, D.G., Emmerich, C.H. et al. (2013). The anti-inflammatory drug BAY 11-7082 suppresses the MyD88-dependent signalling network by targeting the ubiquitin system. *Biochem. J.* 451: 427–437.

75 Cheng, J., Fan, Y.-H., Xu, X. et al. (2014). A small-molecule inhibitor of UBE2N induces neuroblastoma cell death via activation of p53 and JNK pathways. *Cell Death Dis.* 5: e1079.

76 Huang, H., Ceccarelli, D.F., Orlicky, S. et al. (2014). E2 enzyme inhibition by stabilization of a low affinity interface with ubiquitin. *Nat. Chem. Biol.* 10: 156–163.

77 Ceccarelli, D.F., Tang, X., Pelletier, B. et al. (2011). An allosteric inhibitor of the human Cdc34 ubiquitin-conjugating enzyme. *Cell* 145: 1075–1087.

78 Ivanov, A.A., Khuri, F.R., and Fu, H. (2013). Targeting protein–protein interactions as an anticancer strategy. *Trends Pharmacol. Sci.* 34: 393–400.

79 Sadowski, M., Suryadinata, R., Lai, X. et al. (2010). Molecular basis for lysine specificity in the yeast ubiquitin-conjugating enzyme Cdc34. *Mol. Cell. Biol.* 30: 2316–2329.

80 Camilloni, C., Cavalli, A., and Vendruscolo, M. (2013). Assessment of the use of NMR chemical shifts as replica-averaged structural restraints in molecular dynamics simulations to characterise the dynamics of proteins. *J. Phys. Chem. B.*

81 Kannan, A., Camilloni, C., Sahakyan, A.B. et al. (2014). A conformational ensemble derived using NMR methyl chemical shifts reveals a mechanical clamping transition that gates the binding of the HU protein to DNA. *J. Am. Chem. Soc.* 136: 2204–2207.

82 Bonomi, M., Camilloni, C., Cavalli, A. et al. (2016). Metainference: a Bayesian inference method for heterogeneous systems. *Sci. Adv.* 2: e1501177–e1501177.

83 Piana, S., Lindorff-Larsen, K., and Shaw, D.E. (2012). Protein folding kinetics and thermodynamics from atomistic simulation. *Proc. Natl. Acad. Sci. U.S.A.* 109: 17845–17850.

Index

a

Ab initio quantum chemical methods 185–186
accelerated molecular dynamics (aMD) 32, 49, 52
ACE model 94
adaptive biasing force (ABF) 52, 252, 253, 268, 271
adiabatic bias metadynamics 238–241
adiabatic bias molecular dynamics (ABMD) 238
adio-ligand-binding assay 254
Alchemical methods 15, 16, 50–51, 88
Aliskiren 182
allosteric effect 118, 282–284, 287, 288
allosteric two-state model (ATSM) 283, 287
allostery
- behavior and application 288–289
- chaperones 291–293
- classic view of 283
- defined 281
- ensemble allostery model 287–288
- GPCRs 294–296
- Protein kinases 293–294
- structural and dynamic approaches 289–291
- thermodynamics to protein structure and dynamics 285–287
- thermodynamic two-state model of 283–285

Amprenavir 180
amyloid beta (Abeta) 34
anaplastic lymphoma kinase (ALK) 174, 183, 184
angiotensin-converting enzyme (ACE) 94, 167, 169
angiotensin II 169, 170, 238
angiotensin receptor II 170
angle bending 250
anisotropic network model (ANM) 91–93
antimicrobial peptides 34
antiparallel packing (APP) 34, 304
antiparallel sheets (APS) 304–306, 308, 313, 317
Arrhenius-like formula 59
asparagine (N) and glutamine(Q) polypeptides
- aggregate formation 318–319
- initial structures 313–314
- N-rich oligomers 315–316
- polyQ β-hairpins 314–315
- polyQ β structures 316–318

atomic solvation factor (SF) 239
atomistic representations 206, 226
average path length 109

b

backbone-oxygen atoms 257
Bayes factor 77
Bayesian agglomerative clustering engine (BACE) algorithm 77
Bennett acceptance ratio 16
benzamidine-trypsin system 189
β-arrestin-biased ligands 211
β-lactamase enzyme inhibitor 179
betrixaban 178
bias exchange metadynamics 13, 35, 264

Biomolecular Simulations in Structure-Based Drug Discovery,
First Edition. Edited by Francesco L. Gervasio and Vojtech Spiwok.

biasing potential 52, 252
BILN-2061 177, 178
binding free energy 12, 14–17, 35–37, 55, 56, 176, 187, 189, 260, 286
binding kinetics 6, 20, 57–59, 190, 211, 226, 234
binding pocket 37, 55, 75, 77, 118, 169, 173, 176, 184, 185, 190, 210, 211, 214, 227, 232, 236, 238, 239
biochemical assays 331
biochemical pathway regulation 281
Biomolecular and Organic Simulation System (BOSS) 88, 89
biomolecular recognition 29, 30
biomolecular simulations
  accuracy of 8–10
  binding free energy 14–16
  design of 4–6
  equilibrium design 5
  evaluative design 4
  free energy stimates 16–20
  refinement design 5
  sampling 10–14
  trajectory clustering 6–8
biomolecular systems 10, 11, 17, 141, 166, 250
Boltzmann constant 7, 49, 53, 69, 235, 251, 262, 286
Boltzmann distribution 54, 69–71, 87, 137, 236
Brigatinib 183, 184

c

Ca^{2+} accumulating region (CAR) 269
calcium-saturated calmodulin (CaM) 287
Captopril 167, 169
carbonic anhydrase (CA) II 185
carboxypeptidase A 167
casein kinase 2 (CK2) 329–331
Cdc34
  acidic loop of 332–333
  catalytic activity 334
  different strategies to target 333–334
  heterogeneous conformational ensemble 329–330
  leucine 329
  overexpression and increased ubiquitination 326
  from phenotype to structure 331–332
  protein sequence and structure 328–329
  structure and dynamics 326
Cdc34-like proteins 330
cell cycle progression 325, 328, 333
centrality measures 109–114, 116, 118
chaperones 119, 291–293
Chapman–Kolmogorov test 73, 210
CHARMM-GUI 9
chemoinformaticians 20
Chou–Fasman method 182, 184
CK2 kinase 330
classical metadynamics 13, 19
clustering coefficient 109
coarse-grained (CG) models of proteins 106
coarse-grained representations 121, 206
coarse grained elastic network models (CG ENMs) 122–125, 130, 136, 137, 139, 141
collective variables (CVs) 6–8, 32, 51, 52, 56, 59, 240, 250, 252
computational alanine scanning (CAS) 34
computer-aided drug discovery (CADD) 30, 45, 47, 50, 59, 165, 166, 190, 296
conformational flooding 32, 52
continuous tempering MD (CTMD) 48
corticotropin-releasing factor 239
corticotropin-releasing factor receptor 1 (CRF_1R) 239
Coulomb potential 132, 250
Coulomb's law 179
Crizotinib 174, 184
cryo-electron microscopy 173, 248

cyclic adenosine monophosphate (cAMP) 282, 286, 287, 291
cyclin-dependent kinase 5 (CDK5) 56, 190
cyclin-dependent kinase (CDK) inhibitor 333
cyclophilin A (CypA) 189
cyclosporin A 78, 79

d

(S,S,S)-decahydro-isoquinoline-3-carbonyl (DIQ) 168
degree of complexity 30
degrees of freedom 12, 13, 30, 32, 54, 120, 137, 169, 175, 206, 240, 252, 287, 318, 319
delta opioid receptor (DOR) 190
de novo drug design 175, 180–181
dentatorubral-pallidoluysian atrophy (DRPLA) 301
diarylpyrimidine (DAPY) compound 176
digraphs 107
dimethylphosphine oxine (DMPO) 183, 184
directed graphs 107
divalent cation sensor (DCS) 265, 266
docking algorithms 30, 179, 254
dominant eigenspace 70–72
dopamine receptor D_2 (D_2R) 212
dopamine receptor D_3 (D_3R) 211
Dorzolamide 185–186
double decoupling method 36
drug (un)binding 209
drug design, Markov state models 67
drug design process 165, 180
drug development
  classical approaches of 281
  targets and strategies for 281
drug discovery
  allostery for 289
  exploiting allostery in 288–289
  partnerships 281
  pipeline 57
  target-centric 281
drug discovery process
  Captopril 167
  early days 166–167
  ligand-based methods 169–173
  pharmacophore modeling 171–172
  quantitative structure-activity relationships (QSAR) 172–173
  Ritonavir 168
  Saquinavir 167–168
  structure-based methods 173–185
  use of computer-aided methods in 168–186
drug resistance mechanisms 281

e

E2 catalysis 327
E2 enzymes 325
  families of 327–328
  phosphorylation 332
elastic network models (ENMs) 106, 124
  general principles 123–124
  residue interaction networks (RINs) 124–127
electrostatic ENM (eENM) 137, 138, 253, 308
end-point methods 36, 47, 50–51
energy minimization 89, 92, 94, 175, 179
enfuvirtide 184
enhanced sampling method 38, 46, 47, 57, 59, 88, 180, 294
ensemble allostery model 287–288
enzyme characterization 96–97
equilibrium constant 17, 18, 235, 284
E3 recognition site 330–331
Eyring's equation 235

f

fibroblast growth factor receptor (FGFR) 13, 190
flexible receptor molecular docking 179
fluorescence microscopy technique 20

flying Gaussian method, 11force field 8–10, 16, 20, 21, 31, 32, 37–39, 45–48, 55, 59, 87, 88, 94, 121, 166, 175–180, 182, 188, 206, 230–232, 241, 250, 251, 263, 299, 303, 304
free energy
  ABF 252–253
  barrier 286
  estimates 16–20
  metadynamics 252
  methods 264–270
  perturbation 88, 251, 257–260
  umbrella sampling 251–252, 257–260
  of zipper formation 318
free energy perturbation (FEP) 16, 32, 51, 81, 88, 187, 230, 251, 257
  FEP thermodynamic cycle 187, 188
free-energy simulation methods 32
free energy surface (FES) 7, 8, 11–15, 19, 20, 31, 48, 52, 53, 240, 253
FTMAP algorithm 296
funnel metadynamics 15

g
gamma aminobutyric acid (GABA) receptors 262
Gartner hype cycle 4
gastrointestinal stromal tumor (GIST) 183
Gaussian accelerated MD (GaMD) 52
Gaussian function 132, 252
Gaussian network models 294
General Amber Force Field (GAFF) 9
Generic GPCR Residue Numbering 227
genetic algorithm for graph similarity (GAGS) search 134
genetic algorithms 175
Gibbs' free energy 235
glucokinase activation 282
GPCR ligand design
  adiabatic bias metadynamics 238–240
  application of MD simulations 226
  ligand binding free energy 230–233
  ligand binding kinetics 233–240
  role of water 226–230
  supervised molecular dynamics (SuMD) 235–237
  WaterMap and WaterFlap 228–230
G protein-coupled receptors (GPCRs) 96, 118, 294, 296
  activation or inactivation 208
  drug design 210–214
  modulation 213
  plasticity of 207
  sampling problem and simulation timescales 208–209
  sharing dynamic data on 216
  simplified energy landscape of 208
  simulation data 209–210
  subtypes 212
  timescales 214
graphical processing units (GPU) 11, 30, 31, 67, 166, 180, 215, 232, 237
graphs, of protein structures 107
Grazoprevir 177, 178
GRID 181, 227–230, 241
Gromos clustering algorithm 7

h
hepatitis C virus (HCV) 177
high-throughput screening (HTS) 165, 183, 295
Hilbert space 70
HOLE program 257
homology modeling 33, 169, 175, 182, 183, 248–249, 254, 264, 269, 271, 331
Hsp90 34, 59, 119, 291–293
  C-terminal domain (CTD) 292, 330, 335
  C-terminal targeted drugs 292
human immunodeficiency virus-1 protease (HIV-1 PR) 35, 96, 120, 167, 175, 180, 290
Huntington's disease (HD) 301

hydrogen bond 10, 34, 78, 121, 125, 167, 171–173, 176, 180, 183, 227, 230, 302, 304, 306, 308, 315, 318, 334
hydrophobic gate 264–270
hydroxyethylamine compounds 168
5-hydroxytryptamine (5-HT) 171, 211, 266
5-hydroxytryptamine receptor 2A (5-HT_{2A}) 211
5-hydroxytryptamine receptor 3 (5-HT_3) receptor 266

i

ICM Docking module 179
indinavir 175
INFOMap algorithm 128, 139, 140
instantaneous force 252, 253
ion channel
  calculation of free energy 251–253
  drug discovery studies 247
  force field 250
  free energy perturbation and umbrella sampling 257–260
  homology modeling 248–249
  ion conductance calculations 260–263
  malfunctioning of 247
  melatonin receptors 254
  molecular dynamics simulations 249–250
  properties of 253–264
  Trk1p 253–254
  TRP channels 263–264
  voltage-gated sodium channels (Nav) 254–256
ion conductance calculations
  low conductance GLIC channel 261–263
  VDACs 261–262
isothermal titration calorimetry (ITC) 33, 187

j

Jacobean matrix 253

k

protein kinases 118, 293, 294
kinetics rate constants 234, 235
knock-on mechanism 258

l

λ windows 187, 233
Lennard–Jones potentials 3, 51, 179, 250
ligand-based methods 169–173
ligand binding free energy 230–233
ligand binding kinetics 233–240
ligand perturbation 91, 92
ligand–protein complex 36, 176, 229, 235
ligand–receptor complex 29, 237
ligand-receptor interactions 211, 229
local-elevation method 32
lock-and-key model 29, 67
long-lived conformations 75, 77–79, 81
losartan 169–171
low conductance GLIC channel 261–263

m

mapping protein ligand 90, 94–96
Markov chain model 8, 20
Markov chain Monte Carlo (MCMC) 3, 48, 87
Markovian dynamics 70, 71
Markov state models (MSM) 57, 72–75, 210
  dominant eigenspace 70–72
  long-lived conformations 77–79
  MD simulations 68
  microstates 75–77
  molecular ensemble 69
  propagator 69–70
  transition paths 79
MC for proteins (MCPRO) 88, 89
MC-minimization technique 89
MC software 90
MDM2 76
MD simulations
  advantage of 179
  of cAMP-free (apo) 291

MD simulations (*contd.*)
  Markov state models 68
  multi-replicate approach 330
  technique 68
mean first passage time (MFPT) 80, 81
melatonin receptors 254, 270
metadynamics (MetaD) 12, 52–57
  accuracy of 12
  application of 12
  biased simulation 12
  potential 53
microstates, drug design 75–77
migration inhibitory factor (MIF) 232
MK-927 185, 186
muscarinic (M_3) receptor 238
molecular docking-virtual screening 175–178
molecular dynamics (MD)
  alchemical methods 50–51
  binding kinetics 57–59
  collective variables (CVs)-based methods 51–52
  vs. docking 30–31
  end-point methods 50
  limits of 46–47
  metadynamics (MetaD) 52–57
  methods 225
  multiple replica methods 48–50
  non-Markovian 32
  protein-ligand binding 36–38
  protein-peptide binding 34–36
  protein-protein binding 32–34
  simulations 6, 30, 45, 179–180
    application of 210–214
    GPCR functionality 205
    ion channel 249–250
    robustness and stability 249
    timescales 214
  state of art 31–32
  tempering methods 47–48
  tools 234
metadynamics
  bias exchange 13, 35, 264
  classical 13, 19
  funnel 15
molecular ensemble 69
molecular graphs 105, 141, 173
molecular mechanics (MM) 8, 9, 16, 35, 36, 50, 81, 88, 94–97, 172, 176, 177, 229, 238
molecular mechanics/generalized born surface area (MM/GBSA) 16, 50
molecular mechanics/Poisson-Boltzmann surface area (MM/PBSA) 16, 36, 50, 51, 81
molecular modeling techniques 165
Monte Carlo (MC)
  method 3, 11, 18
  simulations 179, 230
  techniques
    free energy calculations 88
    and MD combinations 89–90
    optimization 88–89
    sampling procedure 91–92
Monte Carlo *de novo* ligand generator (MCDNLG) 181
MSMs. *See* Markov state models (MSMs)
multi-grained (MG) models 131
multiple replica methods 14, 48–50
MWC model 283

n

negative allosteric modulators (NAMs) 239
Nelfinavir 180–181
nerve signal transmission 247
network diameter 109, 113
network sciences
  development of 106
  modularization techniques 107
  rise of 105–107
Newton's equations of motion 87, 250
Newton's laws of motion 31, 179
nicotinic acetylcholine receptors (nAChRs) 262, 266
NMR spectroscopy 29, 36, 117, 169, 173, 286
nonequilibrium candidate Monte Carlo (NCMC) 89
non-nucleoside reverse-transcriptase inhibitor (NNRTI) 176

nonpeptide compound 169
non-small cell lung cancer (NSCLC) 174, 183, 184
norfloxacin 173
nuclear magnetic resonance (NMR) 17, 29, 33, 79, 117, 169, 248, 286, 315
nucleation-growth polymerization 301

o
optimization algorithm 48

p
parallel artificial membrane permeability assays (PAMPA) 12
parallel packing (PP) 304
parallel sheets (PS) 304, 313
parallel tempering (PT) 14, 31, 48, 49, 189
partial-path transition interface sampling (PPTIS) 58
path-based collective variables (PathCVs) 56
peptides and proteins
  design of multiscale ENMs 135–140
  graph reduction of three-dimensional molecular fields of 132–135
periodic boundary conditions (PBC) 8, 237, 260
Perron-cluster cluster analysis+ (PCCA+) 77
pharmacokinetics (PK) 172, 178, 183, 184, 233, 236, 239
pharmacophore modeling 20, 171, 172
Philadelphia chromosome–positive chronic myeloid leukemia (Ph+ CML) 175
phosphoglycerate dehydrogenase (PGDH) 285
Planck's constant 235
platelet-derived growth factor receptors (PDGFRs) 183
Poisson–Boltzmann surface area (PBSA) 81, 187
polar zipper mechanism 308
polar zipper model 308
polyglutamine protofibrils and aggregates
  monomeric Q40 protofibrils 308
  oligomeric Q8 structures 303–306
  time evolution, steric zippers and crystal structures 306–308
poly(ADP-ribose) polymerase (PARP) 184
population shift 29, 285
potential energy 49
  function 250
  surface 71
potential of mean force (PMF) 189, 252, 256, 262, 264
principal component analysis (PCA) 7, 34, 91, 209
Prion diseases 302
probability distribution 7, 54, 72, 74, 137
Protein Data Bank (PDB) 33, 116, 126, 249, 267, 285
protein-drug interactions 96
protein dynamics, networks
  coarse-graining 120–122
  ENMs 123–124
  molecular simulations 117–119
  RINs design 124–127
  software 119–120
protein energy landscape exploration (PELE) method 89
  coordinate exploration 93–94
  energy function 94
  enzyme characterization 96–97
  ligand perturbation 91
  mapping protein ligand and biomedical studies 94–96
  MC sampling procedure 91–92
  minimization 93
  process flow in 92
  receptor perturbation 91–92
  side chain readjustment 93
  web server 95
protein folding 7, 9, 253
protein kinases 118, 293, 294
  A (PKA) 35, 79, 291

protein kinases (*contd.*)
  casein kinase 2 (CK2) 329–331
  cyclin-dependent kinase 5 (CDK5) 56, 190
  receptor tyrosine (RTK) 174, 182, 183
protein–ligand binding 15, 30, 32, 33, 36–38, 87, 175, 227, 230
protein–ligand complex 4, 6, 10, 17, 33, 87, 175, 176, 227, 238
protein–ligand docking 4, 20
protein–ligand interaction 12, 15, 30, 38, 90, 97, 174, 235, 238
protein–ligand unbinding 87
protein–peptide binding 34–36
protein–peptide complex 34, 35
protein–peptide recognition 34
protein–protein binding 32–34, 129
protein–protein complexes 33, 34
protein–protein docking 33, 34, 107
protein–protein interactions (PPIs) 32, 77, 111, 113, 170, 226, 325, 326, 333, 334
protein–protein interfaces 33, 34
protein–protein system 33
protein stability 109, 129
protein structure network 107, 331
protein structure prediction 89, 181–184
protein structures, networks
  centrality measures 110–114
  software 114–117
  topological features and analysis 107–110
protofibrils 301, 303–310, 319, 320

q
quantitative structure-activity relationships (QSAR) 172–173, 190, 292
quantum mechanics and molecular mechanics (QM/MM) 8, 96, 97

r
raltegravir 46, 290
Really Interesting New Gene (RING) 117, 328, 333
receptor perturbation 91–92
receptor tyrosine kinase (RTK) 174, 182, 183
reduced point charge models (RPCMs) 121, 137
replica-exchange accelerated molecular dynamics (REXAMD) 49
replica exchange algorithm 49
replica-exchange molecular dynamics (REMD) 31, 32, 49
replica exchange thermodynamic integration (RETI) 89
replica exchange with solute tempering (REST) 49, 89, 190, 232
residue interaction networks (RINs) 107, 124–130
  centrality measures of 118
  design of mesoscale protein models 130–131
  modularization of 128–130
reweighting formula 12, 19
rilpivirine 176, 177
Ritonavir 168
root mean square distance (RMSD) 5–7, 56, 77, 93, 95, 134, 209
Rucaparib 184–185

s
Saccharomyces cerevisiae Cdc34 (ScCdc34) 328, 331, 334
  acidic loop 328
  casein kinase 2 (CK2) 331
sampling water interfaces through scaled Hamiltonians (SWISH) 50
Saquinavir 167–168
selectivity filter 253, 255–260, 264, 268, 269, 271
side chain readjustment 93
simulated annealing (SA) 31, 48, 88, 177, 181
simulated tempering (ST) 48
single-molecule fluorescence detection 29
single nucleotide polymorphisms (SNPs) 212
six spinocerebellar ataxias (SCA) 301

solvent accessible surface area (SASA) 91
spinal-bulbarmuscular atrophy (SBMA) 301
standard deviation 16, 18, 20, 59
standard error 16, 18
steered molecular dynamics (SMD) 11, 267
stromal interaction molecules (STIMs) 267, 268
structure-activity relationships (SAR) 168, 229, 234, 235
structure-based drug design (SBDD) 3, 37, 67, 75, 82, 165, 169, 174, 175, 180, 182, 208, 230, 235, 248, 271, 295
structure-based methods 173, 185
  Brigatinib 183–184
  Crizotinib 174–175
  de novo drug design 180
  enfuvirtide 184
  flexible receptor molecular docking 179
  Grazoprevir 177–178
  molecular docking–virtual screening 175–176
  molecular dynamics simulations 179–180
  Nelfinavir 180–181
  protein structure prediction 181–183
  Rucaparib 184–185
structure-kinetics relationships (SKR) 234
Sunitinib 182, 183
supervised molecular dynamics (SuMD) 235–238, 241
surface generalized born (SGB) model 94

t

teleportation 128
tempering methods 47–48
thermodynamic ATSM 287
thermodynamic cycle 18, 187, 231, 284
thermodynamic integration 16, 32, 51, 88, 89, 187
thermodynamic two-state model 283–285
3D Quantitative Structure Activity Relationship (3D-QSAR) 292
time-dependent independent component analysis (tICA) 81, 209
time-homogeneous dynamics 69
time-lagged independent component analysis (tICA) 81
tirofiban 170, 171
TmCorA protein 265
total dipole (TD) 304
trajectory clustering 6–8
transduction mechanism 109, 292
transient receptor potential (TRP) channels 263–264
transient receptor potential vanilloid type 1 (TRPV1) 264
transition paths 52, 56–59, 79–81, 291
transition path sampling (TPS) 57–59
transition state partial-path transition interface sampling (TS-PPTIS) 58
transition state theory 32, 58, 235
transmembrane (TM)
  helices 255–257
  potential 261
  protein 184
7-transmembrane (7TM) receptors 294
trinucleotide repeat expansion diseases (TREDs) 319
Trk1p 253–254
truncated Newton algorithm 92, 93
tyrosine kinase (Syk) inhibitors 189
Tyr-Tyr interactions 315

u

ubiquitination assay 334
ubiquitin conjugating (UBC) 327, 329–332, 335

ultra-CG models 122
umbrella sampling 11, 12, 19, 32, 52, 189, 190, 251–252, 257–260, 262, 266, 267, 271
Food and Drug Administration (FDA) 166, 290

V
Vaborbactam 179
vacancy diffusion mechanism 258
valsartan 170, 171
van der Waals interaction 250, 303, 315
vascular endothelial growth factor receptor (VEGFR) 182
venous thromboembolism (VTE) 178
virtual-library screening (VS) 30
visual molecular dynamics (VMD) 10
voltage-dependent anion channels (VDACs) 261, 262
voltage-gated sodium channels (Nav) 254–256
Voronoi cells 126

W
Wang–Landau sampling 32, 89
WaterFlap 228–230, 241
WaterMap 228–230
water network 30, 227, 230, 241, 242
weighted histogram analysis method (WHAM) 12, 19, 20, 32, 52, 252
weighted histogram techniques 32, 189
well-tempered metadynamics 13, 19, 252

X
X-ray crystallography 29, 33, 36, 117, 169, 173, 238, 262, 286, 290
X-ray diffraction patterns 303, 308, 315

Z
Zanamivir 181
zero dipole (ZD) 304, 305
zolmitriptan 171, 172
Zwanzig relationship 51